MICROBIOLOGY

An Illustrated Dictionary

New India Publishing Agency

New □elhi□□□□ □□□

About the Authors

Dr. Sunil V. Patel, is working as Professor at the Department of Biochemistry, Junagadh Agricultural University, Junagadh. He obtained his M.Sc. and Ph.D. in Department of Chemistry from Bhavnagar University, Bhavnagar (Gujarat). He has 31 years of experience in research and teaching. He has published number of Research Papers and Eight Books and also presented papers in National and International Seminars. His present interest is quality assessment of food grains, vegetables and fruit crops.

Dr. B.A. Golakiya is working as Professor and Head, Department of Biochemistry and Biotechnology at Junagadh Agricultural University, Gujarat, India. He has been national merit scholar in HSC and ICAR. He received young scientist award of Indian Science Congress (1989) and Zonal Awards (1986, 1989) of Indian Soil Science Society for excellent post graduate research. He also received best poster award from International Potash Institute, Switzerland (2001) and a momento of appreciation for excellent project work from Dead Sea Works, Israel. He had been a fellow of the Academy of Plant Sciences, India. He has some seven books, 45 research papers and over hundred popular articles to his credit. He has been PI of half a dozen National and International Projects. He guided many P.G Students up till date.

Dr. L.F. Akbari, Professor, Junagadh Agricultural University, Junagadh, has good background and 31 years of experience in field of Research in Plant Pathology and Agricultural Microbiology. He had completed three Research Projects as co-investigator. Agrochemical Manufacturer Association of India has recognized his contribution on biological control of stem/pod roots of groundnut through *Trichoderma* and awarded AMAI Plant Protection Research Award for same. He has published many research papers in International and National Journals. He attended 11 National Symposium/Seminars and one Global conference. He participated in training on Molecular Biology and Bioinformatics (GSBT, Gandhinagar), Alfatoxin Management in Groundnut (NRCG, Junagadh), Detection on Bacterial Plant Pathogens: Symptomatology to Advance Technique (IARI, New Delhi) and Advances in Agricultural Research Project Management (NAARM, Hyderabad). Junagadh Agricultural University had also appointed him as a member of Project Monitor and Evaluation Cell.

Dr. Valentina V. Umrania possesses M. Sc. with Microbiology and her doctoral research was on biogeotechnological applications of extremophiles. At present she is in faculty of Microbiology at MVM Sc & HSc College, Saurashtra University, Rajkot since 1986.

For last several years she has actively pursued research programs in Environmental Biotechnology and has obtained promising results appreciated by peers and experts. The results have been published in National and International Journals of repute.

Dr Umrania had advanced research associations abroad in her specialized areas of activity. She has visited Germany, United Kingdom and USA and was invited as a Post Doctoral Fellow at University Blaise Pascal, France, during 2005-06.

She has been active as a social activist by running a philanthropic trust named *Nishtha Trust*, Rajkot in capacity as an honorary chairperson. Apart from her academic pursuit Indian, Classical dance and Music are her pastime.

MICROBIOLOGY
An Illustrated Dictionary

S. V. Patel, M.Sc., Ph.D.
Professor
Department of Biochemistry
College of Agriculture, Junagadh Agricultural University
Junagadh - 362 001 (Gujarat)

B. A. Golakiya, M.Sc., Ph.D.
Professor and Head
Department of Biochemistry
College of Agriculture, Junagadh Agricultural University
Junagadh - 362 001 (Gujarat)

L. F. Akbari, M.Sc., Ph.D.
Professor
Department of Plant Pathology
College of Agriculture, Junagadh Agricultural University
Junagadh - 362 001 (Gujarat)

Valentina V. Umrania, M.Sc., Ph.D.
Lecturer in Microbiology
M.V. Mahilla Science & Home Science College
Saurashtra University, Rajkot - 360 001

New India Publishing Agency
New Delhi- 110 034

NEW INDIA PUBLISHING AGENCY
101, Vikas Surya Plaza, CU Block, LSC Market
Pitam Pura, New Delhi – 110 034, India
Email: info@nipabooks.com
Web: www.nipabooks.com

For customer assistance, please contact
Phone: + 91-11-27 34 17 17 Fax: + 91-11- 27 34 16 16
E-Mail: feedbacks@nipabooks.com

ISBN: 978-81-89422-95-0

Composed and Designed by NIPA.

Dr. B. K. Kikani
Vice-Chancellor

JUNAGADH AGRICULTURAL UNIVERSITY
Junagadh-362 001

Foreword

Education is the process of development from every angle of human personality. The journey of life on this planet is amazing. The study of nature, through the multiple disciplines of science, makes it more exciting and fascinating. Every student can experience this while satisfying their thirst by getting perfect answers regarding the theme under study. Awareness in all facets of life including environmental issues is becoming a global talk. People world wide are realizing its importance as they are able to smell a 'polluted tomorrow.' Cautious usage of today's environment would only provide as a legacy for tomorrow's generation.

While understanding the microbes we will be wiser about their significance in sustaining life on the earth. Seemingly invisible world of microbes is full of wonders and bizarre interrelations among their millions of associates. We cannot imagine our life and comfort without microbes. The over specialization of this branch in to mycology, bacteriology, virology, microbe-biotechnology led to an isolation of mindsets of the respective disciplines. Glossary of microbiology is to provide a bridge among all these faculties.

Advances made in the field of microbiology and biotechnology can be best understood through integrated discipline including information from microbiology, mycology, parasitological, interrelationship among microbes with plants and animals, cytology *etc*. Keeping up with its major mandate of disseminating scientific information to large masses, an effort has been made for understanding the basic essential concepts of this interesting multi disciplinary subject. This is truly useful from the view point of excellence in biology and its relevance to mankind.

-Sd/-

Place: Junagadh

(B. K. Kikani)

Dr. D. B. Kuchhadiya
Director of Research and
Dean of P. G. Faculty

JUNAGADH AGRICULTURAL UNIVERSITY
Junagadh

Foreword

Nature of life is very complex. Man had always been interested (curious) to know more and more about creation of life. Questioning mind of scientists led them to unveil many secrets of life. The success history started with the discovery of the cell in the 19th century and then continued with marvelous information and knowledge till date. Today, much is known about life at the molecular level.

Microbes are primordial life on this earth. Their course of life went through harsh climatic conditions. Their sustained life embodied the secret of sustainability under uncomfortable situations. Microbes can provide us the blue print of survival. We need to understand them holistically. The glossary of microbiology consists of terminologies along with abbreviations of various branches bifurcated originally from it.

A glossary normally takes the uses of words as such for its subject matter. This glossary includes about 6,900 terms. We have made sincere effort to present up-to-date knowledge of the subjects like Microbiology, Mycology, Cytology, Parasitological, Environmental and Microbial Biotechnology, and Virology, Soil Microbiology and many more similar concepts of biological sciences.

The group of scientists working at the department of biochemistry and biotechnology is putting their rigorous efforts in compiling this sort of information. Since they are directly connected with research and education they know it better what is the requirement of the concerned scientist, students and teachers affiliated with microbiology field.

The attempt has been made for all concerned with biological sciences including all branches like plant, animal and microbial biotechnology as well as molecular biology.

We hope this collection will be useful, interesting and even entertaining.

-Sd/-

Place: Junagadh **(D.B. Kuchhadiya)**

Preface

We have been teaching UG and PG students of biosciences since last 20 years. Students of various disciplines have completed their UG and PG programmes. While screening them at the common entrance examination or common competition examinations we found a poor interface between and among the various faculties and even students. The main barrier was the terminology. The concern faculties and students were many a times unaware of the sister branches of the same discipline. Microbiology is not an exception in it. When we seat together with pathological microbiologists, environmental microbiologists, microbial biotechnologists, marine microbiologists, soil microbiologists, virologists and so on, our group discussion looses its consistence with logical proliferation. And the cause is diversity in the terminology each one of us using. We are unaware of the terms being used in a single discipline of microbiology.

Teaching or research assumes a logical, interesting and useful band only when we are comprehensively aware of the terms of the at least sister concerned branches say for instance microbiology here. It is our humble effort in this collection to break the terminological barrier among the various segment of microbiology discipline. We hope this will bridge the knowledge and know-how to microbiologist working in various facets of microbiology.

Authors

Contents

Important Symbols

α: In the context of immunoglobulins, α is the type of heavy chain in IgA.
α: β T cell : See T-cell.
α: β T-cell recept : See T-cell receptor.
β : The light chain of the MHC class I proteins is called β2-microglobulin. It binds noncovalently to the heavy or (α–chain).
β barrel : See β sheet.
β sheet, β strands : A β sheet is one of the fundamental
β barrel, β sandwich : Structural building blocks of proteins, consisting of adjacent, extended strands of amino acids (β strands) that are bonded together by interactions between backbone amide and carbonyl groups. B Sheets can be parallel, in which case the adjacent strands run in the same direction, or antiparallel, where adjacent strands run in opposite directions. All immunoglobulin domains are made up of antiparallel β-sheet structures. A β barrel or a β sandwich is another way of describing the structure of the immunoglobulin domain.
γ : In the context of immunoglobulins, γ is the heavy chain of IgG.
γ globulins, gamma globulins: Plasma proteins can be separated on the basis of electrophoretic mobility into albumin and the α, β, and γ globulins. Most antibodies migrate in electrophoresis as g globulins (or gamma globulins), and patients who lack antibodies are said to have agammaglobulinemia.
γ:δ T-cell receptor : Most T lymphocytes have α:β heterodimeric
γ: δ T cells: T-cell receptors, but some bear a distinct γ:δ T-cell receptor composed of different antigenrecognition chains, γ and δ, assembled in a γ:δ heterodimer. Cells bearing these receptors are called γ:δ T cells and their specificity and function are not yet clear.
ε : In the context of immunoglobulins, ε (epsilon) is the heavy chain of IgE.
κ : In the context of immunoglobulins, κ is one of the two classes of lightchain.
λ : In the context of immunoglobulins, λ. is one of the two is one of the two classes of light-chain.
λ5 : See pre-B-cell receptor.
μ : In the context of immunoglobulins, μ is the heavy chain of IgM.
μ : Micro (10^{-6}m). *Or* Micromolar (μ mol/L)
μ mol : Micromole.

A

(+) sense RNA (plus-sense RNA) : A virus with a single-stranded RNA genome of the same polarity ('sense') as mRNA.

(Os)Coxa : Collectively,the hip bones.

12/23 rule : The 12/23 rule states that gene segments of immunoglobulin or T-cell receptors can be joined only if one has a recognition signal sequence with a 12 base pair spacer, and the other has a 23 base pair spacer.

16S rRNA : Large polynucleotide (about 1,500 bases) that functions as a part of the small subunit of the ribosome of prokaryotes and from whose sequence evolutionary information can be obtained; the eukaryotic counterpart is 18s rRNA.

2PG : 2-phosphoglycerate.

3PG : 3-phosphoglycerate.

4-1BB ligand (4-1BBL) : 4-1BB ligand (4-1BBL) is a member of the TNF family that binds to 4-1BB.

A : Adenine.

A&P : Anterior & posterior; auscultation & palpation; auscultation & percussion.

A form : A duplex DNA structure with right-handed twisting in which the planes of the base pairs are tilted about 70° with respect to the helix axis.

AAV : Adeno-associated viral.

Ab : Antibody.

AB toxins : The structure and activity of many exotoxins are based on the AB model. In this model, the B portion of the toxin is responsible for toxin binding to a cell but does not directly harm it. The A portion enters the cell and disrupts its function.

ABC : ATP-binding cassette.

Abduct : A muscle which moves a body part away from the midline of the body.

Abiotic : A term used to describe a process which occurs outside of the body of a living organism,occurring without involvement of a living organism.

Abiotic factors : Non living; moisture, soil, nutrients, fire, wind, temperature, climate.

ABO blood group system : The ABO blood group system antigens are expressed on red blood cells. They are used for typing human blood for transfusion. Individuals who do not express A or B antigens on their red blood cells naturally form antibodies that interact with them.

Abtive Infection : When a virus infects a cell (or host), but cannot complete the full replication cycle, i.e. A non-productive infection.

Abruptly bulbous (shape of stipe, i.e. stem, base) : A bulbous stipe base with a distinct margin and an upper surface either attached at a right angle to the stipe or sloping upward to the stipe; the upper surface is not shaped like the edge of a saucer as in a marginate-depressed bulb (when the upper surface is attached at a right angle, abruptly bulbous marginate bulb).

Abscess : Pocket of pus; pusfilled lesion.

Absidia (Zygomycetes) : A common contaminant (fungus) isolated from soil, air, stored grains, foods, and the indoor environment. Reported to be allergenic and a rare cause of human infection called zygomycosis.

Absidia sp. **:** A zygomycete fungus. Reported to be allergenic. May cause mucorosis in immune compromised individuals. The sites of infection are the lung, nasal sinus, brain, eye and skin. Infection may have multiple sites.

Absorption : Movement of ions and water into as organism as a result of metabolic processes, frequently against an electrochemical potential gradient (active) or as a result of diffusion along an activity gradient (passive). ***OR*** The removal of antibodies specific for one antigen from an antiserum to render it specific for another antigen or antigens is called absorption. ***OR*** the taking in or soaking up of one substance into the body of another by molecular or chemical action (as tree roots absorb dissolved nutrients in the soil).

Absorption field : A system of properly sized and constructed narrow trenches partially filled with a bed of washed gravel or crushed stone into which perforated or open joint pipe is placed. The discharge from the septic tank is distributed through these pipes into trenches and surrounding soil. While seepage pits normally require less land area to install, they should be used only where absorption fields are not suitable and wellwater supplies are not endangered.

Acanthocytes : Needle-like crystalline deposits on the mycelium of certain mushrooms.

ACAT : Acyl-coa:cholesterol acyl transferase.

Accessy effect cells : Accessory effector cells in adaptive immunity are cells that aid in the response but do not directly mediate specific antigen recognition. They include phagocytes, mast cells, and NK cells.

Accessy pigments : Photosynthetic pigments such as carotenoids and phycobiliproteins that aid chlorophyll in trapping light energy.

Acetal : The product formed by the successive condensation of two alcohols with a single aldehyde. It contains two ether-linked oxygens attached to a central carbon atom.

Acetogenic bacterium : Prokaryotic organism that uses carbonate as a terminal electron acceptor and produces acetic acid as a waste product.

Acetyl CoA (Acetyl coenzyme) : Acetyl-coenzyme A, a high-energy ester of acetic acid that is important both in the tricarboxylic acid cycle and in fatty acid biosynthesis. ***OR*** A combination of acetic acid and coenzyme A that is energy rich; it is produced by many catabolic pathways and is the substrate for the tricarboxylic acid cycle, fatty acid biosynthesis, and other pathways.

Acetylene block assay : Estimates denitrification by determining release of nitrous oxide (N_2O) from acetylenetreated soil.

Acetylene reduction assay : Estimates nitrogenase activity by measuring the rate of acetylene reducted to ethylene.

ACh : Acetylcholine.

Achene : A type of fruit, such as a dandelion fruit,which "looks like" a seed and does not burst open when ripe.

Achilles Tendon : The tendon that ends the gastro cnemius.

Achlhydria (achlhydric) : No stomach acid; condition may be physiological or pharmacological (e.g., omeprazole (prilosec(r)) or other antisecretory protonpump blocking drugs prescribed commonly for gastric ulcers).

Achondroplasia : A dominant genetic disorder where the homozygote (AA) dies before birth, resulting in a miscarriage, and the heterozygote (Aa) is a dwarf.

ACI : Acute coronary insufficiency.

Acid : A substance that dissolves in water with the formation of hydrogen ions, contains hydrogen which may be replaced by metals to form salt, and/or is corrosive.

Acid dyes : Dyes that are anionic or have negatively charged groups such as carboxyls.

Acid fast : Refers to bacteria like the mycobacteria that cannot be easily decolorized with acid alcohol after being stained with dyes such as basic fuchsin.

Acid soil : Soil with a pH value < 6.6.

Acid-fast staining : A staining procedure that differentiates between bacteria based on their ability to retain a dye when washed with an acid alcohol solution.

Acidity : The capacity of water or wastewater to neutralize bases. Acidity is expressed in milligrams per liter of equivalent calcium carbonate. Acidity is not the same as pH.

Acidoduric : An organism capable of surviving under acidic conditions, e.g. *Streptococcus mutans*

Acidophile : Organism that grows best under acid conditions (down to a pH of 1). ***OR*** A microorganism that has its growth optimum between about pH 0 and 5.5.

ACJT : Acromioclavicular joint.

ACL : Anterior clavicular line.

ACO : Alert, cooperative, oriented.

ACP : Acyl carrier protein.

Acquired enamel pellicle : A membranous layer on the tooth enamel surface formed by selectively adsorbing glycoproteins (mucins) from saliva. This pellicle confers a net negative charge to the tooth surface.

Acquired immuno deficiency syndrome (AIDS) : An infectious disease syndrome caused by the human immunodeficiency virus and is characterized by the loss of a normal immune response, followed by increased susceptibility to opportunistic infections and an increased risk of some cancers.

Acquired immuno response : Acquired immune response: see adaptive immune response.

Acquired immune tolerance : The ability to produce antibodies against nonself antigens while "tolerating" (not producing antibodies against) self-antigens.

Acquired immunity : Refers to the type of specific immunity that develops after exposure to a suitable antigen or is produced after antibodies are transferred from one individual to another.

Acremonium (Hyphomycetes) : Fungus naturally found in soils, decaying organic matter, and plant debris. Can be found in food and the indoor environment. Acremonium is a common allergen, can produce a trichothecene mycotoxin, and volatile organic compounds (vocs).

Acremonium sp. (Cephalospium sp.) **:** Reported to be allergenic. Can produce a trichothecene toxin which is toxic if ingested. It was the primary fungus identified in at least two houses where the occupant complaints were nausea, vomiting and diarrhea. Asexual state of *Emericellopsis* sp., *Chaetomium* sp., and *Nectripsis* sp. It can produce mycetomas, infections of the cornea and nails.

Acrid : Intensely sharp and burning (peppery).

Acrogenous : Conidia born at the tip of the conidiophore.

Acropetal : A chain of conidia in which the youngest conidium is at the tip and the oldest is at the base.

Acropleurogenos : Conidia developing at the tip and along the sides of the conidiophore.

ACS : Acute cervical strain; acute cervical sprain.

ACT : Anticoagulant therapy; active motion.

ACTH : Adrenocorticotropic hormone.

Actin : i) protein involved in smooth muscle contraction and relaxation; ii) protein that forms microfilaments at the cellular level that enable the movement of certain intracellular materials ***OR*** A protein found in combination with myosin in muscle and also found as filaments constituting an important part of the cytoskeleton in many eukaryotic cells.

Actinobacteria : A group of gram-positive bacteria containing the actinomycetes and their high G 1 C relatives.

Actinomycete : Nontaxonomic term applied to a group of high G + C base composition, Gram positive bacteria that have a superficial resemblance to fungi. Includes many but not all organisms belonging to the order Actinomycetales. ***OR*** An aerobic, gram-positive bacterium that forms branching filaments (hyphae) and asexual spores.

Actinomycin D : An antibiotic that binds to DNA and inhibits RNA chain elongation.

Actinorhizae : Associations between actinomycetes and plant roots

Activated complex : The highest free energy state of a complex in going from reactants to products.

Activated sludge : Sludge particles produced in raw or settled wastewater (primary effluent) by the growth of organisms (including zoogleal bacteria)

in aeration tanks in the presence of dissolved oxygen. The term "activated" comes from the fact that the particles are teeming with fungi, bacteria, and protozoa. Activated sludge is different from primary sludge in that the sludge particles contain many living organisms which can feed on the incoming wastewater. ***OR*** Solid matter or sediment composed of actively growing microorganisms that participate in the aerobic portion of a biological sewage treatment process. The microbes readily use dissolved organic substrates and transform them into additional microbial cells and carbon dioxide.

Activation enery : Amount of energy required to bring all molecules in one mole of a substance to their reactive state at a given temperature. ***OR*** The energy required to bring reacting molecules together to reach the transition state in a chemical reaction.

Active carrier : An individual who has an overt clinical case of a disease and who can transmit the infection to others.

Active immunization : The induction of active immunity by natural exposure to a pathogen or by vaccination.

Active site : The region of an enzyme molecule that contains the substrate binding site and the catalytic site for converting the substrate(s) into product(s). ***OR*** Region of an enzyme where substrates bind.

Active transport : The transport of solute molecules across a membrane against an electrochemical gradient; it requires a carrier protein and the input of energy. ***OR*** The energy-dependent transport of a substance across a membrane.

ACTS : Acute cervical trauma syndrome.

Aculeate : Having narrow spines; spores of *Laccaria* which are round and have narrow spines on the outer wall.

Acute : Course of a disease that is of sudden onset, marked intensity (relatively severe) and short duration (see chronic) ***OR*** sharp or pointed.

Acute carrier : See casual carrier.

Acute Infection : Relatively brief infections, i.e. A few days to a few weeks, following which the virus is usually eliminated completely from the body by the immune system.

Acute lymphoblastic leukemia : Acute lymphoblastic leukemia is a highly aggressive, undifferentiated form of lymphoid malignancy derived from a progenitor cell that is thought to be able to give rise to both T and B lineages of lymphoid cells. Most of these leukemias show partial differentiation toward the B-cell lineage (so called B-ALL) whereas a minority show features of T cells (T-ALL).

Acute viral gastroenteritis : An inflammation of the stomach and intestines, normally caused by Norwalk and Norwalklike viruses, other caliciviruses, rotaviruses, and astroviruses.

Acute-phase proteins : Acute-phase proteins are found in the blood shortly after the onset of an infection. These proteins participate in early phases of host defense against infection. An example is the mannan-binding lectin.

Acute-phase response : The acute-phase response is a change in the blood that occurs during early phases of an infection. It includes the production of acute-phase proteins and also of cellular elements.

ACVD : Acute cardiovascular disease.

Acyclovir : A synthetic purine nucleoside derivative with antiviral activity against herpes simplex virus.

AD : Right ear; admitting diagnosis; adult disease; admission and discharge; alternating days.

AD LIB : As desired.

ADA : Anterior descending artery.

Adaptation : Changes in the genetics of individuals within a population that make them more fit for their environment.

Adaptive immune response, adaptive immunity, acquired immune responses : The adaptive immune response or adaptive immunity is the response of antigen-specific lymphocytes to antigen, including the development of immunological memory. Adaptive immune responses are generated by clonal selection of lymphocytes. Adaptive immune responses are distinct from innate and nonadaptive phases of immunity, which are not mediated by clonal selection of antigen-specific lymphocytes. Adaptive immune responses are also known as acquired immune responses.

Adaptive radiation : The process of a species spreading into and making use of a new environment.

Adapt proteins : The adaptor proteins are key linkers between receptors and downstream members of signaling pathways. These proteins are functionally heterogeneous, but all use a similar domain, known as an SH2 domain, as the means of interacting with the phosphotyrosine residues generated by the receptor-associated tyrosine kinases. The protein known as Vav is an adaptor protein of this type.

ADCC : Antibody-dependent cell-mediated cytotoxicity.

ADD : Addiction; adduct; adduction.

Adenine : A purine derivative, 6-aminopurine, found in nucleosides, nucleotides, coenzymes, and nucleic acids. ***OR*** A purine base found in DNA or RNA.

Adenoids : The adenoids are mucosal-associated lymphoid tissues located in the nasal cavity.

Adenosine 5-triphosphate (ATP) : The triphosphate of the nucleoside adenosine, which is a high energy molecule or has high phosphate group transfer potential and serves as the cell's major form of energy currency.

Adenosine deaminase deficiency : The enzyme defect adenosine deaminase deficiency leads to the accumulation of toxic purine nucleosides and nucleotides, resulting in the death of most developing lymphocytes within the thymus. It is a cause of severe combined immunodeficiency.

Adenosine diphosphate (ADP) : The nucleotide formed by adding a pyrophosphate group to the 5'-OH group of adenosine. ***OR*** The nucleoside diphosphate usually formed upon the breakdown of ATP when it provides energy for work.

Adenosine triphosphate (ATP) : Common energy donating molecule in biochemical reactions. Also an important compound in transfer of phosphate groups. ***OR*** The nucleotide formed by adding yet another phosphate group to the pyrophosphate group on ADP. ***OR*** A molecule used by cells to store and transport energy.

Adenosine : A purine nucleoside found in DNA, RNA, and many cofactors.

Adenylate cyclase : The enzyme that catalyzes the formation of cyclic 3',5' adenosine monophosphate (camp) from ATP.

ADH : Alcohol dehydrogenase. ***OR*** Antidiuretic hormone; adhesions.

Adherence (attachment) : Close association of bacterial cells and host cells generally characterized by receptors and target sites.

Adhesin : Structure or macromolecule located on the surface of a cell or extracellularly that facilitates the adherence of a cell to a surface or to another cell; site of attachment is often a specific receptor, but the adherence may also be nonspecific; host cell receptors are often sugar moieties (see lectin).

Adhesion molecules : Adhesion molecules mediate the binding of one cell to other cells or to extracellular matrix proteins. Integrins, selecting, members of the immunoglobulin gene superfamily, and CD44 and related proteins are all adhesion molecules important in the operation of the immune system.

Adhesion/attachment : A stable interaction of a cell with respect to a surface. Living cells actively excrete holdfast chemicals from their surface to anchor themselves to a substratum. This is referred to as adhesion or attachment.

Adiaconidia : A large, globose, thick-walled conidium, usually produced by *Emmonsia* (*Chrysosporium*) *parvum*, in the lungs of humans and animals.

Adipocyte : A specialized cell that functions as a storage depo for lipid.

ADJ : Adjacent; adjoining; adjunct; adjust.

Adjuvant : Material added to an antigen to increase its immunogenicity. Common examples are alum, killed *Bordetella pertussis*, and an oil emulsion of the antigen, either alone (Freund's incomplete adjuvant) or with killed mycobacteria (Freund's complete adjuvant).

ADL : Activities of daily living.

ADM : Admit; admission.

ADME : Absorption, distribution, metabolism, and excretion

Adnate : Used to describe gills attached to the stem for their total width.

Adnexed : Gills tapering toward the stem so that their attachment is narrow.

Adomet : S-adenosylmethionine.

Adoptive immunity, adoptive transfer, adoptive immunization : Adoptive immunity is immunity conferred on a naive or irradiated recipient by transfer of lymphoid cells from an actively immunized donor. This is called adoptive transfer or adoptive immunization.

ADP : Adenosine diphosphate. See ATP.

ADR : Adverse drug reaction.

ADR : Adverse drug reaction.

Adrenal glands : Endocrine glands which are located on top of the kidney sand which secrete hormones such as cortisone and adrenaline (epinephrine).

Adrenaline : Hormone,also known a sepin ephrine, secreted by adrenal glands which triggers the "fight-or-flight" response.

Adsption : The gathering of a gas, liquid, or dissolved substance on the surface or interface zone of another substance. ***OR*** Process by which atoms, molecules, or ions are taken up and retained on the surfaces of solids by chemical or physical binding.

Adult stem cell : A stem cell in adult tissue, normally found in the brain, liver and bone marrow.

Adult T-cell leukemia : A type of white blood cell cancer caused by the HTLV-1 virus.

ADV : Arterial deep venous; advised.

AE : Above the elbow; active and equal.

Aequi-hymeniiferous : Hymenium having basidia which mature evenly all over the surface of each lamella (from Ainsworth and Bisby, "Dictionary of the Fungi"); in opposition to inaequi-hymeniiferous.

AER : Auditory evoked response; average evoked response.

Aeration : The process of adding air to water. In wastewater treatment, air is added to freshen wastewater and to keep solids in suspension.

Aeration tank : The tank where raw or settled wastewater is mixed with return sludge and aerated. This is the same as an aeration bay, aerator, or reactor.

Aerenchyma : Plant tissue incorporating large, gasfilled spaces.

Aerial mycelium : Hyphal elements growing above the agar surface.

Aerobe : An organism that utilizes oxygen for growth. ***OR*** An organism that requires free oxygen for growth.

Aerobic : (i) Having molecular oxygen as a part of the environment. (ii) Growing only in the presence of molecular oxygen, as in aerobic organisms. (iii) Occurring only in the presence of molecular oxygen, as in certain chemical or biochemical processes such as aerobic respiration.

Aerobic anoxygenic photosynthesis : Photosynthetic process in which electron donors such as organic matter or sulfide, which do not result in oxygen evolution, are used under aerobic conditions.

Aerobic anoxygenic photosynthesis : Photosynthetic process in which electron donors such as organic matter or sulfide, which do not result in oxygen evolution, are used under aerobic conditions.

Aerobic respiration : A metabolic process in which molecules, often organic, are oxidized with oxygen as the final electron acceptor.

Aerotolerant anaerobes : Microbes that grow under both aerobic and anaerobic conditions, but do not shift from one mode of metabolism to another as conditions change. They obtain energy exclusively by fermentation.

Aestivation : State of dormancy during the summer.

Aether : The incorruptible,eternal matter which Aristotle thought composed the universe outside of the moon's orbit, including the sun,the stars,and the other planets.

AF : Adult female; amniotic fluid; aortic flow; atrial fibrillation; atrial flutter.

Afferent lymphatic vessels : Afferent lymphatic vessels drain fluid from the tissues and carry macrophages and dendritic cells from sites of infection in most parts of the body to the lymph nodes.

Affinity : Affinity is the strength of binding of one molecule to another at a single site, such as the binding of a monovalent Fab fragment of antibody to a monovalent antigen. See also avidity.

Affinity chromatography : Affinity chromatography is the purification of a substance by means of its affinity for another substance immobilized on a solid support. For example, an antigen can be purified by affinity chromatography on a column of beads to which specific antibody molecules are covalently linked.

Affinity maturation : Affinity maturation refers to the increase in the affinity for the specific antigen of the antibodies produced during the course of a humoral immune response. It is particularly prominent in secondary and subsequent immunizations.

AFL : Artificial limb.

Aflatoxin : A polyketide secondary fungal metabolite that can cause cancer.

AFM : Atomic force microscopy.

AFP : Amplified fragment length polymorphism.

AFTR : Atrophy, fasciculation; tremor; rigidity.

AG : Antigen.

AGA : Appropriate for gestational age.

Agammaglobulinemia, X-linked agammaglobulinemia (XLA): Agammaglobulinemia: see X-linked agammaglobulinemia (XLA).

Agar : Complex polysaccharide derived from certain marine algae that is a gelling agent for solid or semisolid microbiological media. Agar consists of about 70% agarose and 30% agaropectin. Agar can be melted at temperature above 100ºC; gelling temperature is 40-50 ºC.

Agar plate : A sterile dish that contains agar mixed with nutrients, which is used to grow microorganisms.

Agaric : A member of the Agaricaceae (*Sensu lato*); often considered a synonym of the mushroom.

Agarose : Nonsulfated linear polymer consisting of alternating residues of D-galactose and 3,6anhydrolgalactose. Agarose is extracted from seaweed, and agarose gels are often used as the resolving medium in electrophoresis.

AGE : Angle of greatest extension.

AGF : Angle of greatest flexion.

AGG : Aggrevation.

Agglutinates : The visible aggregates or clumps formed by an agglutination reaction.

Agglutination : Clumping of RBCs due to an antigen-antibody interaction.

Agglutination reaction : The formation of an insoluble immune complex by the crosslinking of cells or particles

Agglutination, agglutinate : Agglutination is the clumping together of particles, usually by antibody molecules binding to antigens on the surfaces of adjacent particles. Such particles are said to agglutinate. When the particles are red blood cells, the phenomenon is called hemagglutination.

Agglutinin : The antibody responsible for an agglutination reaction.

AGL : Acute granulocytic leukemia.

AGN : Acute glomerulonephritis.

Agonist peptides : Agonist peptides are peptide antigens that activate their specific T cells, inducing them to make cytokines and to proliferate. They are thought to differ from antagonist peptides by their ability to induce T-cell receptor dimerization.

Agrocybe (Basidiomycetes) : Medium to large mushrooms growing in grassy areas, on wood mulch, and pine needles. Some species are edible.

AGTT : Abnormal glucose tolerance test.

AH : Abdominal hysterectomy; arterial hypertension.

AHG : Antihaemophilic globulin.

AI : Artificial intelligence.

AI : Aortic insufficiency; artificial insemination.

AID : Acute infectious disease; artificial insemination by donor.

AIDS : Acquired immuno deficiency syndrome.

AIDS-related complex (ARC) : A collection of symptoms such as lymphadenopathy (swollen lymph glands), fever, malaise, fatigue, loss of appetite, and weight loss. It results from an HIV infection and may progress to frank AIDS.

AIF : Apoptosis-inducing factor.

AIIS : Anterior inferior iliac spine.

Airborne transmission : The type of infectious organism transmission in which the pathogen is truly suspended in the air and travels over a meter or more from the source to the host.

AIT : Acute intensive treatment.

AJ : Ankle jerk.

AK : Above the knee.

Akinetes : Specialized, nonmotile, dormant, thick-walled resting cells formed by some cyanobacteria.

AL : Left ear.

Ala : Alanine.

ALA : Aminolevulinic acid.

Alae : Flattened, wing-like expansions of the cuticle of nematodes (cervical, caudal or lateral).

Albumin : The main protein in albumen, found in egg white.

ALC : Approximate lethal concentration.

Alcohol : An organic molecule with at least one hydroxyl (–OH)group attached.

Alcohol : A molecule with a hydroxyl group attached to a carbon atom.

Alcoholic fermentation : A fermentation process that produces ethanol and CO_2 from sugars.

ALD : Alcoholic liver disease.

Aldehyde : An organic molecule with a carbonyl group (–C=O, but note that the C needs a fourth bond, too) attached at one *end* of the molecule

Aldehyde : A molecule containing a doubly bonded oxygen and a hydrogen attached to the same carbon atom.

ALDH : Aldehyde dehydrogenase.

Aleurioconidium : A thallic conidium released by lysis or fracture of the supporting cell.

Alexander Ivanovich Oparin : Person who, in 1936, published *The Origins of Life* in which he described. Hypothetical conditions necessary for life to have arisen on early earth.

Alfred Russell Wallace : Person who lived about the same time as Charles Darwin and came up with a very similar theory of evolution.

Algae : A type of plant that grows in freshwater or seawater, examples include pond scum and seaweed. ***OR*** Phototrophic eukaryotic microorganism. Algae could be unicellular or multicellular. Bluegreen algae are not true algae; they belong to a group of bacteria called cyanobacteria.

Algicide : An agent that kills algae.

Alginate : Component of the cell walls of many rhodophytes and kelps. Alginates have an affinity for water, and so help to slow dessication when the algae are exposed to the air; they are commercially important in the production of paper, toothpaste, beer, and frozen foods.

Algology : The scientific study of algae.

Alimentary canal : The large muscular tube from the mouth to the anus approx 8.3 metres long where food is digest.

Aliphatic : Organic compound in which the main carbon structure is a straight chain.

Aliquot : A measured sub-volume of a solution taken used to prepare a dilution of that solution.

Alkaline soil : Soil having a pH value >7.3.

Alkaline substance : Chemical compounds in which the basic hydroxide (OH-) ion is united with a metallic ion, such as sodium hydroxide (NaOH) or potassium hydroxide (KOH). These substances impart alkalinity to water and are employed for neutralization of acids. Lime is the most commonly used alkaline material in wastewater treatment.

Alkalophile : A microorganism that grows best at pHs from about 8.5 to 11.5.

Alkane : Straight chain or branched organic structure that lacks double bonds.

ALL : Acute lymphoblastic leukemia; acute lymphocytic leukemia; allergy.

Allele : One of a number of alternative forms of the same gene, for

Allele frequencies : The percentage of each allele in a population.

Alleles : Alleles are variants of a single genetic locus. ***OR*** Alternative forms of a gene.

Allelic exclusion : Allelic exclusion refers to the fact that in a heterozygous individual, only one of the alternative C-region alleles of the heavy or light chain is expressed in a single B cell and in an immunoglobulin molecule. The term has come to be used more generally to describe the expression of a single receptor specificity in cells with the potential to express two or more receptors.

Allelopathy : When plants secrete chemicals to retard / inhibit the growth of other plants.

Allergen/allergec : A substance capable of inducing allergy or specific susceptibility. ***OR*** Allergens are antigens that elicit hypersensitivity or allergic reactions. An allergen is an antigen, principally a protein, which can elicit symptoms of allergic disease in a previously sensitized individual. This antigen is specifically recognized by the individual's immune system, with subsequent development of specific antibody and/or cell mediated immunity. Fungi can elicit an allergic reaction ranging from mild to severe, anywhere from a stuffy nose, through hay fever and asthma to pneumonitis. In most cases, the physical condition of the host, the amount of allergen the host is exposed to spores, fungal hyphae, dust, pollen, etc. and the degree of sensitization of the individual determines the severity of the reaction. In general, common environmental or indoor air contaminating fungi most often affect humans as irritants that elicit an immune response that we generally associate with allergies (hypersensitivity).

Allergic asthma : Allergic asthma is constriction of the bronchial tree due to an allergic reaction to inhaled antigen.

Allergic conjunctivitis : The lining of the eye, called the conjunctiva, manifests allergic conjunctivitis in sensitized individuals exposed to allergens.

Allergic contact dermatitis : An allergic reaction caused by haptens that combine with proteins in the skin to form the allergen that produces the immune response.

Allergic reaction : An allergic reaction is a response to innocuous environmental antigens, or allergens, due to preexisting antibody or primed T cells. There are various immune mechanisms of allergic reactions, but the most common is the binding of allergen to IgE antibody on mast cells, which causes asthma, hay fever, and other common allergic reactions.

Allergic rhinitis : Allergic rhinitis is an allergic reaction in the nasal mucosa, also known as hay fever, that causes runny nose, sneezing, and tears.

Allergy : Allergy is a symptomatic reaction to a normally innocuous environmental antigen. It results from the interaction between the antigen and antibody or primed T cells produced by earlier exposure to the same antigen.

Alloantigens : Alloantigens are classically defined as polymorphisms at the MHC locus; they stimulate intense reactions to allografted tissues, hence their naming.

Allochthonous flora : Organisms that are not indigenous to the soil but that enter soil by precipitation, diseased tissues, manure, and sewage. They may persist for some time but do not contribute in a significant way to ecologically significant transformations or interactions.

Allogeneic : Two individuals or two mouse strains that differ at the MHC are (Major Histo Compibility Complex) said to be allogeneic. The term can also be used for allelic differences at other loci. Rejection of grafted tissues from unrelated donors usually results from T-cell responses to allogeneic MHC molecules expressed by the grafted tissues. See also syngeneic; xenogeneic.

Allogenic facts : Causes of succession due to abiotic components.

Allograft : An allograft is a graft of tissue from an allogeneic or nonself donor of the same species; such grafts are invariably rejected unless the recipient is immunosuppressed. ***OR*** A transplant between genetically different individuals of the same species

Allopatric species : Species which occupy areas separated by space or time,therefore can not come into contact with each other.

Alleactivity, alleactions : Alloreactivity describes the stimulation of T cells by MHC molecules other than self; it marks the recognition of allogeneic MHC molecules. Such responses are also called alloreactions.

Allosteric enzyme : An enzyme whose active site can be altered by the binding of a small molecule at a nonoverlapping site. ***OR*** An enzyme whose activity is altered by the binding of a small effector or modulator molecule at a regulatory site separate from the catalytic site; effector binding causes a conformational change in the enzyme and its catalytic site, which leads to enzyme activation or inhibition.

Allosteric site : Site on the enzyme other than the active site to which a nonsubstate compound binds. This may result in a conformational change at the active site so that the normal substrate cannot bind to it.

Allotypes : Allotypes are allelic polymorphisms that can be detected by antibodies specific for the polymorphic gene products; in immunology, allotypic differences in immunoglobulin molecules were important in deciphering the genetics of antibodies.

Alluvial soil : Water-deposited soil.

ALMI : Anterior lateral myocardial infarction.

Alpha hemolysis : A greenish zone of partial clearing around a bacterial colony growing on blood agar.

Alpha-proteobacteria : One of the five subgroups of proteobacteria, each with distinctive 16s rRNA sequences. This group contains most of the oligotrophic proteobacteria; some have unusual metabolic modes such as methylotrophy, chemolithotrophy, and nitrogen fixing ability. Many have distinctive morphological features.

ALS : Amyotrophic lateral sclerosis. ***OR*** Acute lumbar strain; anterolateral sclerosis; amytrophic lateral sclerosis.

ALT : Alternate.

Altered peptide ligand : An altered peptide ligand, or partial agonist, is a peptide, usually closely related to an agonist peptide in amino acid sequence, that induces only a partial response from T cells specific for the agonist peptide.

Alternaria (Hyphomycetes) : A common saprobe found on decaying wood, decaying plants, food, soil, and outdoor air. Some species are plant pathogens. Indoors, it can be found in house dust, carpet, damp areas around showers and window frames, and anywhere condensation occurs. Because of its abundance and ubiquity, Alternaria is one of the most important fungal allergens and is recognized as the chief fungal cause of hay fever. Infection is extremely rare.

Alternaria sp. **:** Aw-0.89. Conidia dimensions 18-83 x 7-18 microns (5). A very common allergen with an IgE mediated response . It is often found in carpets, textiles, and on horizontal surfaces in building interiors. Often found on window frames. Outdoors it may be isolated from samples of soil, seeds and plants. It is commonly found in outdoor samples. The large spore size 20-200 microns in length and 7-18 microns in sizes, suggests that the spores from this fungi will deposited in the nose, mouth and upper respiratory tract. It may be related to bakers asthma. It has been associated with hypersensitivity pneumonitis. The species *Alternaria alternata* is capable of producing tenuazonic acid and other toxic metabolites which may be associated with disease in humans or animals. Common cause of extrinsic asthma (immeadiate-type hypersensitivity: type I. Acute symptoms include edema and bronchiospasms, chronic cases may develop pulmonary emphysema.

Alternative complement pathway : An antibody-independent pathway of complement activation that includes the C_3-C_9 components of the classical pathway and several other serum protein factors (e.g., factor B and properdin).

ALTS : Acute lumbar trauma syndrome.

Alum : Astringent crystalline double sulfate of an alkali. $K_2SO_4Al_2(SO_4)_3$ $24H_2O$. Used in the processing of pickles and as a flocking agent. Excess aluminum in the environment can be hazardous.

Alveolar macrophage : A vigorously phagocytic macrophage located on the epithelial surface of the lung alveoli where it ingests inhaled particulate matter and microorganisms.

Alveolate : Surface of spore or cap (pileus) with broad pits .

Alveolus : The tiny air sacs in the lungs in which gases are exchanged with the blood. ***OR*** Air sac in lung; Alveoli (plural of alveolus).

ALVF : Acute left ventricular failure.

AM : Adult male; aerospace medicine; before noon.

AMA : Against medical advice; American Medical Association.

Amantadine : An antiviral compound used to prevent type A influenza infections.

AMB : Ambulate; ambulatory.

Ambient temperature : Temperature of the surroundings.

AMD : Age-related macular degeneration.

Amebiasis (amebic dysentery) : An infection with amoebae, often resulting in dysentery; usually it refers to an infection by *Entamoeba histolytica*.

Amensalism : A relationship in which the product of one organism has a negative effect on another organism.

Amensalism (antagonism) : Production of a substance by one organism that is inhibitory to one or more other organisms. The terms antibiosis and allelopathy also describe cases of chemical inhibition. ***OR*** Interaction between organisms where one organism is adversely affected and the other organism is unaffected, like antibiosis and allelopathy.

Ameroconidium : A one-celled conidium. ***OR*** A test that uses a special Salmonella strain to test chemicals for mutagenicity and potential carcinogenicity.

Ames test : A test that uses a special Salmonella strain to test chemicals for mutagenicity and potential carcinogenicity.

AMI : Acute myocardial infarction.

Amine : An organic molecule with at least one amino($-NH_2$) group attached.

Amino acid : A carbon bonded to four different things: an amino group($-NH_2$), a carboxyl group ($-COOH$), a hydrogen ($-H$), and a side chain (symbolized by"$-R$").

Amino acid activation : The initial stage of protein synthesis in which amino acids are attached to transfer RNA molecules.

Amino acid : Unit molecule from which proteins are constructed by polymerization. ***OR*** The basic building block of proteins. A short chain of amino acids is called a peptide and a long chain of amino acids (normally more than fifty) is called a protein.

Amino group : An NH_2 group attached to a carbon skeleton as in the amines and amino acids.

Aminoacyl accept site (A site) : The site on the ribosome that contains an aminoacyl-trna at the beginning of the elongation cycle during protein synthesis; the growing peptide chain is transferred to the aminoacyl-tRNA and lengthens by an amino acid.

Aminoglycoside antibiotics : A group of antibiotics synthesized by *Streptomyces* and *Micromonospora*, which contain a cyclohexane ring and amino sugars; all aminoglycoside antibiotics bind to the small ribosomal subunit and inhibit protein synthesis.

AML : Acute monocytic leukemia; acute myeloblastic leukemia.

Ammonia oxidation : Test drawn during manufacturing process to evaluate the ammonia oxidation rate for the nitrifiers.

Ammonification : Liberation of ammonium (ammonia) from organic nitrogenous compounds by the action of microorganisms.

Amnesic shellfish : The disease arising in humans and animals that eat seafood such as mussels contaminated with domoic acid from diatoms. The disease produces short-term memory loss in its victims.

Amniocentesis : Removal of about 10 ml of amniotic fluid from around a pre-born baby, often used to determine the child's sex or check for chromosomal abnormalities.

Amniotic fluid : Liquid that surrounds a developing foetus.

Amoeba **:** Genus name for one-celled organisms which have pseudopodia and belong to Kingdom Protista; Division Rhizopoda.

Amoeba (plural, amoebae) : Protozoa that can alter their cell shape, usually by the extrusion of one or more pseudopodia.

Amoebic dysentery : Disease caused by the parasite, *Entamoeba histolytica*, a member of Division Rhizopoda.

Amoeboid movement : Moving by means of cytoplasmic flow and the formation of pseudopodia (temporary cytoplasmic protrusions of the cytoplasm).

AMP : Adenosine monophosphate

Amphibolic pathways : Metabolic pathways that function both catabolically and anabolically

Amphiphilic : Description of a molecule that has both a hydrophilic region and a hydrophobic region (e.g., lipopolysaccharide)

Amphitrichous : A cell with a single flagellum at each end

Amphotericin B : An antibiotic from a strain of *Streptomyces nodosus* that is used to treat systemic fungal infections; it also is used topically to treat candidiasis.

Ampicillin : A type of antibiotic, which is related to penicillin. It is commonly used in laboratories to test for uptake of genes by bacteria.

Amylase : An enzyme which converts starch to maltose.

Amyloid : Blue black when stained with Meltzers iodine

ANA : Antinuclear antibodies; anesthesia.

Anabolism (anabolic) : Energy consuming metabolic process of combining smaller molecules into larger molecules that is driven by ATP (see also catabolism). ***OR*** Metabolic processes involved in the synthesis of cell constituents from simpler molecules. An anabolic process usually requires energy.

Anaerobe/Anaerobic : An organism normally growing only in the absence of oxygen. ***OR*** An organism that lives and reproduces in the absence of dissolved oxygen, instead deriving oxygen from the breakdown of complex substances. ***OR*** (i) Absence of molecular oxygen. (ii) Growing in the absence of molecular oxygen, such as anaerobic bacteria. (iii) Occurring in the absence of molecular oxygen, as a biochemical process. ***OR*** An organism that grows in the absence of free oxygen

Anaerobic bacteria : Bacteria which grow in the absence of oxygen.

Anaerobic digestion : The microbiological treatment of sewage wastes under anaerobic conditions to produce methane.

Anaerobic respiration : An energy-yielding process in which the electron transport chain acceptor is an inorganic molecule other than oxygen. ***OR*** Metabolic process whereby electrons are transferred from an organic, or in some cases, inorganic compounds to an inorganic acceptor molecule other than oxygen. The most common acceptors are nitrate, sulfate, and carbonate.

Anaesthetic : A drug used to prevent people from feeling pain. It can affect the whole body (general anaesthetic) or a particular part (local anaesthetic).

ANAL : Analgesic.

Anammox process : The coupled use of nitrite as an oxidant and ammonium ion as a reductant under anaerobic conditions to yield nitrogen gas.

Anamnestic response : The recall, or the remembering, by the immune system of a prior response to a given antigen.

Anamorph : An asexual state of a fungus. ***OR*** Asexual stage of fungal reproduction in which cells are formed by the process of mitosis. ***OR*** The imperfect stage or asexual state of the fungus produced by mitosis.

Anaphase : The third stage of mitosis or meiosis in which the sister chromatids or chromosomes separate and move to the poles of the cell.

Anaphylactic shock : Anaphylactic shock or systemic anaphylaxis is an allergic reaction to systemically administered antigen that causes circulatory collapse and suffocation due to tracheal swelling. It results from binding of antigen to IgE antibody on connective tissue mast cells throughout the body, leading to the disseminated release of inflammatory mediators.

Anaphylatoxins : Anaphylatoxins are small fragments of complement proteins, released by cleavage during complement activation. These small fragments are recognized by specific receptors, and they recruit fluid and inflammatory cells to sites of their release. The fragments C_{5a}, C_{3a}, and C_{4a} are all anaphylatoxins, listed in order of decreasing potency in vivo.

Anaphylaxis : An immediate (type I) hypersensitivity reaction following exposure of a sensitized individual to the appropriate antigen. Mediated by reagin antibodies, chiefly IgE.

Anaplerotic reactions : Reactions that replenish depleted tricarboxylic acid cycle intermediates.

Anastomosing : Gills connected by cross-veins.

Anatomy : Structure of biological tissue. ***OR*** The study of body parts,their names,and locations.

Anaximander : Person who lived from 611 to 546 BC and is credited with the first written work on natural science, a classical poem entitled *"On Nature"*.

Anch residues : Peptide fragments of antigens are bound to specific MHC (Major Histocompatibility Complex) class I molecules by anchor residues. These are residues of the peptide that have amino acid side chains that bind into pockets lining the peptidebinding groove of the MHC class I molecule.

Each MHC class I molecule binds different patterns of anchor residues, called anchor motifs, giving some specificity to peptide binding. Anchor residues exist but are less obvious for peptides that bind to MHC class II molecules.

Androgens : The collective name for male sex hormones, including testosterone.

Anergy : A state of unresponsiveness to antigens. Absence of the ability to generate a sensitivity reaction to substances that are expected to be antigenic.

Anergy, anergic : Anergy is a state of nonresponsiveness to antigen. People are said to be anergic when they cannot mount delayed-type hyper-sensitivity reactions to challenge antigens, whereas T cells and B cells are said to be anergic when they cannot respond to their specific antigen under optimal conditions of stimulation.

ANES : Anesthesia.

Aneuploidy : Having an abnormal number of chromosomes.

ANG : Angle.

Angiogenesis : The growth of new blood vessels.

Angiospermae : Subdivision of plants within Division Spermatophyta which produce flowers and bear seeds enclosed in an ovary.

Angstrom (Å) : A unit of length equal to 10^{-10} m.

Anhydrobiosis : Life history strategy to tolerate desiccation in which organisms lose water, initiating lower physiological activity; upon rehydration the organism becomes active again.

Anhydrous : Very dry. No water or dampness is present.

ANI : Acute nerve irritation.

Anion : A negatively charged ion in an electrolyte solution, attracted to the anode under the influence of a difference in electrical potential. Chloride is an anion.

Anion exchange capacity : Sum total of exchangeable anions that a soil can adsorb. Expressed as centimoles of negative charge per kilogram of soil.

Annelid : Phylum of segmented aquatic and terrestrial worms. Each segment has its own excretory and sensory elements, with specialized functions (e.g. Reproduction) associated with specific segments. ***OR*** A specialized conidiogenous cell producing conidia in basipetal succession by a series of short percurrent proliferations (annellations). The tip of an annellide increases in length and becomes narrower as each subsequent conidium is formed.

Annelloconidium : A conidium produced by an annellide.

Annotation : The process of determining the location of specific genes in a genome map after it has been produced by nucleic acid sequencing.

Annulus : Ring of tissue on the stem left from a torn partial veil; collar

Anogenital condylomata : Warts that are sexually transmitted and caused by types 6, 11, and 42 human papillomavirus. Usually occur around the cervix, vulva, perineum, anus, anal canal, urethra, or glans penis.

Anomers : The sugar isomers that differ in configuration about the carbonyl carbon atom. This carbon atom is called the anomeric carbon atom of the sugar.

Anoxic : Literally "without oxygen." An adjective describing a microbial habitat devoid of oxygen.

Anoxygenic photosynthesis : Type of photosynthesis in green and purple bacteria in which oxygen is not produced. ***OR*** Type of photosynthesis in green and purple bacteria in which oxygen is not produced. ***OR*** Photosynthesis that does not oxidize water to produce oxygen; the form of photosynthesis characteristic of purple and green photosynthetic bacteria.

ANP : Atrial natriuretic peptide

ANRI : Acute nerve root irritation.

ANS : Autonomic nervous system.

ANT : Anterior.

Antacids : Chemicals consumed by people to react with stomach acid to change the pH of the stomach contents to a neutral pH.

Antagonist : Biological agent that reduces the number or disease producing activities of a pathogen.

Antagonist peptides : Antagonist peptides are peptides, usually closely related in sequence to an agonist peptide, that inhibit the response of a cloned T-cell line specific for the agonist peptide.

Anteri : Front or head end of an organism.

Anthelmintic : A chemical compound that kills or expels certain intestinal worm parasites.

Anther : "Male" plant structure in which male gametophytes(pollen) are produced and mature.

Antheridium : A male gamete producing organ, which may be unicellular or multicellular. ***OR*** Male gametangium found in the phylum Oomycota (Kingdom Stramenopila) and phylum Ascomycota (Kingdom Fungi).

Antheridiumanthe : A male gamete-producing organ, which may be unicellular or multicellular.

Anthrax : An infectious disease of animals caused by ingesting *Bacillus anthracis* spores. Can also occur in humans and is sometimes called woolsorter's disease.

Anthropogenic : Derived from human activities.

Anti diuretic hormone : A hormone secreted by the pituitary gland which affects kidney function and prevents water loss, the production of which is inhibited by alcohol consumption, potentially resulting in dehydration.

Antibacterial : A substance that can kill or prevent the growth of bacteria.

Antibiosis : Inhibition or lysis of an organism mediated by metabolic products of the antagonist; these products include lytic agents, enzymes, volatile compounds, and other toxic substances.

Antibiotic : Chemicals that kill bacteria and are used to treat infectious diseases. ***OR*** A chemical agent that is produced by one organism that is harmful to another organism. ***OR*** Organic substance produced by one species of organism that in low concentrations will kill or inhibit growth of certain other organisms. ***OR*** A chemical agent which directly or indirectly kills bacteria in a person's body.

Antibiotic resistance : When antibiotics are incorrectly used (to treat bacteria that are not sensitive to the specific antibiotic or to treat viruses, which never respond to antibiotics) and are not taken for the full term prescribed (usually from 5 to 21 days, depending on the specific antibiotic and disease being treated), surviving pathogenic organisms develop immunity to the antibiotic and pass it along to descendants and might choose to pass the trait along to unrelated bacteria via a process known as horizontal gene transfer.

Antibodies : Special microbe-fighting proteins in your blood. By attaching to invading microbes, they help white blood cells spot invaders and 'eat' them up. ***OR*** Protein that is produced by animals in response to the presence of an antigen and that can combine specifically with that antigen. A glycoprotein produced in response to the introduction of an antigen; it has the ability to combine with the antigen that stimulated its production. Also known as an immunoglobulin (Ig).

Antibody : A chemical produced by B-cells to "tag" foreign invaders.

Antibody combining site : Antibody combining site: see antigen-binding site.

Antibody repertoire : The antibody repertoire or immunoglobulin repertoire describes the total variety of antibodies in the body of an individual.

Antibody-dependent cell-mediated cytotoxicity (ADCC) : The killing of antibody-coated target cells by cells with Fc receptors that recognize the Fc region of the bound antibody. Most ADCC is mediated by NK cells that have the Fc receptor or CD16 on their surface.

Antibody mediated immunity : See Humoral immunity.

Anticodon triplet : The base triplet on a tRNA that is complementary to the triplet codon on mRNA.

Anticodon : A sequence of three bases on the transfer RNA that pair with the bases in the corresponding codon on th messenger RNA.

Antigen : Substance that can incite the production of a specific antibody and that can combine with that antibody. A foreign (nonself) substance (such as a protein, nucleoprotein, polysaccharide, or sometimes a glycolipid) to which lymphocytes respond; also known as an immunogen because it induces the immune response. ***OR*** An antigen is any molecule that can bind specifically to an antibody. Their name arises from their ability to generate antibodies. However, some antigens do not, by themselves, elicit antibody production; those antigens that can induce antibody production are called immunogens. ***OR*** Substance that can incite the production of a specific antibody and that can combine with that antibody

Antigen display libraries : Both libraries of cDNA clones in expression vectors and bacteriophage libraries encoding random peptide sequences have been used to identify the targets of specific antibodies and, in some cases, of T cells. Such libraries are termed antigen display libraries.

Antigen presentation : Antigen presentation describes the display of antigen as peptide fragments bound to Major Histocompitibility Complex (MHC) molecules on the surface of a cell; T cells recognize antigen when it is presented in this way.

Antigen receptors : T and B lymphocytes collectively bear on their surface highly diverse antigen receptors capable of recognizing a wide diversity of antigens. Each individual lymphocyte bears receptors of a single antigen specificity.

Antigen:antibody complexes : Antigen:antibody complexes are noncovalently associated groups of antigen and antibody molecules that can vary in size from small soluble complexes to large insoluble complexes that precipitate out of solution; they are also known as immune complexes.

Antigenbinding fragment (Fab) : "Fragment antigen binding." A monovalent antigenbinding fragment of an immuno-globulin molecule that consists of one light chain and part of one heavy chain, linked by interchain disulfide bonds.

Antigen-binding site, antibody combining site : The antigen-binding site of an antibody, or antibody combining site, is found at the surface of the antibody molecule that makes physical contact with the antigen. Antigen-binding sites are made up of six hypervariable loops, three from the light-chain V region and three from the heavy-chain V region.

Antigen-capture assay : In an antigen-capture assay, the antigen binds to a specific antibody, and its presence is detected by a second antibody that must be labeled and directed at a different epitope.

Antigenic determinant : An antigenic determinant is the portion of an antigenic molecule that is bound by a given antibody or antigen receptor; it is also known as an epitope.

Antigenic drift, antigenic shift : Influenza virus varies from year to year by a process of antigenic drift in which point mutations of viral genes cause small differences in the structure of viral surface antigens. Periodically, influenza viruses undergo an antigenic shift through reassortment of their segmented genome with another influenza virus, changing their surface antigens radically. Such antigenic shift variants are not recognized by individuals immune to influenza, so when antigenic shift ***OR*** A small change in the antigenic character of an organism that allows it to avoid attack by the immune system variants arise, there is widespread and serious disease.

Antigenic variation : Many pathogens evade the adaptive immune response by antigenic variation in which new antigens are displayed that are not recognized by antibodies or T cells elicited in earlier infections.

Antigen-presenting cells : Antigen presenting cells (apcs) are cells that take in protein antigens, process them, and present antigen fragments to B cells and T cells in conjunction with class II MHC molecules so that the cells are activated. Macrophages, B cells, dendritic cells, and Langerhans cells may act as apcs.

Antigens : Foreign substances (eg bacteria or viruses) in the body that are capable of causing disease. The presence of antigens in the body triggers an immune response, usually the production of antibodies.

Antigens : Substances on the surfaces "foreign" cells that our immune systems use to distinguish "me"from a foreign invader.

Anti-immunoglobulin antibodies, anti-isotype antibodies, anti-allotype antibodies, anti-idiotype antibodies : Anti-immunoglobulin antibodies are antibodies against immunoglobulin constant domains, useful for detecting bound antibody molecules in immunoassays and other applications. These can be divided into anti-isotype antibodies made in a different species, anti-allotype antibodies made in the same species against allotypic variants, and anti-idiotype antibodies, made against unique determinants to a single antibody.

Anti-inflammaty : A substance that reduces or prevents inflammation.

Anti-lymphocyte globulin : Anti-lymphocyte globulin is antibody raised in another species against human T cells.

Antimetabolite : A compound that blocks metabolic pathway function by competitively inhibiting a key enzyme's use of a metabolite because it closely resembles the normal enzyme substrate.

Antimetabolite : A compound that blocks metabolic pathway function by competitively inhibiting a key enzyme's use of a metabolite because it closely resembles the normal enzyme substrate.

Antimicrobial : A substance that stops the growth of microorganisms including bacteria, viruses, and fungi.

Antimicrobial agent : An agent that kills microorganisms or inhibits their growth.

Antioxidant : A chemical that prevents oxidation, including the damage caused by reactive oxygen species (ROS). Examples of antioxidants are Vitamin C, Vitamin E and selenium. Some enzymes, like catalase, glutathione peroxidase and superoxide dismutase, are also antioxidants.

Antiparallel ?-pleated sheet (?-sheet) : A hydrogen bonded secondary structure formed between two or more extended polypeptide chains.

Antisense RNA : A single-stranded RNA with a base sequence complementary to a segment of another RNA molecule that can specifically bind to the target RNA and inhibit its activity.

Antisepsis : The prevention of infection or sepsis.

Antiseptic : Agent that kills or inhibits microbial growth but is not harmful to human tissue.

Antiserum, antisera : An antiserum (plural: antisera) is the fluid component of clotted blood from an immune individual that contains antibodies against the molecule used for immuni-zation. Antisera contain heterogeneous collections of antibodies, which bind the antigen used for immunization, but each has its own structure, its own epitope on the antigen, and its own set of cross-reactions. This heterogeneity makes each antiserum unique.

Antitoxin : An antibody to a microbial toxin, usually a bacterial exotoxin, that combines specifically with the toxin, *in vivo* and *in vitro*, neutralizing the toxin.

Antivenin : Sites by poisonous snakes can be treated by identification of the snake and injection of an antivenin specific for that snake's venom.

Anton van Leeuwenhoek : Person who lived in the 1600s and was the first person to see sperm cells, in which he thought he could see a homunculus.

ANX : Anxiety.

AO : Angle of; aorta.

AOB : Ammonia-oxidizing bacteria

AOB : Alcohol on breath.

AOD : Arteriosclerotic occlusive disease.

AOH : Anterior pituitary hormone; aphasia.

AOM : Acute otitis media.

Aotearoa : Translates as Land of the Long White Cloud; an alternative name for New Zealand.

AP : After parturition; anteroposterior; angina pectoris.

AP & L : Anteroposterior and lateral.

APB : Atrial premature beat.

APC : Antigen-presenting cell.

APC : Aspirin; phenacetin; caffeine.

API : Active pharmaceutical ingredient (dict.) ***OR*** annual parasite incidence ***OR*** application program (ming) interface.

API separat : A facility developed by the Committee on Disposal or Refinery Wastes of the American Petroleum Institute for separation of oil from wastewater in a gravity differential and equipped with means for recovering the separated oil and removing sludge.

Apical complex : A complex set of organelles found at the anterior end of the protozoans of the Phylum Apicomplexa. The apical complex has a role in the penetration of host cells.

Apical pore : In certain Gastromycetes, the mouth at the top of the spore case through which the spores are released; in spores, a germ pore.

Apicomplexa : Division of Protista which includes many parasites such as *Plasmodium vivax* which causes malaria.

Apicomplexan : A sporozoan protist that lacks special locomotor organelles but has an apical complex and a spore-forming stage. It is either an intra- or extracellular parasite of animals; a member of the phylum Apicomplexa.

Apiculus : The projection on a spore by which it is attached to the sterigma (spore stalk).

Aplanospe : A nonflagellated, nonmotile spore that is involved in asexual reproduction.

Aplastic anemia : Aplastic anemia is a failure of bone marrow stem cells so that formation of all cellular elements of the blood ceases; it can be treated by bone marrow transplantation.

Apoactivat : A regulatory protein that stimulates transcription from one or more genes in the presence of a coactivator molecule.

Apoenzyme : The protein part of an enzyme that also has a nonprotein component.

Apophysis : A swelling. The term is primarily applied to the funnel-shaped swelling of a sporangiophore, immediately below the columella, seen in some zygomycetes.

Apoptosis : Programmed cell death. The fragmentation of a cell into membranebound particles that are eliminated by phagocytosis. Apoptosis is a physiological suicide mechanism that preserves homeostasis and occurs during normal tissue turnover. It is responsible for cell death in pathological circumstances, such as exposure to low concentrations of xenobiotics and infections by HIV and various other viruses.

Apepress : An inactive form of the repressor protein, which becomes the active repressor when the corepressor binds to it.

Apothecium : Open ascoma of fungi in the phylum Ascomycota.

APP : Amyloid β-protein precursor

APP : Appears; appendix; appointment; approximate.

AP-PCR : Arbitrarily primed polymerase chain reaction.

Appendiculate : Used to describe the cap margin when hung with veil fragments.

Appendix : The appendix is a gut-associated lymphoid tissue located at the beginning of the colon.

Appressed : Flattened down on a surface

Aquatic : Growing or living in or near water (either fresh water or marine).

AR : Active resistance; admitting room; apical rate; artificial respiration; assisted respiration; at risk.

Arable : Land that is used for farming.

Araceae : A family of monocots whose flowers are in the form of a spathe and spadix.

Arboviruses : A large a diverse group of viruses, taxonomically unrelated which are classically transmitted by arthropod vectors, e.g. Mosquitoes, ticks, etc.

Arbuscular mycorrhiza (AM) : Mycorrhizal type that forms highly branched arbuscules within root cortical cells.

Arbuscule : Special "treeshaped" structure formed within root cortical cells by arbuscular mycorrhizal fungi.

ARC : Aids-related complex.

Archaea : The domain that contains procaryotes with isoprenoid glycerol diether or diglycerol tetraether lipids in their membranes and archaeal rRNA (among many differences).

Archaea : Evolutionarily distinct group (domain) of prokaryotes consisting of the methanogens, most extreme halophiles and hyperthermo-philes, and Thermoplasma.

Archaebacteria : Older term for the Archaea.

ARD : Acute respiratory disease.

ARE : Active-resistance exercise.

Areolate : Cap surface cracked into plaques or blocks (like dried mud).

ARF : Acute respiratory failure; acute rheumatic fever.

Arg : Arginine.

ARI : Anxiety reaction, intense.

Aristotle : Person who lived from 343 to 322 BC,and contributed ideas such as the *Scala Naturae* and the idea that the Earth was round and surrounded by the other spheres of matter.

ARM : Anxiety reaction, mild.

Armed effect T cells : In this book, we have termed primed effector T cells armed effector T cells, because these cells can be triggered to perform their effector functions immediately on contact with cells bearing the peptide: MHC complex for which they are specific. They contrast with memory T cells, which need to be activated by antigen-presenting cells to differentiate into effector T cells before they can mediate effector responses.

ARMG : Antibiotic resistance marker gene

Arnium (Ascomycetes) : Fungus commonly isolated from dung. Not generally associated with human disease and is most often considered benign.

AROM : Active range of motion.

Aromatic : Organic compounds which contain a benzene ring, or a ring with similar chemical characteristics.

Arrested development : See hypobiosis.

ARS : Autonomously replicating sequence

ARS : Autonomously replicating segment ***OR*** autonomously replicating sequence.

ART : Achilles tendon reflex test; artery; articulation; artificial.

Arteries : The blood vessels that carry blood from the heart to the rest of the body.

ARTH : Arthritic.

Arthric : A type of conidial ontogeny involving the conversion and subsequent disarticulation of a determinant conidiogenous hypha.

Arthrinium (Hyphomycetes) : A cosmopolitan fungus isolated from plant debris and soil. Often isolated from air near grassy places, but rarely found indoors. Not reported to cause human or animal infection.

Arthritis : Inflammation of joint, causing pain, swelling, and stiffness.

Arthroconidium : A thallic conidium released by the fragmentation or lysis of hypha. It is not notably larger than the parental hypha, and separation occurs at a septum.

Arthropod : Invertebrate with jointed body and limbs (includes insects, arachnids, and crustaceans).

Arthrospore : A spore resulting from the fragmentation of a hypha.

Arthus reaction : The Arthus reaction is a skin reaction in which antigen is injected into the dermis and reacts with IgG antibodies in the extracellular spaces, activating complement and phagocytic cells to produce a local inflammatory response.

Artificially acquired active immunity : The type of immunity that results from immunizing an animal with a vaccine. The immunized animal now produces its own antibodies and activated lymphocytes.

Artificially acquired passive immunity : The type of immunity that results from introducing into an animal antibodies that have been produced either in another animal or by *in vitro* methods. Immunity is only temporary.

AS : Antisense.

AS : Anal sphincter; auris sinistra (left ear); arteriosclerosis; astigmatism.

ASA : Allele-specific amplification

ASA : Acetylsalicylic acid; aspirin.

ASC : Arteriosclerosis; ascending.

ASCAD : Arteriosclerotic coronary artery disease.

Ascertainment artifact : Ascertainment artifact refers to data that seem to demonstrate some finding, but fail to do so because they are collected from a population that is selected in a biased fashion.

Asclepiadaceae : Plant family with milky juice and flowers with a corona and reflexed petals.

Ascocarp : A multicellular structure in ascomycetes lined with specialized cells called asci in which nuclear fusion and meiosis produce ascospores. An ascocarp can be open or closed and may be referred to as a fruiting body.

Ascogenous hypha : A specialized hypha that gives rise to one or more asci.

Ascogonium : The receiving (female) organ in ascomycetous fungi which, after fertilization, gives rise to ascogenous hyphae and later to asci and ascospores.

Ascoma (plural, ascomata) : Fungal fruiting body that contains ascospores; also termed an ascocarp.

Ascomycete : Name of fungal group that all bear spores in sac-like cells called asci (see ascus).

Ascomycetes : Constitutes the largest class of fungi characterized by the production of sexual spores in structures called asci. This includes plant pathogens, saprobes, and decomposers. With a few exceptions, most Ascomycetes do not grow in buildings and are seldom agents of wood rot. Ascomycetes are the perfect stages of molds like *Aspergillus* and *Penicillium*. At high levels, Ascomycetes spores may cause allergies. Since most Ascomycetes are plant pathogens, ascospores are common during the growing season of plants and rare during winter, such as those of the Ascomycetes genera: *Daldinia*, *Hypoxylon*, *Paraphaeosphaeria*, *Phaeosphaeria*, and *Leptosphaeria*.

Ascomycetous : Referring to the Ascomycetes.

Ascorbic acid : A chemical name for vitamin C—when present in adequate amounts,this prevents scurvy.

Ascospore : Spores resulting from karyogamy and meiosis that are formed within an ascus. Sexual spore of the Ascomycota.

Ascospores : Sexual spores produced by Ascomycetes.

Ascus : The cylindrical or club-shaped spore-producing cell that characterises the Ascomycetes. Contains (usually 8) ascospores.

ASCVD : Arteriosclerotic cardiovascular disease; atherosclerotic coronary artery disease.

ASD : Atrial septal defect.

Aseptate : Lacking septa, often pertaining to the hyphae seen in zygomycetes (also see coenocytic).

Aseptic : Free from living germs of disease, fermentation or putrefaction.

Aseptic meningitis syndrome : See meningitis.

Aseptic technique : Manipulating sterile instruments or culture media in such a way as to maintain sterility.

Aseptical : With out bacteria,sterile,under sterile conditions.

Aserotolerant anaerobes : Microbes that grow under both aerobic and anaerobic conditions, but do not shift from one mode of metabolism to another as conditions change. They obtain energy exclusively by fermentation.

Asexual : Reproduction without cell division.

Asexual reproduction : Growth and cell duplication that does not involve the union of nuclei from cells of opposite mating types.

ASHD : Arteriosclerotic heart disease.

ASIC : Acid-sensing ion channel.

ASIS : Anterior superior iliac spine.

ASL : Alternative splicing library.

Asn : Asparagine.

ASO : Antisense oligonucleotide.

Asp : Aspartic acid.

AS-PCR : Allele-specific PCR.

Aspect : The overall shape of a fruiting body as determined by the portions of its parts.

Aspergillus (Hyphomycetes) : A fungal disease caused by species of *Aspergillus*. ***OR*** Teleomorph: *Emericella* (Ascomycetes), *Eurotium* (Ascomycetes) Found in soil, compost piles, decaying vegetation, stored grain, and other kinds of organic matter. Can be found indoors in water-damaged buildings. A few species can cause aspergillosis in humans with compromised or defective immune systems. Most people are naturally immune to this infection of the lung. *Aspergillus fumigatus* is the most common cause of aspergillosis, followed by *A. flavus* and *A. niger*. Some species are able to produce mycotoxins, depending on the strain, substrate, and/or food source. Others species are used in the manufacture of food, such as *A. oryzae* or *A. soyae* for soy sauce.

Aspergillus caesiellus **:** This species is only occasionally pathogenic.

Aspergillus candidus **:** Aw 0.75. Conidia dimensions 2.5-4 microns. Found in warm soils, grain and in the secondary decay of vegetation. Associated with respiratory complaints in a recent house investigation. Can produce the toxin petulin which may be associated with disease in humans and other animals.

Aspergillus carneus **:** This species is only occasionally pathogenic.

Aspergillus clavatus **:** Conidia dimensions 3-4.5 x 2.5-4.5 microns. Found in soils and animal manure. Can produce the toxin petulin which may be associated with disease in humans and other animals. This species is only occasionally pathogenic.

Aspergillus deflectus **:** This species is only occasionally pathogenic.

Aspergillus flavus **:** Aw 0.78. Conidia dimensions 3-6 microns or 3-5 microns. It grows on moldy corn and peanuts. It can be found in warm soil, foods and dairy products. Some strains are capable of producing a group of mycotoxins - in the aflatoxin group. Aflatoxins are known animal carcinogen. There is limited evidence to suggest that this toxin is a human carcinogen. The toxin is a poisonous to humans by ingestion. It may also result in occupational disease via inhalation. Experiments have indicated that it is teratogenic and mutagenic. It is toxic to the liver. It is reported to be allergenic. Its presence is associated with reports of asthma. It can be found in water damaged carpets. The production of the fungal toxin is dependent on the growth conditions and on the substrate used as a food source. This fungus is associated with aspergillosis of the lungs andor disseminated aspergillosis. This fungus is occasionally identified as the cause of corneal, otomycotic and nasoorbital infections.

Aspergillus fumigatus : Aw 0.82; Optimum > 0.97. Conidia dimensions 2-3.5 microns. Major cause of aspergillosis. Both invasive and allergic aspergillosis are caused by this organism. Aspergillosis affects individuals who are immune compromised. It is considered a human pathogen. It grows well at 35 degrees C. It is commonly found outdoors in compost piles with temperatures higher than 40 degrees C, in mild to warm soils and on cereals.

Aspergillus glaucus : Conidia dimensions 5-6.5 microns. Common outdoor fungus in the winter. It is reported to be allergenic. This species is only occasionally pathogenic. It can grow on leather. This fungus can grow at low moisture levels on grains, sugary food products, meat and wool. The ascomycetous state is Eurotium sp.

Aspergillus nidulans : Aw 0.78. Conidia dimensions 2-4 microns. Found in mild to warm soils and on slowly decaying plants. Can produce the mycotoxin sterigmatocystin. This toxin has been shown to produce liver and kidney damage in lab animals. This fungus is associated with aspergillosis of the lungs andor disseminated aspergillosis. This species is only occasionally pathogenic.

Aspergillus niger : Aw 0.77; Optimum > 0.97. Conidia dimensions 3.5 - 5 microns or 4 to 5 microns. Less common cause of aspergillosis. Fungus has a musty odor. It is commonly found in the environment on textiles, in soils, grains, fruits and vegetables. It has been reported to cause skin and pulmonary infections. It is a common cause of fungal related ear infections-otomycosis.

Aspergillus ochraceus : Aw 0.77. Conidia dimensions 2.5 - 3 microns. Found in grains, soil and salted food products. It is not usually associated with decaying vegetation. Can produce a kidney toxin ochratoxin A which may produce ochratoxicosis in humans. This is also known as Balkan nephropathy. The toxin is produced at optimum growth conditions at 25 degrees C and high moisture conditions. The ochratoxin may also be produced by other Aspergillus sp. And Penicillium sp. Other toxins which can be produced by this fungus include penicillic acid, xanthomegnin and viomellein. These are all reported to be kidney and liver toxins.

Aspergillus yzae : This species is only occasionally pathogenic.

Aspergillus parasiticus : Some strains are capable of producing a group of mycotoxins - in the aflatoxin group. Aflatoxins are known animal carcinogen. There is limited evidence to suggest that this toxin is a human carcinogen. The toxin is a poisonous to humans by ingestion. Experiments have indicated that it is teratogenic and mutagenic. It is toxic to the liver. The production of the fungal toxin is dependent on the growth conditions and on the substrate used as a food source.

Aspergillus penicilloides **:** Conidia dimensions 3-3.5 x 4-5 microns. Can grow in areas with low water activity. It is found in house dust and food.

Aspergillus restrictus **:** This species is only occasionally pathogenic.

Aspergillus sp. **:** Aw 0.75 - 0.82 Reported to be allergenic. Members of this genus are reported to cause ear infections. Many species produce mycotoxins which may be associated with disease in humans and other animals. Toxin production is dependent on the species or a strain within a species and on the food source for the fungus. Some of these toxins have been found to be carcinogenic in animal species. Several toxins are considered potential human carcinogens. Common cause of extrinsic asthma (immeadiate-type hypersensitivity: type I). Acute symptoms include edema and bronchiospasms, chronic cases may develop pulmonary emphysema .

Aspergillus sydowi **:** This species is only occasionally pathogenic.

Aspergillus terreus **:** Aw 0.78. Conidia dimensions 1.8-2.4 microns or 2-2.5 microns. Aleurospores 6-7 microns in diameter are also produced. Found in warmer soil and in grains, straw. Cotton and decomposing vegetation. Can produce the toxin patulin and citrinin which may be associated with disease in humans and other animals. This fungus is associated with aspergillosis of the lungs and or disseminated aspergillosis. Found as an isolate from otomycosis-ear infection and onychomycosis-infection of finger or toe nails.

Aspergillus ustus **:** This species is only occasionally pathogenic.

Aspergillus versicol **:** Aw 0.78. Conidia dimensions 2-3.5 microns. It is commonly found in soil, hay. Cotton and dairy products, It can produce a mycotoxin sterigmatocystin and cyclopiaxonic acid. These toxins can cause diarrhea and upset stomach. It is reported to be a kidney and liver carcinogen. This species is only occasionally pathogenic.

ASR : Analyte specific reagent.

ASS : Anterior superior spine.

Assay : A test, usually carried out in a laboratory, to determine the concentration of a particular substance. In biology, this often refers to a test to determine the activity of a biological molecule, for example, an enzyme.

Assembly : The stage of replication during which all the structural components come together at one site in the cell and the basic structure of the virus particle is formed.

ASSET : Alternative splicing sequence enriched tag

Assimilable ganic carbon (aoc) : The fraction of total organic carbon in water that can be used for microbiological growth. It is measured by a growth potential assay.

Assimilate : To take in, similar to eating food.

Assimilaty nitrate reduction : Conversion of nitrate to reduced forms of nitrogen, generally ammonium, for the synthesis of amino acids and proteins.

Assimilaty reduction : The reduction of an inorganic molecule to incorporate it into organic material. No energy is made available during this process.

Associative dinitrogen fixation : Close interaction between a freeliving diazotrophic organism and a higher plant that results in an enhanced rate of dinitrogen fixation.

Associative symbiosis : Close but relatively casual interaction between two dissimilar organisms or biological systems. The association may be mutually beneficial but is not required for accomplishment of a particular function.

AST : Astigmatism.

Aster : Microtubules originating around centrioles in dividing animal cells, from which the mitotic spindle eventually forms.

Asteroid : Smaller, planet-like objects whose orbits lie between Mars and Jupiter.

Asthma : An allergy-induced constriction of the bronchiole muscles, thereby reducing the air passages.

Asymmetric carbon : A carbon that is covalently bonded to four different groups.

AT : Anterior tibial; atlas.

Ataxia telangiectasia : Ataxia telangiectasia is a disease characterized by staggering, multiple disorganized blood vessels, and an immuno deficiency in a protein called ATM, which contains a kinase thought to be important in signaling of double-stranded DNA breaks.

ATCase : Aspartate transcarbamoylase.

ATCC : American type culture collection.

Atherosclerosis : A hardening of arteries caused by a build up of fatty substances, which may limit blood flow.

Athlete's foot : See tinea pedis.

Atm : Atmosphere.

Atmosphere : The gaseous mass(air) surrounding a planet.

ATN : Acute tubular necrosis.

Atom : The smallest unit of matter. ***OR*** One particle,one piece of an element.

Atomic force microscope : A type of scanning probe microscope that images a surface by moving a sharp probe over the surface at a constant distance; a very small amount of force is exerted on the tip and probe movement is followed with a laser.

Atomic structure : The way in which atoms are arranged in a molecule.

Atopic allergy, atopy : Atopic allergy, or atopy, is the increased tendency seen in some people to produce immediate hypersensitivity reactions (usually mediated by IgE antibodies) against innocuous substances.

ATP : Adenosine triphosphate. ***OR*** Adenosine triphosphate. Chemical energy generated by substrate oxidations is conserved by formation of high-energy compounds such as adenosine diphosphate (ADP) and adenosine triphosphate (ATP) or compounds containing the thioester bond. ***OR*** "Adenosine triphosphate". A relatively stable, high energy molecule used to fuel chemical reactions within cells.

ATP : Attending physician.

ATP-binding cassette transpters (ABC transpters) : Membrane protein complexes that use ATP energy to move substances across membranes without modifying the compound being transported. They require an extracytoplasmic substrate-binding protein for proper function.

ATR : Achilles tendon reflex; atrophy.

Atrophy : Wasting away of a tissue or organ.

ATS : Anxiety tension state; arteriosclerosis.

Attached growth processes : Wastewater treatment processes in which the microorganisms and bacteria treating the wastes are attached to the media in the reactor. The wastes being treated flow over the media. Trickling filters, bio-towers, and rbcs are attached growth reactors. These reactors can be used for removal of BOD, nitrification, and denitrification.

Attachment : The binding of a virus particle to a specific receptor on the surface of a host cell.

Attenuation : 1. A mechanism for the regulation of transcription of some bacterial operons by aminoacyl-trnas. 2. A procedure that reduces or abolishes the virulence of a pathogen without altering its immuno-genicity. ***OR*** Pathogens are said to be attenuated when they can grow in their host and induce immunity without producing serious clinical disease.

Attenuat : A provisional transcription stop signal.

AU : Both ears.

AUC : Analytical ultracentrifugation.

Aureobasidium (Hyphomycetes) : A common saprobe frequently isolated from soil, plant surfaces, seeds, grains, fruits and other food, human skin, and nails. Common indoors in humid areas such as bathrooms, kitchens, poorly maintained HVAC systems, and window frames. Allergies to Aureobasidium are common but infections are rarely reported.

Australian realm : The biogeographical realm including the continent of Australia and some of the surrounding islands.

Autecology : The study of the relationship of an individual or a species to its environment.

Autism : Autism is a disorder of brain function that appears early in life, generally before the age of three. Children with autism have problems with social interaction, communication, imagination and behaviour. Autistic traits persist into adulthood, but vary in severity.

Autoantibodies : Antibodies specific for self antigens are called autoantibodies.

Autodigestion : Self digesting or liquefying. Characteristic of the genus *coprinus*.

Autogenic factors : Causes of succession due to the plants, themselves

Autogenous Infection : An infection that results from a patient's own microbiota, regardless of whether the infecting organism became part of the patient's microbiota subsequent to admission to a clinical care facility

Autograft : A graft of tissue from one site to another on the same individual is called an autograft.

Autoimmune : When a person's immune system mistakenly attacks normal, healthy tissues in their body.

Autoimmune disease : Diseases in which the pathology is caused by adaptive immune responses to self antigens are called autoimmune diseases.

Autoimmune hemolytic anemia : Autoimmune hemolytic anemia is a pathological condition with low levels of red blood cells (anemia), which is caused by autoantibodies that bind red blood cell surface antigens and target the red blood cell for destruction.

Autoimmune response, autoimmunity : An adaptive immune response directed at self antigens is called an autoimmune response; likewise, adaptive immunity specific for self antigens is called autoimmunity.

Autoimmune thrombocytopenic purpura : In the disease autoimmune thrombo-cytopenic purpura, antibodies against a patient's platelets are made. Antibody binding to platelets causes them to be taken up by cells with Fc receptors and complement receptors, causing a fall in platelet counts that leads to purpura (bleeding).

Autoimmunity : Autoimmunity is a condition characterized by the presence of serum autoantibodies and selfreactive lymphocytes. It may be benign or pathogenic. Autoimmunity is a normal consequence of aging; is readily inducible by infectious agents, organisms, or drugs; and is potentially reversible in that it disappears when the offending "agent" is removed or eradicated.

Autolysin : Autolytic enzyme important in enzymatically catabolizing peptidoglycan in preexisting cell walls so that new cell wall could be "laid down" during bacterial cell division; also important component in cell lysis and resulting cell death during betalactam antibiotic inhibition of cell wall synthesis

Autolysins : Enzymes that partially digest peptido-glycan in growing bacteria so that the peptidoglycan can be enlarged

Autonomic NS : The functional part of the NS which controls visceral and cardiac muscles ;includes sympathetic and parasympathetic NS.

Autonomy : The ability to make decisions for oneself. A person acts autonomously when such decisions can be put into effect.

Autoradiography : The technique of exposing film in the presence of disintegrating radioactive particles. Used to obtain information on the distribution of radioactivity in a gel or a thin cell section. ***OR*** Detecting radioactivity in a sample, such as a cell or gel, by placing it in contact with a photographic film.

Autoregulation : The process in which a gene regulates its own expression.

Autosome : All the chromosomes other than the X and Y chromosomes. ***OR*** A chromosome that is not directly involved in determining the sex of the organism. Humans have 23 pairs of chromosomes in their cells; 22 pairs are autosomal chromosomes and one pair is the sex chromosomes (XX in females, XY in males).

Autotroph : An organism,such as a plant,which makes its own food. ***OR*** Organism which uses carbon dioxide as the sole carbon source.

Autotrophic nitrification : Oxidation of ammonium to nitrate through the combined action of two chemoauto-trophic organisms, one forming nitrite from ammonium and the other oxidizing nitrite to nitrate.

Autotrophy : A unique form of metabolism found only in bacteria. Inorganic compounds (e.g., NH_3, NO_2^-, S_2, and Fe^{2+}) are oxidized directly (without using sunlight) to yield energy. This metabolic mode also requires energy for CO_2 reduction, like photosynthesis, but no lipid-mediated processes are involved. This metabolic mode has also been called chemotrophy, chemoautotrophy, or chemolithotrophy.

Auxin : A plant growth hormone usually concentrated in the apical bud.

Auxotroph : A mutated prototroph that lacks the ability to synthesize an essential nutrient and therefore must obtain it or a precursor from its surroundings.

Auxotroph : A mutated prototroph that lacks the ability to synthesize an essential nutrient and therefore must obtain it or a precursor from its surroundings.

AV : Accurate volume.

AV : Aortic valve; avulsion.

AVG : Average.

AVH : Acute viral hepatitis.

Avidity : Avidity is the sum total of the strength of binding of two molecules or cells to one another at multiple sites. It is distinct from affinity, which is the strength of binding of one site on a molecule to its ligand.

Avidity hypothesis : The avidity hypothesis of T-cell selection in the thymus states that T cells must have a measurable affinity for self MHC molecules in order to mature, but not so great an affinity as to cause activation of the cell when it matures, as this would require that the cell be deleted to maintain self tolerance.

Avogadro's number : The number of molecules in a gram molecular weight of any compound (6.023×10^{23}).

AW : Alive and well; able to work.

AWA : As well as.

AWT : Advanced Waste Treatment - any process of water renovation that upgrades treated wastewater to meet reuse requirements.

AX : Axial: axillary.

Axenic : Literally "without strangers." A system in which all biological populations are defined, such as a pure culture.

Axial filament : The organ of motility in spirochetes. It is made of axial fibrils or periplasmic flagella that extend from each end of the protoplasmic cylinder and overlap in the middle of the cell. The outer sheath lies outside the axial filament.

Azathioprine : Azathioprine is a potent immuno-suppressive drug that is converted to its active form *in vivo* and then kills rapidly proliferating cells, including lymphocytes responding to grafted tissues.

Azimuth : The distance in angular degrees in a clock wise direction from north.

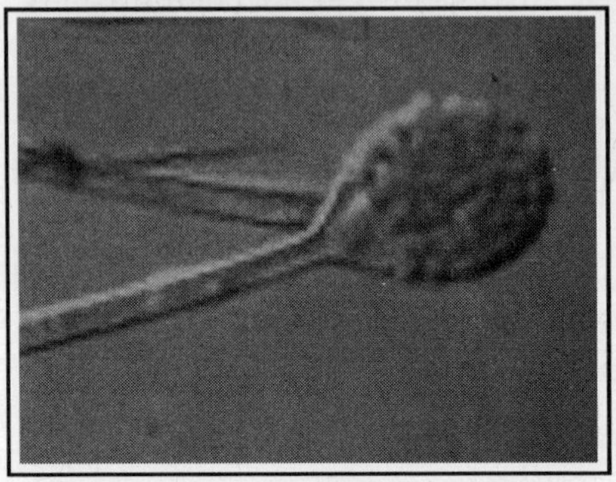

Absidia

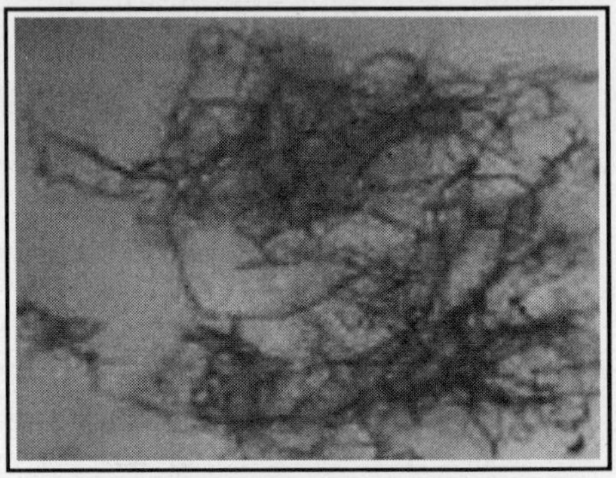

Acid fast bacteria

Acromonium

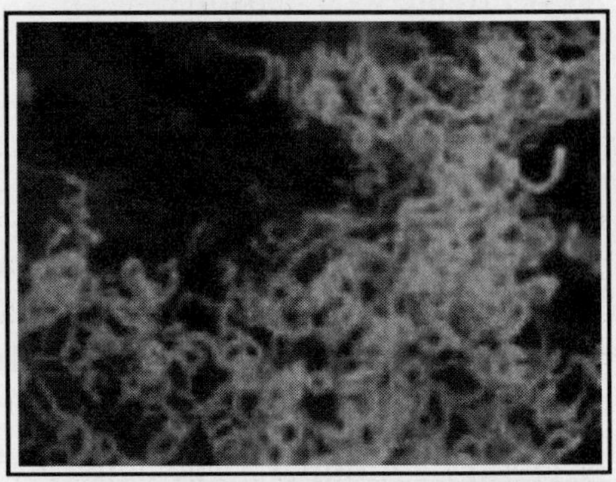

Actinomycetes

Agaric

Algae

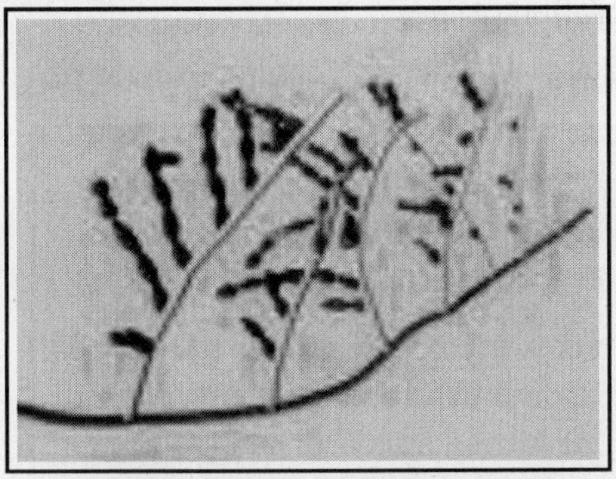

Alternaria

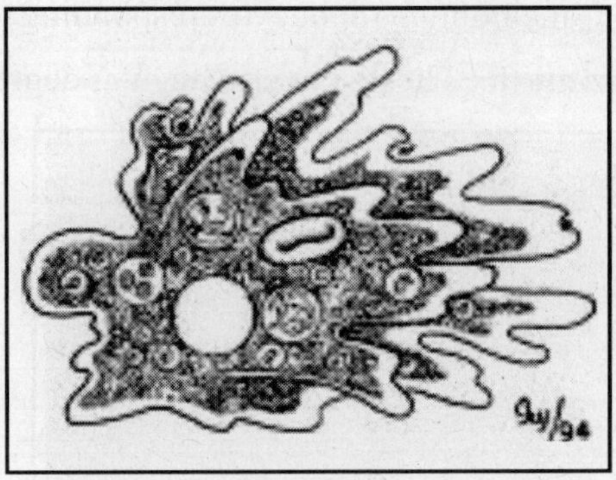

Amoeba

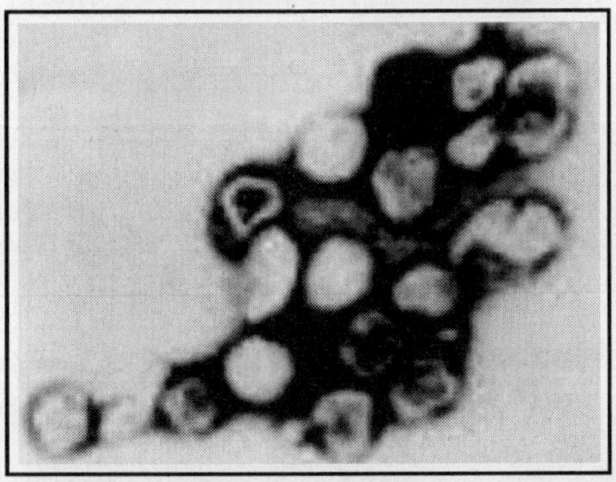

Arbovirus

Arbuscular mycorrhiza

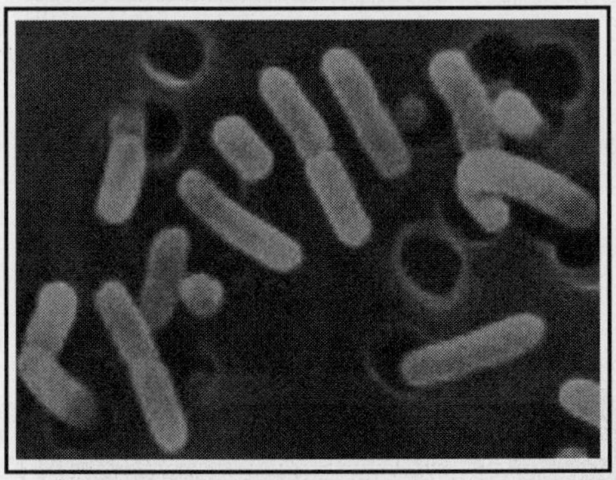

Archaea

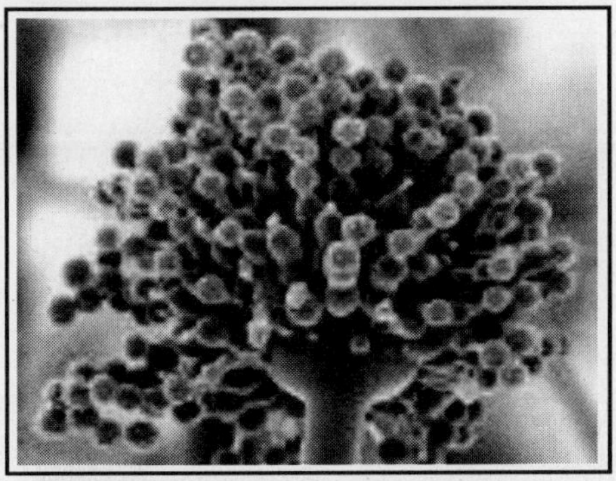

Aspergillus

B

B : Born.

B cell, also known as a B lymphocyte : One of the major types of cells in the immune system. B cells can differentiate to form memory cells or antibody-forming cells. ***OR*** A type of lymphocyte derived from bone marrow stem cells that matures into an immunologically competent cell under the influence of the bursa of Fabricius in the chicken and bone marrow in nonavian species. Following interaction with antigen, it becomes a plasma cell, which synthesizes and secretes antibody molecules involved in humoral immunity.

B cell, B lymphocyte, B-1 cells, B-2 cells : A B cell, or B lymphocyte, is one of the two major types of lymphocyte. The antigen receptor on B lymphocytes, usually called the B-cell receptor, is a cell-surface immunoglobulin. On activation by antigen, B cells differentiate into cells producing antibody molecules of the same antigen specificity as this receptor. B cells are divided into two classes. B-1 cells, also known as CD5 B cells, are a class of atypical, self-renewing B cells found mainly in the peritoneal and pleural cavities in adults. They have a far less diverse repertoire of receptors than do B-2 cells, also known as conventional B cells, which are generated in the bone marrow throughout life, emerging to populate the blood and lymphoid tissues.

B form : The most common form of duplex DNA, containing a right-handed helix and about 10 (10.5 exactly) base pairs per turn of the helix axis.

b.i.d. : Twice daily (Latin, bis in die) (see also q.d.; q.i.d.; t.i.d.)

B7 molecules, B7.1, B7.2 : The major T-cell co-stimulatory molecules are the B7 molecules, B7.1 (CD80) and B7.2 (CD86). They are closely related members of the immunoglobulin gene superfamily and both bind to the CD28 molecule on T cells. They are expressed differentially on various antigen-presenting cell types. We use the term B7 molecules to refer to both B7.1 and B7.2.

Ba : Barium.

BA : Blood alcohol; bone age; brachial artery; bronchial asthma;

BAB : Babinski's sign.

BABS : Biosynthetic antibody binding site.

BAC : Bacterial artificial chromosome. ***OR*** Blood alcohol concentration.

Bacillary dysentery : Dysentery caused by bacterial infection with invasion of host cells/tissues and/or production of exotoxins.

Bacillus : Bacterium with an elongated, rod shape. ***OR*** Bacterium with an elongated, rod shape. ***OR*** any rodshaped bacterial cell; bacilli (plural) (see also *coccus; spirillum*).

Back crosses : An individual with a dominant phenotype is deliberately mated with a homozygous recessive individual to see what offspring they produce. The aim is to determine whether the parent is homozygous or heterozygous for the dominant characteristic.

Bacteremia : The presence of viable bacteria in the blood.

Bacteria : All prokaryotes that are not members of the domain Archaea.The domain that contains procaryotic cells with primarily diacyl glycerol diesters in their membranes and with bacterial rRNA. Bacteria also is a general term for organisms that are composed of procaryotic cells and are not multicellular.

Bacterial artificial chromosome (BAC) : A cloning vector constructed from the *coli* F-factor plasmid that is used to clone foreign DNA fragments in *E. coli*.

Bacterial photosynthesis : A lightdependent, anaerobic mode of metabolism. Carbon dioxide is reduced to glucose, which is used for both biosynthesis and energy production. Depending on the hydrogen source used to reduce CO_2, both photolithotrophic and photoorganotrophic reactions exist in bacteria.

Bacterial vaginosis : Bacterial vaginosis is a sexually transmitted disease caused by *Gardnerella vaginalis*, *Mobiluncus* spp., *Mycoplasma hominis*, and various anaerobic bacteria. Although a mild disease it is a risk factor for obstetric infections and pelvic inflammatory disease.

Bactericidal : Able to kill bacteria; e.g., serum has innate ability to kill many microorganisms.

Bacteriochlophyll : Lightabsorbing pigment found in green sulfur and purple sulfur bacteria.

Bacteriocin : Agent produced by certain bacteria that inhibits or kills closely related isolates and species.

Bacteriocin : A protein produced by a bacterial strain that kills other closely related strains.

Bacteriophage : Virus that infects bacteria, often with destruction or lysis of the host cell.

Bacteriophage : A type of virus that invades,lives in,and kills a host species of bacterium.

Bacteriophage (phage) typing : A technique in which strains of bacteria are identified based on their susceptibility to a variety of bacteriophages.

Bacterihodopsin : A protein containing retinal found in the membranes of certain extremely halophilic Archaea and which is involved in lightmediated ATP synthesis.

Bacteriostatic : Inhibiting the growth and reproduction of bacteria.

Bacteriovous : Feeding on bacteria.

Bacterium : The singular form of bacteria.

Bacteroid : Altered form of cells of certain bacteria. Refers particularly to the swollen, irregular vacuolated cells of rhizobia in nodules of legumes.A modified, often pleomorphic, bacterial cell within the root nodule cells of legumes; after transformation into a symbiosome it carries out nitrogen fixation.

Baeocytes : Small, spherical, reproductive cells produced by pleurocapsalean cyano-bacteria through multiple fission.

BAF : B-cell activating factor.

Balanced growth : Microbial growth in which all cellular constituents are synthesized at constant rates relative to each other.

Balanitis : Inflammation of the glans penis usually associated with Candida fungi; a sexually transmitted disease. ***OR*** Microbial growth in which all cellular constituents are synthesized at constant rates relative to each other.

Ballistoconidium : A conidium that is forcible discharged.

BALT : See bronchial-associated lymphoid tissue.

Bare lymphocyte syndrome : Bare lymphocyte syndrome is an immunodeficiency disease in which MHC class II molecules are not expressed on cells as a result of one of several different regulatory gene defects. Patients with bare lymphocyte syndrome are severely immunodeficient and have few CD4 T cells. In MHC class I deficiency, or bare lymphocyte syndrome (MHC class I), the most common defect is mutation in either TAP1 or TAP2.

Barometer : An instrument for measuring air pressure.

Barophile : An organism able to live optimally at high hydrostatic pressure.

Barotolerant : An organism able to tolerate high hydrostatic pressure, although growing better at normal pressures.

Bas : Basal.

Basal body : The cylindrical structure at the base of procaryotic and eucaryotic flagella that attaches them to the cell.

Basal medium : A (unsupplemented) medium which allows the growth of many types of micro-organisms which do not require any special nutrient supplements, e.g. Nutrient broth.

Base : A substance which dissociates (separates) in aqueous solution to yield hydroxyl ions, or one containing hydroxyl ions (OH-) which reacts with an acid to form a salt or which may react with metal to form a precipitate.

Base : The adenine, guanine, cytosine or thymine group attached to a nucleotide or nucleoside. Also may be used to refer to a nucleic acid unit within a polynucleotide chain, as when a gene is said to be 2000 bases long.

Base analog : A compound, usually a purine or a pyrimidine, that differs somewhat from a normal nucleic acid base. ***OR*** Molecules that resemble normal DNA nucleotides and can substitute for them during DNA replication, leading to mutations.

Base composition : Proportion of the total bases consisting of guanine plus cytosine or thymine plus adenine base pairs. Usually expressed as a guanine + cytosine (G + C) value, e.g. 60% G+C.

Base pairs : The unit of double-stranded DNA. Bases pair up so that A joins with T and C joins with G.

Base stacking : The close packing of the planes of base pairs, commonly found in DNA and RNA structures.

Basic dyes : Dyes that are cationic, or have positively charged groups, and bind to negatively charged cell structures. Usually sold as chloride salts.

Basidiocarp : Fruiting body in which basidia and basidiospores are produced.

Basidioma (plural, basidiomata) : Fruiting body that produces basidia; also termed a basidiocarp.

Basidiomycete : Name of fungal group that all bear spores on basidia (see basidium).

Basidiomycetes **:** Fungal spores which are from mushrooms. The specific mushroom species can not be identified on the culture plate. Many mushroom spores are reported to be allergenic. ***OR*** A class of fungi characterized by spores formed on basidia. Includes the mushrooms, toadstools, boletes, wood bracket fungi, and puffballs. Some species are edible, such as *Agaricus bisporus*, the commercially cultivated mushroom. A few species cause wood brown rot, white rot, and dry rot in buildings.

Basidiospe : Spore resulting from karyogamy and meiosis that are formed on a basidium that usually is formed on a basidium. Sexual spore of the Basidiomycota. ***OR*** Sexual spores produced by Basidiomycetes.

Beauveria (Hyphomycetes) : Fungus found in plant debris and soil. Some species are well known parasites of insects. It is also isolated from food materials and indoor environments.

Basidium : A structure that bears on its surface a definite number of basidiospores (typically four) that are formed following karyogamy and meiosis. Basidia are found in the basidiomycetes and are usually club-shaped sexual spore producing cell (or cells) that characterises the Basidiomycetes.

Basipetal : A chain of conidia, the oldest conidium is at the apex and the youngest is at the base.

Basocatenulate : A chain of conidia having the youngest cell at the base.

Basophils : Basophils are white blood cells containing granules that stain with basic dyes, and which are thought to have a function similar to mast cells. ***OR*** A phagocytic leukocyte whose granules stain bluish-black with a basic dye. It has a segmented nucleus. The granules contain histamine and heparin.

Batch culture : A culture of microorganisms produced by inoculating a closed culture vessel containing a single batch of medium.

Batch process : A treatment process in which a tank or reactor is filled, the wastewater (or solution) is treated or a chemical solution is prepared and the tank is emptied. The tank may then be filled and the process repeated. Batch processes are also used to cleanse, stabilize or condition chemical solutions for use in industrial manufacturing and treatment processes.

Bathypelagic zone : Equivalent of hypo limnion,bottom layer of water in the ocean.

Bb : Bb is the large active fragment of complement component factor B. It is produced when factor B is captured by bound C3b and cleaved by factor D. Bb remains associated with C3b and is the serine protease component of the alternative pathway 03 convertase.

BB : Bed board; blood bank; both bones; breakthrough bleeding.

BB/FS : Breast biopsy/frozen section.

BBB : Blood brain barrier.

BBT : Basal body temperature.

BC : Back care; bone conduction; birth control.

BC/BS : Blue cross/blue shield.

BCC : Basal cell carcinoma.

B-cell antigen receptor, B-cell receptor (BCR) : The B-cell antigen or receptor, or B-cell receptor (BCR), is the cellsurface receptor of B cells for specific antigen. It is composed of a transmembrane immunoglobulin molecule associated with the invariant IgA and IgA chains in a noncovalent complex.

B-cell corona : The B-cell corona in the spleen is the zone of the white pulp primarily made up of B cells.

B-cell mitogens : B-cell mitogens are substances that cause B cells to proliferate.

BChE : Butyrylcholinesterase.

BChl : Bacteriochlorophyll.

BCP : Birth control pills.

BD : Birth date.

BDNF : Brain-derived neurotrophic factor .

BE : Barium enema; below the elbow; brisk and equal.

Bench scale analysis : Also known as "bench test". A method of studying different ways of treating wastewater and solids on a small scale in a laboratory.

Benign : Referring to a nonlife or nonhealth threatening condition = commensalism.

Benthic : Pertaining to the bottom of the sea or another body of water.

Benthic zone : The bottom mudina lake or ocean.

Benzene : An aromatic hydrocarbon which is a colorless, volatile, flammable liquid. Benzene is obtained chiefly from coal tar and is used as a solvent for resins and fats in dye manufacture.

Beta hemolysis : A zone of complete clearing around a bacterial colony growing on blood agar. The zone does not change significantly in color.

Beta-bend : A characteristic way of turning an extended polypeptide chain in a different direction, involving the minimum number of residues, and held together by hydrogen bonding.

Beta-glucans : A type of sugar molecule.

Betalactam : A class of antibiotics characterized by a betalactam ring structure (e.g., penicillins, cephalosporins).

Betalactamase : Bacterial enzyme that degrades betalactam antibiotics by breaking betalactam ring.

Beta-oxidation : Oxidative degradation of fatty acids that occurs by the successive oxidation of the carbon atom.

Beta-proteobacteria : One of the five subgroups of proteobacteria, each with distinctive 16S rRNA sequences. Members of this subgroup are similar to the alpha-proteobacteria metabolically, but tend to use substances that diffuse from organic matter decomposition in anaerobic zones.

Beta-sheet : A sheetlike structure formed by the interaction between two or more extended polypeptide chains.

BEV : Baculovirus expression vector.

BF : Blood flow; bone fragment.

BFGF : Basic fibroblast growth factor.

BG : Bone graft.

BGG : Bovine gamma globulin.

BGS : Balance, gate, station.

BGTT : Borderline glucose tolerance test.

BHP : Benign hypertrophic prostate.

BHS : Beta hemolytic streptococcus.

Bias : An error caused by favouring some outcomes more than others.

Bic : Biceps.

Biceps brachii : The flex or muscle on the front side of the upper arm.

Biceps femis : The outer muscle on the back of the thigh.

BID : Twice a day; brought in dead (b.i.d.).

Bidirectional replication : Replication in both directions away from the origin, as opposed to replication in one direction only (unidirectional replication).

Bifidobacterium : A good bacteria which lives in our large intestine and helps fight infection.

Bifurcate : Divided into two branches of equal or unequal length; often used to describe the branching pattern of gills (particularly in the genus *Russula*).

BIL : Bilateral.

Bilabiate : Having two lips(used to refer to the shape of flowers on plants in the mint family).

Bilayer : A double layer of lipid molecules with the hydrophilic ends oriented outward, in contact with water, and the hydrophobic parts oriented inward.

Bile salts : Derivatives of cholesterol with detergent properties that aid in the solubilization of lipid molecules in the digestive tract.

Binal symmetry : The symmetry of some virus capsids (e.g., those of complex phages) that is a combination of icosahedral and helical symmetry.

Binary fission : Asexual reproduction in which a cell or an organism separates into two cells. ***OR*** Division of one cell into two cells by the formation of a septum. It is the most common form of cell division in bacteria. ***OR*** During binary fission, a single cell divides transversely to form two new cells called daughter cells. Both daughter cells contain an exact copy of th geneticin-formation contained in the parent cell. ***OR*** Division of one cell into two cells by the formation of a septum. It is the most common form of cell division in bacteria.

Binocular vision : Being able to focus both eyes,together,on an object and process the information to allow for depth perception.

Binomial nomenclature : System of having two names, genus and specific epithet, for each organism.

Binomial system : The nomenclature system in which an organism is given two names; the first is the capitalized generic name, and the second is the uncapitalized specific epithet.

Bioaccumulation : Intracellular accumulation of environmental pollutants such as organic materials by living organisms. Accumulation of a chemical substance in living tissue.

Bioactive : A substance that has an effect on living tissue.

Bioassay : The determination of kinds, quantities or concentrations, and in some cases, the locations of radioactive material in the human body, whether by direct measurement (*in-vivo* counting) or by analysis and evaluation of materials excreted or removed from the human body. ***OR*** Use of the reaction of live plants in testing for virus, in contrast to methods using serology.

Bioaugmentation : The addition to the environment of microorganisms that can metabolize and grow on specific organic compounds.

Bioavailability : The availability of chemicals to potentially biodegradative microorganisms. ***OR*** The availability of a nutrient for microbially mediated degradation or transformation. Typically, for a nutrient to be bioavailable, it must (1) be within physical proximity to the cell and (2) be dissolved in the aqueous phase, to be transported across the cell membrane. Thus, nutrients that are adsorbed to surfaces or sequestered into micropores that physically exclude cells are generally considered unavailable. Bioavailability should not be confused with recalcitrance, although distinguishing the two in the environment can be very difficult.

Biocatalysis : Chemical reactions mediated by biological systems (microbial communities, whole organisms or cells, cell-free extracts, or purified enzymes aka catalytic proteins).

Biochemical oxygen demand (BOD) : Amount of dissolved oxygen consumed in five days by biological processes breaking down organic matter. In particular: The requirement for molecular oxygen by microbes during oxidation of biological substances in sewage. The BOD test measures the oxygen consumed (in mg/L) over 5 days at 20 degrees C. ***OR*** The amount of oxygen used by organisms in water under certain standard conditions; it provides an index of the amount of microbially oxidizable organic matter present.

Biochemical pathway : A series of enzyme-catalyzed reactions that results in the conversion of a precursor molecule into a product molecule.

Biochemistry : The study of those molecules used and manufactured by living things.

Biocide / disinfectant : An agent that is capable of killing microorganisms; however, it may not be 100% effective. In biofilm usage A chemical used to disinfect or remove biofilm in order to control a detrimental effect of the biofilm.

Biocontrol : The use of a living organism, or the product of a living organism to control the population size of another organism, usually a pest.

Biodegradable : Substance capable of being decomposed by biological processes.

Biodegradation : The breakdown of a complex chemical through biological processes that can result in minor loss of functional groups, fragmentation into larger constitutents, or complete breakdown to carbon dioxide and minerals. Often the term refers to the undesired microbial-mediated destruction of materials such as paper, paint, and textiles. ***OR*** The breakdown of organic substances by microorganisms.

Biodiesel : An alternative to diesel fuel that is made from renewable resources, such as animal fats or vegetable oils.

Biodiversity : The variety of living organisms, including plants, animals and microorganisms, living in a particular area.

Bioelectric effect : The undulating expansion and contraction of the charged biofilm matrix exopolymers created by the application of a DC field current which reverses polarity on a frequent basis. The net effect is to enhance the efficacy of antimicrobial agents against biofilm microorganisms.

Biofermentation : The use of a microorganism's ability to survive by fermentation and in so doing to produce products of use to humans. For example, a gene that codes for a particular protein may be inserted into a microorganism, and the required protein is made as part of the cell's regular life activities. The energy for this process would in this case come from fermentation.

Biofilm : Microbial cells encased in an adhesive, usually a polysaccharide material, and attached to a surface. ***OR*** ganized microbial systems consisting of layers of microbial cells associated with surfaces, often with complex structural and functional characteristics. Biofilms have physical/chemical gradients that influence microbial metabolic processes. They can form on inanimate devices (catheters, medical prosthetic devices) and also cause fouling (e.g., of ships' hulls, water pipes, cooling towers).

Bioflocculation : The clumping together of fine, dispersed organic particles by the action of certain bacteria and algae.

Biofuel : A fuel made from biological materials, such as wood, fish oil, vegetable oil or animal waste.

Biogeochemical cycling : The oxidation and reduction of substances carried out by living organisms and/or abiotic processes that results in the cycling of elements within and between different parts of the ecosystem (the soil, aquatic environment, and atmosphere).

Biogeochemistry : Study of microbially mediated chemical transformations of geochemical interest, such as nitrogen or sulfur cycling.

Bioinformatics : The use of information technology and computers to capture, record and interpret complex biological data, for example DNA sequences and patterns of gene expression.

Bioinsecticide : A pathogen that is used to kill or disable unwanted insect pests. Bacteria, fungi, or viruses are used, either directly or after manipulation, to control insect populations.

Biologic transmission : A type of vector borne transmission in which a pathogen goes through some morphological or physiological change within the vector.

Biological Oxygen Demand (BOD) : The amount of oxygen needed by the organisms living in an aquatic environment. In cases of high biological oxygen demand (BOD), less oxygen is available for other organisms to be able to grow. Oxygen is needed by living organisms for cellular respiration. This is a process in cells that releases energy for cell reactions.

Biologically active : Describes a specific substance which has an effect on the metabolic activity of living cells.

Biology : The study of life.

Bioluminescence : The production of light by living cells, often through the oxidation of molecules by the enzyme luciferase. ***OR*** Production of light by living organisms. ***OR*** The production of light by a chemical reaction within an organism. The process occurs in many bacteria and protists, as well as certain animals and fungi. ***OR*** The production of light by a biochemical system.

Biomagnification : The increase in concentration of a substance in higher-level consumer organisms. ***OR*** Increase in the concentration of a chemical substance as it is progresses to higher trophic levels of a food chain.

Biomass : A mass or clump of living organisms feeding on the wastes in wastewater, dead organisms and other debris.

Biome : A major regional biotic community of plants and animals.

Biomedical : Using biological science for medical developments.

Biomolecules : A biomolecule is a chemical compound that naturally occurs in living organisms. Biomolecules consist primarily of carbon and hydrogen, along with nitrogen, oxygen, phosphorus and sulfur.

Biopesticide : A pesticide that is biological in origin (ie, viruses, bacteria, pheromones, natural plant compounds) in contrast to synthetic chemicals. For example, transgenic Bt cotton and corn are biopesticides because *Bacillus thuringiensis* (Bt) is a naturally occurring soil bacterium that has been genetically engineered into the plants.

Biopolymer : Any large molecule in an organism that is made up of a repeating number of smaller components. Examples include proteins (made up of amino acids), starch and cellulose (made up of sugars) and DNA (made up of nucleic acids).

Bioreact : A vessel for growing live organisms or cells, which can be used for the production of substances such as pharmaceuticals, antibodies or vaccines, or for breaking down organic waste.

Bioremediation : The use of biologically mediated processes to remove or degrade pollutants from specific environments. Bioremediation can be carried out by modification of the environment to accelerate biological processes, either with or without the addition of specific microorganisms.

Biosecurity : The protection of a geographical area from invasion by an unwanted organism. In New Zealand, action is taken to prevent unwanted organisms from getting into the country, and to get rid of or control unwanted organisms when they do occur.

Biosensor : The coupling of a biological process with production of an electrical signal or light to detect the presence of particular substances.

Biosolid : The resides of wastewater treatment. Formerly called sewage sludge.

Biosphere : Zone incorporating all forms of life on earth. The biosphere extends from deep in sediment below the ocean to several thousand meters elevation in high mountains.

Biosphere : All the living organisms on Earth.

Biostimulation : Any process that increases the rates of biological degradation, usually by the addition of nutrients,oxygen, or other electron donors and acceptors so as to increase the number of indigenous microorganisms available for degradation of contaminants.

Biosynthesis : Production of needed cellular constituents from other, usually simpler, molecules.

Biotechnology : Any technique that uses living organisms, or parts of organisms, to make or modify products, improve plants or animals, or to develop microorganisms for specific uses.

Bioterrism : The intentional or threatened use of viruses, bacteria, fungi, or toxins from living organisms to produce death or disease in humans, animals, and plants.

Biotic potential : All the factors that contribute to a species increasing its number. Reproduction, migration, adaptation etc.

Biotin : A "B" vitamin related to vitamin B_2 and formerly called vitamin H.

Biotower : An attached culture system. A tower filled with a media similar to rachet or plastic rings in which air and water are forced up a counterflow movement in the tower.

Biotransformation : Alteration of the structure of a compound by a living organism or enzyme.

Biotrophic : Nutritional relationship between two organisms in which one or both must associate with the other to obtain nutrients and grow.

Bioventing : The process of supplying oxygen *in situ* to oxygen deprived soil microbes by forcing air through unsaturated contaminated soil at low flow rates. This stimulates biodegradation and minimizes stripping volatiles into the atmosphere. Frequently used to remediate soil under structures since it is relatively noninvasive.

BiP : Binding protein

Bipolar budding : Blastoconidia developing at the opposite poles of a parent cell.

Bipolaris (Hyphomycetes) : A common saprobe fungus and plant pathogen frequently isolated from plant debris and soil. It is also a common cause of leaf spot on golf course turf. A few species are capable of causing disease in humans.

Bipolaris sp. **:** A fungus with large spores which would be expected to be deposited in the upper respiratory tract. This fungus can produce the mycotoxin - sterigmatocystin which has been shown to produce liver and kidney damage when ingested by laboratory animals.

Biscoctifm : Biscuit-shaped.

Biseriate : Phialides arising from metulae as in the genus *Aspergillus*.

Bivalent : A pair of homologous chromosomes during meiosis,also known as a tetrad.

Biverticillate : Two or rarely three levels of branching directly below the phialides as in the genus Penicillium.

BJ : Biceps jerk; bone and joint.

BJE : Bones, joints, extremities.

BJM : Bones, joints, muscles.

BK : Back; below the knee.

BL : Bilateral lower lung fields; blood loss.

Black peidra : A fungal infection caused by *Piedraia hortae* that forms hard black nodules on the hairs of the scalp.

Black smoker : Thermal vent emitting very hot (270-380 °C) water and minerals.

Blast : Basic local alignment search tool.

Blastic : A form of conidial development where there is a recognizable enlargement or "blowing out" of a conidial initial before being delimited by a septum.

Blastocatenate : A chain of conidia having the youngest cell at the tip.

Blastoconidium (pl. blastoconidia) : An asexual conidium that forms by a blowing out or budding process.

Blastocyst : An early human embryo, about 4-5 days after fertilisation, when the embryo consists of a hollow ball of cells filled with fluid.

Blastocyst : The hollow ball stage of embryonic development.

Blastoderm : The stage in embryogenesis when a unicellular layer at the surface surrounds the yolk mass.

Blastomyces sp. : Human pathogen. The fungus is commonly found in soil. It is a dimorphic fungus which has filamentous fungus when grown at 25ºC and a yeast form at 37ºC.

Blastomycosis : An acute or chronic mycosis which usually affects man and animals (e.g. Dogs). Blastomycosis is caused by a fungus called *Blastomyces dermatitidis* and occurs in North America, Africa and Israel. Infection apparently occurs by inhalation of spores from the fungus although *B. dermatitidis* has proved difficult to isolate from environmental habitats. ***OR*** A systemic fungal infection caused by *Blastomyces dermatitidis* and marked by suppurating tumors in the skin or by lesions in the lungs.

Blastospore : A spore formed by budding from a hypha.

BLE : Both lower extremities.

Blight : A disease symptom that describes withering of above ground plant parts that usually progresses rapidly. Blight may be used for specific plant parts, for example leaf blight.

Blinding : The clogging of the filtering medium of a microscreen or a vacuum filter when the holes or spaces in the media become sealed off due to a buildup of grease or the material being filtered.

Blk : See tyrosine kinase.

BLNK : The adaptor protein BLNK operates in B cells in the same way as LAT does in T cells; it has multiple tyrosine residues that are targets for phosphorylation, and recruits signaling molecules to membrane lipid rafts.

Blood group antigens : Blood group antigens are surface molecules on red blood cells that are detectable with antibodies from other individuals. The major blood group antigens are called ABO and Rh (Rhesus), and are used in routine blood banking to type blood. There are many other blood group antigens that can be detected in cross-matching.

Blood typing : In transfusion medicine, blood typing is used to determine whether donor and recipient have the same ABO and Rh blood group antigens. A cross-match, in which serum from the donor is tested on the cells of the recipient, and *vice versa*, is used to rule out other incompatibilities. Transfusion of incompatible blood causes a transfusion reaction, in which red blood cells are destroyed and the released hemoglobin causes toxicity.

Bloom's syndrome : Bloom's syndrome is a disease characterized by low T-cell numbers, reduced antibody levels, and an increased susceptibility to respiratory infections, cancer, and radiation damage. It is caused by mutations in a DNA helicase.

BLQ : Both lower quadrants.

Blue green alga : See cyanobacterium.

B-lymphocyte chemokine, BLC : B-lymphocyte chemokine (BLC) is a CXC chemokine that attracts B cells and activated T cells into the follicles of peripheral lymphoid tissues by binding to the CXCR5 receptor.

BLYS : Blys is a secreted member of the TNF family of cytokines. It is secreted by T cells and plays critical roles in germinal center and plasma-cell formation, and possibly dendritic-cell maturation.

BM : Basal metabolism; body mass; bone marrow; bowel movement.

BMP : Bone morphogentic protein.

BMR : Basal metabolic rate.

BMS : Biomedical sciences.

BND : Barely noticeable difference.

BNO : Bladder neck obstruction.

BNP : Brain natriuretic peptide.

BOD : Biochemical Oxygen Demand the rate at which microorganisms use the oxygen in water or wastewater while stabilizing decomposable organic matter under aerobic conditions. In decomposition, organic matter serves as food for the bacteria and energy results from this oxidation.

BOD test : A procedure that measures the rate of oxygen use under controlled conditions of time and temperature. Standard test conditions include dark incubation at 20 ºC for a specified time (usually 5 days).

BOE : Bilateral otitis externa.

BOM : Bilateral otitis media.

Bond energy : The energy required to break a bond.

Bone marrow : The bone marrow is the site of hematopoiesis, the generation of the cellular elements of blood, including red blood cells, monocytes, polymorphonuclear leukocytes, and platelets. The bone marrow is also the site of B-cell development in mammals and the source of stem cells that give rise to T calls upon migration to the thymus. Thus, bone marrow transplantation can restore all the cellular elements of the blood, including the cells required for adaptive immunity.

Bone marrow chimera : A bone marrow chimera is formed by transferring bone marrow from one mouse to an irradiated recipient mouse, so that all of the lymphocytes and blood cells are of donor genetic origin. Bone marrow chimeras have been crucial in elucidating the development of lymphocytes and other blood cells.

Booster immunization : A booster immunization is commonly given after a primary immunization, to increase the titer of antibodies.

BOP : Blood, ova, parasites.

Botanist : A person who studies plants.

Botany : The study of plants.

Botrytis (Hyphomycetes) : Teleomorph: Sclerotinia (Ascomycetes) Most species are important plant pathogens, such as *B. cinerea*, which can cause gray mold disease on various plant parts. Can be found in food and indoor environment, particularly on plants, fruits, and vegetables.

Botrytis sp. **:** Aw 0.93. Conidia dimensions 7-14 x 5-9 microns (5). Reported to be allergenic. Fungus is parasitic on plants and soft fruits. Found in soil and vegetables (5). Possibly associated with allergic symptoms (skin tests).

Botulism : A form of food poisoning caused by a neurotoxin (botulin) produced by *Clostridium botulinum* serotypes A-G; sometimes found in improperly canned or preserved food.

Bowel : The small and large intestines where digestion of food is completed

Bowel movement : Passing of faeces/stool - solid waste excreted by an animal

B-oxidation pathway : The major pathway of fatty acid oxidation to produce NADH, $FADH_2$, and acetyl coenzyme A.

bp : Base pair.

BP : Blood pressure; body part.

BPC : Blood pressure cuff.

BPG : D-2,3-bisphosphoglycerate.

BPH : Benign prostatic hypertrophy.

BPheo : Bacteriopheophytin.

BPO : Business process outsourcing.

BPP : Blood pressure and pulse.

BPTI : Bovine pancreatic trypsin inhibitor.

BR : Bathroom; bedrest.

Brachial artery : Artery in upper arm used to measure blood pressure.

Brachydactyly : Short fingered ness; genetic disorder in which the homozygous dominant genotype is not viable.

Bradykinin : Bradykinin is a vasoactive peptide that is produced as a result of tissue damage and acts as an inflammatory mediator.

Bradyzoite : A stage in the life cycle of protozoa of the family Sarcocystidae (tissue-cyst-forming coccidia). In particular this is the term used to describe the merozoite which forms within the tissue cyst in the intermediate host (and rarely within the definitive host). This infectious stage rarely infects new cells within the intermediate host, rather it is the infectious stage for the definitive host.

Branchpoint : An intermediate in a biochemical pathway that can follow more than one route in following steps.

Brassica : A genus of plants in the mustard family (Brassicaceae). The members of the genus may be collectively known either as cabbages, or as mustards.

BRB : Bright red blood.

BRBPR : Bright red blood per rectum.

BRET : Biolumiscence resonance energy transfer

Brevitoxin : Neurotoxin produced by the dinoflagellate Ptychodiscus brevis.

Bright-field microscope : A microscope that illuminates the specimen directly with bright light and forms a dark image on a brighter background.

Bright's disease : See glomerulonephritis.

Broad-spectrum drugs : Chemotherapeutic agents that are effective against many different kinds of pathogens.

Bronchial asthma : An example of an atopic allergy involving the lower respiratory tract.

Bronchial-associated lymphoid tissues, BALT : The type of defensive tissue found in the lungs. Part of the nonspecific immune system. ***OR*** The lymphoid cells and organized lymphoid tissues in the respiratory tract have been termed the bronchial-associated lymphoid tissues (BALT). These tissues are very important in the induction of immune responses to inhaled antigens and to respiratory infection.

Bronchus : One of the two main branches into which the trachea splits.

Bronzing : A disease symptom in which plant cells, usually on the upper leaf surface, are damaged and the leaves appear shiny and brownish.

Broodstock : A group of mature adult fish or shellfish that are kept separate and used for breeding.

Brown rot fungus : Fungus that attacks cellulose and hemicellulose in wood, leaving darkcolored lignin and phenolic materials behind.

Brown field : An abandoned, idled, or underused industrial or commercial facility where expansion or redevelopment is complicated by a real or perceived environmental contamination.

BRP : Bathroom privileges.

Bruton's X-linked agammaglobulinemia : See X-linked agammaglobulinemia.

BS : Blood sugar; bowel sounds; breath sounds.

BSA : Bovine serum albumin

BSA : Bowel signs active.

BSE : Bovine spongioform encephalopathy

BSE : Breast self-examination.

BSN : Bowel sounds normal.

BSP : Bilateral salpingo-oophorectomy.

BST : Bovine somatotrophin

Bt : *Bacillus thuringensis*

BTEX : Benzene, toluene, ethylbenzene, and xylenes

BTR : Biceps tendon reflex.

BTW : Back to work.

Bubo : A tender, inflamed, enlarged lymph node that results from a variety of infections.

Bubonic plague : See plague.

Bud : A young conidium. Usually used to denote the young blastoconidia of yeasts.

Budding : A vegetative outgrowth of yeast and some bacteria as a means of asexual reproduction; the daughter cell is smaller than the parent. ***OR*** Asexual reproduction (usually for yeast) beginning as a protuberance from the parent cell that grows and detaches to form a smaller, daughter cell.

BUE : Both upper extremities.

Buff : An indefinite pale color; pale dull yellow or very pale tan

Buffer : A conjugate acid-base pair that is capable of resisting changes in pH when acid or base is added to the system. This tendency will be maximal when the conjugate forms are present in equal amounts.

BUG : Slang (derived from creepy crawlers in the field of entomology) used by microbiologists and others to refer casually to microorganisms such as bacteria, fungi, cyanobacteria, protozoa, microflagellates and other microscopic organisms.

Bulbous : Usually used to describe the swollen base of a stem.

Bulk density : Soil Mass of dry soil per unit bulk volume (combined volume of soil solids and pore space).

Bulking sludge : Sludges produced in sewage treatment that do not settle properly, usually due to the development of filamentous micro-organisms. ***OR*** Clouds of billowing sludge that occur throughout secondary clarifiers and sludge thickeners when sludge becomes too light and will not settle properly. In the activated sludge process, bulking is usually caused by filamentous bacteria.

Bulla : (Singular) fluidfilled blister or vesicle; bullae (plural).

Bulliform : Bubble-shaped, swollen.

BUN : Blood urea nitrogen.

BUQ : Both upper quadrants.

Burkitt's lymphoma : Burkitt's lymphoma is caused by Epstein-Barr virus (EBV) and occurs mainly in sub-Saharan Africa.

Bursa of fabricius : Found in birds; the blind saclike structure located on the posterior wall of the cloaca; it performs a thymuslike function. A primary lymphoid organ where B-cell maturation occurs. Bone marrow is the equivalent in mammals.

Bursa of fabricius, bursectomy : The bursa of fabricius is an outpouching of the cloaca Found in birds. It is an aggregate of epithelial tissue and lymphoid cells and is the site of intense early B-cell proliferation. The bursa of Fabricius is required for B-cell development in birds, as its removal or bursectomy early in life causes an absence of B cells in adult birds. An equivalent structure has not been detected in mammals, where B-cell development follows a different pathway.

Bursiform : Bag-like.

Burst : See rise period.

Burst size : The number of phages released by a host cell during the lytic life cycle.

BUS : Bartholin, urethral, skein's.

Butanediol fermentation : A type of fermentation most often found in the family Enterobacteriaceae in which 2,3-butanediol is a major product; acetoin is an intermediate in the pathway and may be detected by the Voges-Proskauer test.

Button : A young fruiting body before it has opened up.

BV : Blood vessel; blood volume.

BVR : Bureau of Vocational Rehabilitation.

BW : Birth weight; body weight; twice per week.

BX : Biopsy.

Bycatch : Also called incidental catch. The harvest of fish or shellfish other than the species for which the fishing gear was set.

Bypass flow : See preferential flow.

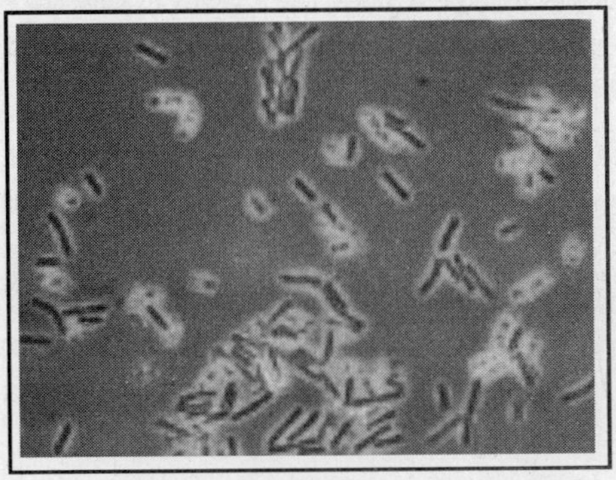

Bacillus

Bacteriophage

Bipolaris

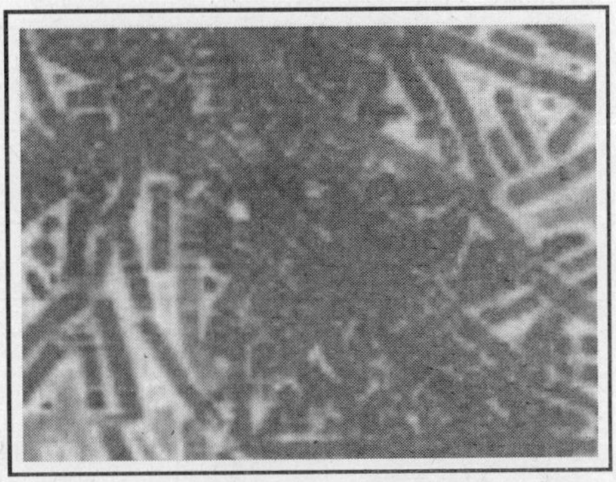

Blue green algae

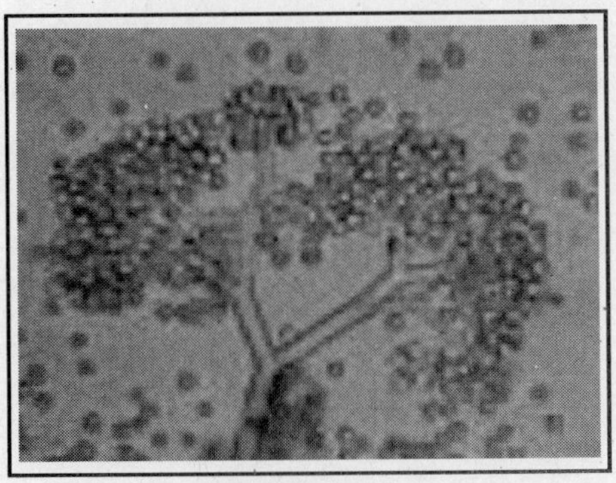

Botrytis

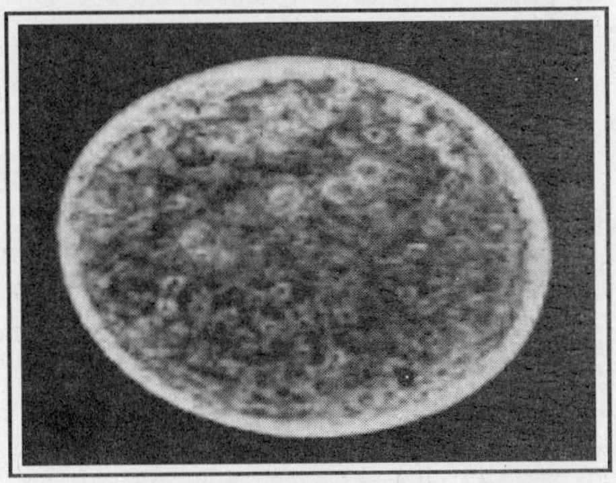

B-cell

C

C : Cytosine.

C domains : The constant regions of the polypeptide chains of immunoglobulin molecules are made up of one or more constant domains or C domains of similar structure; each immunoglobulin chain also has a single variable or V domain.

C region, constant region : See constant region.

C/O : Care of; complains of.

C3 convertase : The generation of the enzyme C3 convertase on the surface of a pathogen or cell is a crucial step in complement activation. The classical pathway C3 convertase is formed from membrane-bound C4b complexed with the protease C2b. The alternative pathway of complement activation uses a different but homologous C3 convertase, formed from membrane-bound C3b complexed with the protease Bb. These C3 convertases have the same activity, catalyzing the deposition of large numbers of C3b molecules that bind covalently to the pathogen surface, leading to opsonization and the activation of the effector cascade that causes membrane lesions.

C3b : The complement fragment C3b is the major product of the C3 convertase, and the principal effector molecule of the complement system. It has a highly reactive thioester bond which allows it to bind covalently to the surface on which it is generated. Once bound, it acts as an opsonin to promote the destruction of pathogens by phagocytes and removal of immune complexes; C3b is bound by the complement receptor CR1, while its proteolytic derivative, iC3b, is bound by the complement receptors CR1, CR2.

C3dg : C3dg is a breakdown product of C3b that remains attached to the microbial surface, where it can bind to CD21, the complement receptor CR2.

C4b-binding protein : C4b-binding protein can inactivate the classical pathway C3 convertase if it forms on host cells, by displacing C2b from the C4b:C2b complex. It binds to C4b attached to host cells, but cannot bind C4b attached to pathogens. This is because it has a second binding site specific for sialic acid, a terminal sugar on vertebrate cell surfaces, but not on pathogens.

C5, C5 convertase, C5a, C5b : C5 is an inactive complement component that is cleaved by the C5 convertase to release the potent inflammatory peptide C5a and a larger fragment, C5b, that initiates the formation of a membraneattack complex from the terminal components of complement.

C5a recept, C3a, C4a : The receptor for the C5a fragment of complement, the C5a receptor, is a seven-transmembrane spanning receptor that couples to a heterotrimeric G protein. Similar receptors bind to C3a and C4a.

C6, C7, C8, C5b, C9 : The complement components C6, C7, and C8 form a complex with the active complement fragment C5b in the late events of complement activation. This complex inserts into the membrane and induces polymerization of C9 to form a pore known as the membrane-attack complex.

CA : Cancer; cardiac arrest; Caucasian adult; chronological age.

CAD : Coronary artery disease.

Cadaver : A dead human body that may be used by physicians and other scientists to study anatomy, identify disease sites, determine causes of death, and provide tissue to repair a defect in a living human being. Students in medical schools study and dissect cadavers as part of their education.

CADD : Computer aided drug design computer assisted drug design.

Caespitose : Crowded together in a cluster but not attached to each other.

CAF : Caucasian adult female.

CAH : Chronic active hepatitis.

CAI : Codon adaptation index.

Cal : Calorie.

Calciferol : Chemical name for vitamin D needed for proper calcium metabolism.

Calcineurin : The cytosolic serine/threonine phosphatase calcineurin has a crucial role in signaling via the T-cell receptor. The immuno-suppressive drugs cyclosporin A and tacrolimus (also known as FK506) form complexes with cellular proteins called immunophilins that bind and inactivate calcineurin, suppressing T-cell responses.

Calcite : A common crystalline form of natural calcium carbonate, caco3, that is the basic constituent of limestone, marble, and chalk. Also called calcspar.

Calcium Carbonate : A calcium ion bonded to a carbonate ion($CaCo_3$). ***OR*** A "salt" used by many marine invertebrates, such as corals and echinoderms, and by protists, such as coccolithophorids, to construct their exoskeletons.

Calnexin : The protein calnexin is an 88 kda protein found in the endoplasmic reticulum. It binds to partly folded members of the immunoglobulin superfamily of proteins and retains them in the endoplasmic reticulum until folding is completed.

Calies : A measure of the energy.

Calreticulin : Calreticulin is the molecular chaperone that binds initially to MHC class I, MHC class II, and other proteins that contain immunoglobulinlike domains, such as the T-cell and B-cell antigen receptors.

Calvin cycle : The main pathway for the fixation (or reduction and incorporation) of CO_2 into organic material by photoautotrophs during photosynthesis; it also is found in chemolithoautotrophs. ***OR*** Biochemical route of carbon dioxide fixation in many autotrophic organisms.

CaM : Calmodulin.

CAM : Caucasian adult male.

CAMM : Computer assisted molecular modeling.

cAMP : Cyclic 3',5'-adenosine monophosphate.

cAMP : Cyclic 3',5' AMP.

cAMP : 3',5' cyclic adenosine monophosphate. The camp molecule plays a key role in metabolic regulation.

Campanulaceae : Plant family commonly known as blue bell shaving bell-shaped flowers with inferior ovaries and alternate leaves.

Campanulate : Bell shaped.

CaMV : Cauliflower mosaic virus.

Cancers (or malignancy) : A disease of the body's cells. Normally, our cells divide and multiply to help our body to grow and heal. Cancer cells, however, divide and grow abnormally and have the ability to invade or damage surrounding tissues. ***OR*** A disease caused by the uncontrolled division and growth of cells. ***OR*** A malignant tumor that expands locally by invasion of surrounding tissues, and systemically by metastasis.

Candida sp. **:** Part of the normal fungal flora of mouth and other mucous membranes in the body. Thrush and other diseases caused by *Candida albicans* usually occur after prolonged treatment with antibiotics or steroids. The environment is not a likely source of exposure for this fungus. Cells from the organism are usually not airborne. Reported to be allergenic.

Candidiasis : An infection caused by *Candida* species of dimorphic fungi, commonly involving the skin.

CAO : Chronic airway obstruction.

CAP : Catabolite activating protein. ***OR*** The catabolite gene activator protein, sometimes incorrectly referred to as the CRP protein. The latter term, in small letters (crp), should be used to refer to the gene but not to the protein.

Cap : Top part of a mushroom which bears the fertile tissue.

Capillary action : The natural upward movement of water in confined areas, such as the spaces between soil particles.

Capillitium : Modified hyphae mixed with spores in the Gasteromycetes.

Capnophile : An organism requiring carbon dioxide at a level exceeding that of air (concentration of carbon dioxide in normal air is ~0.033%) for growth or for enhancement of growth.

Capping : Covalent modification involving the addition of a modified guanidine group in a 5'-5" linkage. It occurs only in eukaryotes, primarily on mrna molecules.

CAPS : Cleaved amplified polymorphic sequence.

Capsid : A protein shell comprising the main structural unit of a virus particle. ***OR*** Protein coat of a virus.

Capsomer : The ring-shaped morphological unit of which icosahedral capsids are constructed.

Capsomere : An individual protein subunit of the virus capsid.

CAPSS : Clone-array pooled shotgun sequencing.

Capsule : Polysaccharide or protein layer surrounding the surface of some bacterial cells and is a major virulence factor that is poorly antigenic and antiphagocytic and functions in adherence and as a protective barrier (see also slime layer and glycocalyx)OR Compact layer of polysaccharide exterior to the cell wall in some bacteria. ***OR*** A hyaline mucopolysaccharide sheath around the cell wall of certain yeasts e.g. *Cryptococcus* and *Rhodotorula*

Capture : Antibodies or antigens can be measured in various capture assays. In these assays, antigens are captured by antibodies bound to plastic (or *vice versa*). Antibody binding to a plate-bound antigen can be measured using labeled antigen or anti-immuno-globulin. Antigen binding to plate-bound antibody can be measured by using an antibody that binds to a different epitope on the antigen.

Capture-release : Where individuals (generally of scarce or endangered species) are captured, and removed to another, more amenable environment and released.

Carbohydrate : Any chemical compound which consists of only carbon (C), oxygen (O), and hydrogen (H) elements, for examples, sugars, starches, and cellulose are carbohydrates. Also the ratio of hydrogen to oxygen atoms in carbohydrates is usually 2:1. ***OR*** A chemical compound made up of carbon, hydrogen and oxygen; forms the main source of energy in our diet. Examples include sugar, starch and cellulose ***OR*** A polyhydroxy aldehyde or ketone. ***OR*** Class of biochemical compounds which includes sugars, starch, chitin, and steroids. ***OR*** One of the three main classes of food and a source of energy. Carbohydrates are the sugars and starches found in breads, cereals, fruits, and vegetables, which, during digestion, are changed into a simple sugar called glucose. Glucose is stored in the liver until cells need it for energy.

Carbon cycle : Sequence where carbon dioxide is converted to organic forms by photosynthesis or chemosynthesis, recycled through the biosphere, with partial incorporation into sediments, and ultimately returned to its original state through respiration or combustion.

Carbon fixation : Conversion of carbon dioxide or other single carbon compounds to organic forms such as carbohydrates.

Carbon nitrogen (C/N) ratio : Ratio of the mass of organic carbon to the mass of nitrogen in soil or organic material.

Carboxyl group : A COOH group attached to a carbon skeleton as in the carboxylic acids and fatty acids.

Carboxylic acid : An organic molecule with a carboxyl group (–COOH) attached. ***OR*** A molecule containing a carbon atom attached to a hydroxyl group and to an oxygen atom by a double bond.

Carboxysomes : Polyhedral inclusion bodies that contain the CO_2 fixation enzyme ribulose 1,5-bisphosphate carboxylase; found in cyanobacteria, nitrifying bacteria, and thiobacilli.

Carbuncle : A coalescence or cluster of furuncles (boils).

Carcinogen : Substance which causes the initiation of tumor formation. Frequently a mutagen.

Carcinogen : Cancer-causing agent. ***OR*** A chemical that can cause cancer. ***OR*** Capable of causing cancer.

Cardiac Sphincter : Ring of muscles between bottom of esophagus and top of stomach to close off stomach.

Cardio vascular system : Another name for the circulatory system, consisting of heart, arteries and veins.

Caries : Tooth decay.

Carnive : Animal which eats other animals.

Carolus Linnaeus : A Latinization of "Karl von Linné".

Carotenoids : Pigment molecules, usually yellowish in color, that are often used to aid chlorophyll in trapping light energy during photosynthesis. ***OR*** Plant pigments that produce a yellow colour ***OR*** Lipid-soluble pigments that are made from isoprene units

Carpels : The wrist bones.

Carpophe : Name given to the stem, cap and gills (fruiting body as a whole).

Carrier (carrier state) : A symptomless (asymptomatic) individual who is host to a pathogenic microorganism and who has the potential to pass (transmit) the pathogen to others. ***OR*** The condition of being a carrier; carriage may be transient or (semi) permanent. ***OR*** Carriers are foreign proteins to which small noni-mmunogenic antigens, or haptens, can be coupled to render the hapten immunogenic. In vivo, self proteins can also serve as carriers if they are correctly modified by the hapten; this is important in allergy to drugs. ***OR*** An infected individual who is a potential source of infection for others and plays an important role in the epidemiology of a disease.

Cartilage : Fibrous, dense connective tissue that acts as a cushion between bone joints.

Cartilaginous : Used to describe consistency of stem tissue; does not bend but breaks with a snap.

Caryophyllaceae : Plant family commonly known as the pinks because the flower petals have Jagged edges as though cut with pinking shears ; typically the stems are swollen at the joints.

Caseation necrosis : Caseation necrosis is a form of necrosis seen in the center of large granulomas, such as the granulomas in tuberculosis. The term comes from the white cheesy appearance of the central necrotic area.

Caseous lesion : A lesion resembling cheese or curd; cheesy. Most caseous lesions are caused by *M. tuberculosis*.

Caspases : Caspases are a family of closely related cysteine proteases that cleave proteins at aspartic acid residues. They have important roles in apoptosis.

Cassical complement pathway : The antibody-dependent pathway of complement activation; it leads to the lysis of pathogens and stimulates phagocytosis and other host defenses.

Casual carrier : An individual who harbors an infectious organism for only a short perio.

CAT : Committee for Advanced Therapies.

CAT : Cataract; computerized axial tomography.

Catabolism : Biochemical processes involved in the breakdown of organic compounds, usually leading to the production of energy. ***OR*** Energy releasing metabolic process of breaking complex compounds into simpler ones (see also anabolism) ***OR*** That part of metabolism in which larger, more complex molecules are broken down into smaller, simpler molecules with the release of energy ***OR*** That part of metabolism that is concerned with degradation reactions.

Catabolite repression : The general repression of transcription of genes associated with catabolism that is seen in the presence of glucose ***OR*** Transcriptionlevel inhibition of a variety of inducible enzymes by glucose or other readily used carbon source.

Catalase : An enzyme that catalyzes the breakdown of hydrogen peroxide to oxygen and water.

$2\ H_2O_2 — O_2 + 2\ H_2O$

Catalyse : The action of enzymes to allow chemical reactions to occur faster than they normally would by lowering the amount of energy required to get the reaction started.

Catalyst : Substance that promotes a chemical reaction by lowering the activation energy without itself being changed in the end. Enzymes are a type of catalyst. ***OR*** A compound that lowers the activation energy of a reaction without itself being consumed.

Catalytic site : The site of an enzyme involved in the catalytic process.

Catenane : An interlocked pair of circular structures, such as covalently closed DNA molecules.

Catenation : The linking of molecules without any direct covalent bonding between them, as when two circular DNA molecules interlock like the links in a chain.

Catenulate : Conidia arranged in chains.

Catheter : A tubular surgical instrument for withdrawing fluids from a cavity of the body, especially one for introduction into the bladder through the urethra for the withdrawal of urine.

Cation exchange capacity (CEC) : Sum of exchangeable cations that a soil can adsorb at a specific pH. Expressed as centimoles of positive charge per kilogram of soil (cmol kg^{-1}). ***OR*** The ability of a soil or other solid to exchange cations (positive ions such as calcium) with a liquid.

Cat-scratch disease (CSD) : A loosely defined syndrome caused by either of the following gram-negative bacilli: *Bartonella (Rochalimaea) henselae* or *Afipia felis*. The typical case of CSD is self-limiting, with abatement of symptoms over a period of days to weeks.

Caulocystidium : A cystidium on the stem.

CB : Chest-back; chronic bronchitis.

CBC : Complete blood count.

CBD : Convention on Biological Diversity.

Cbfa : Core binding factor.

CBI : Codon bias index.

CC : Caucasian child; chief complaint; current complaint.

ccc DNA : Covalently-closed circle DNA.

CCD : Charge coupled device.

CCOT : Cervical compression overloading test.

CD : Circular dichroism.

CD : See clusters of differentiation.

CD : Circular dichroism .

CD : Cardiovascular disease; cause of death.

CD19:CR2:TAPA-1 complex : See B-cell co-receptor.

CD3 complex : The CD3 complex is the complex of a:b or g:d T-cell receptor chains with the invariant subunits CD3g, d, and l, and the dimeric æ chains.

CD4 : The cell-surface protein CD4 is important for recognition by the T-call receptor of antigenic peptides bound to MHC class II molecules. It acts as a co-receptor by binding to the lateral face of the MHC class II molecules.

CD4 T cells : CD4 T cells are T cells that carry the co-receptor protein CD4. They recognize peptides derived from intravesicular sources, which are bound to MHC class II molecules, and differentiate into CD4 TH1 and

CD4 TH2 effector cells that activate macrophages and B-cell responses to antigen.

CD40 ligand, CD154, CD40 : B-cell growth is triggered in part by the binding of CD40 ligand, also known as CD154, expressed on activated helper T cells, to CD40 on the B-cell surface.

CD45 : The leukocyte common antigen, is a transmembrane tyrosine phosphatase found on all leukocytes. It is expressed in different isoforms on different cell types, including the different subtypes of T cells. These isoforms are commonly denoted by the designation of CD45R followed by the exon whose presence gives rise to distinctive antibody-binding patterns.

CD5 B cells : CD5 B cells are a class of a typical, self-renewing B cells found mainly in the peritoneal and pleural cavities in adults. They have a far less diverse receptor repertoire than conventional B cells, and since they are the first B cells to be produced they are also known as B-1 cells.

CD59, protectin : See protectin.

CD8 : The cell-surface protein CD8 is important for recognition by the T-cell receptor of antigenic peptides bound to MHC class I molecules. It acts as a co-receptor by binding to the lateral face of MHC class I molecules.

CD8 T cells : CD8 T cells are T cells that carry the co-receptor CD8. They recognize antigens, for example viral antigens, that are synthesized in the cytoplasm of a cell. Peptides derived from these antigens are transported by TAP, assembled with MHC class I molecules in the endoplasmic reticulum, and displayed as peptide: MHC class I complexes on the cell surface. CD8 T cells differentiate into cytotoxic CD8 T cells.

CD95 pathway : The CD95 receptor is found on many nucleated eucaryotic cells. When the receptor is bound to a specific ligand (CD95L), the CD95CD95L complex activates several cytoplasmic proteins that initiate a cellular suicide cascade leading to apoptosis.

CDH : Congenital dislocation of the hip.

CDK : Cyclin dependent kinase.

cDNA : Complementary DNA, made *in vitro* from the mrna by the enzyme reverse transcriptase using deoxyribonucleotide triphosphates. Unlike mRNA, cDNA can be easily propagated and sequenced.

CDP : Cytidine diphosphate.

CDR : Complimentarity determining region

CDS : Cervico-dorsal syndrome.

CE : Capillary electrophoresis.

CEC : Chicken embryonic cell.

Cecum : Dead-end pouch at the beginning of the large intestine.

Cecum (caecum) : A large, "blind" pouch at the beginning of the large intestine extending from the terminal ileum to the colon, just posterior to the ileocecal junction

Cell cluster : A discrete unit consisting of densely packed (10^{10} cells/ml cluster) bacterial cells (single or multi species) in a polymer matrix. Morphology of clusters can vary.

Cell commitment : That stage in a cell's life when it be comes committed to a certain line of development.

Cell cycle : The sequence of events in a cell's growth division cycle between the end of one division and the end of the next. In eucaryotic cells, it is composed of the G1 period, the S period in which DNA and histones are synthesized, the G2 period, and the M period (mitosis). ***OR*** All of those stages that a cell passes through from one cell generation to the next.

Cell line : An established clone originally derived from a whole organism through a long process of cultivation.

Cell lineage : The pedigree of cells resulting from binary fission.

Cell membrane : See cytoplasmic membrane.

Cell wall : A tough outer coating found in many plant, fungal, and bacterial cells that accounts for their ability to withstand mechanical stress or abrupt changes in osmotic pressure. Cell walls always contain a carbohydrate component and frequently also a peptide and a lipid component. ***OR*** The strong layer or structure that lies outside the plasma membrane; it supports and protects the membrane and gives the cell shape.

Cell-adhesion molecules, CAMS : Cell-adhesion molecules (CAMS) are cell-surface proteins that are involved in binding cells together in tissues and also in less permanent cell-cell interactions.

Cell mediated immunity : The type of immunity that results from T cells coming into close contact with foreign cells or infected cells to destroy them; it can be transferred to a nonimmune individual by the transfer of cells.

Cell-mediated immunity, Cell-mediated immune response : Cell-mediated immunity, or a cell-mediated immune response, describes any adaptive immune response in which antigen-specific T cells have the main role. It is defined operationally as all adaptive immunity that cannot be transferred to a naive recipient with serum antibody. Cf. Humoral immunity.

Cell-surface immunoglobulin : Cell-surface immunoglobulin is the B-cell receptor for antigen. See also B-cell antigen receptor.

Cellular : Where the outermost layer of the cap (cuticle) is made up of spherical cells

Cellular immunology : Cellular immunology is the study of the cellular basis of immunity.

Cellular respiration : The process which occurs within mitochondria in which cells use oxygen to burn sugar for fuel.

Cellular slime molds : Slime molds with a vegetative phase consisting of amoeboid cells that aggregate to form a multicellular pseudoplasmodium; they belong to the division Acrasiomycota.

Cellulitis : A diffuse spreading infection of subcutaneous skin tissue caused by *Streptococci, Staphylococci*, or other organisms. The tissue is inflamed with edema, redness, pain, and interference with function. ***OR*** inflammation of connective tissue underlying the skin

Cellulose : A long, tough carbohydrate molecule that makes up the walls of plant cells. When we eat plants, we find cellulose difficult to digest and call it dietary fibre. ***OR*** Glucose polysaccharide (with beta1,4linkage) that is the main component of plant cell walls. Most abundant polysaccharide on earth. ***OR*** Carbohydrate polymer of the simple sugar glucose. It is found in the cell walls of plants and green algae, as well as dinoflagellates. Cellulose is the most abundant compound on earth that is manufactured by living things.

Central lymphoid organs : Central lymphoid organs are sites of lymphocyte development. In humans, B lymphocytes develop in bone marrow, whereas T lymphocytes develop within the thymus from bone marrow-derived progenitors. They are also sometimes known as the primary lymphoid organs.

Central tolerance : Central tolerance is tolerance that is established in lymphocytes developing in central lymphoid organs. Cf. Peripheral tolerance.

Centres of research excellence : Research centres funded by the New Zealand government. They include: Allan Wilson Centre for Molecular Ecology and Evolution, Maurice Wilkins Centre for Molecular Biodiscovery, The macdiarmid Institute for Advanced Materials and Nanotechnology, National Centre for Advanced Bio-Protection Technologies, National Centre for Growth and Development, New Zealand Institute of Mathematics and its Applications, and Nga Pae o te Maramatanga (Horizons of Insight) - The National Institute of Research Excellence for Maori Development and Advancement.

Centrifuge : (Noun) A machine that uses a high-speed circular motion (centrifugal force) to separate solids from liquids, or different layers in a liquid (e.g., milk and cream). (Verb) To apply a centrifugal force in order to separate solids from liquids, or different layers in a liquid.

Centroblasts : Centroblasts are large, rapidly dividing cells found in germinal centers, and are the cells in which somatic hypermutation is believed to occur. Antibody-secreting and memory B cells derive from these cells.

Centrocytes : Centrocytes are the small B cells in germinal centers that derive from centroblasts. They may mature into antibody-secreting plasma cells or memory B cells, or may undergo apoptosis, depending on their receptor's interaction with antigen.

Centromere : The central region of a chromosome which holds the sister chromatids together.

Cephalosporin : A group of antibiotics derived from the fungus *Cephalosporium*, which share the 7aminocephalosporanic acid nucleus.

Cephalosporium sp. **:** See *Acremonium* sp.

Cercaria : In the trematode life cycle, the cercaria is a larval trematode (which may or may not have a tail, depending on the species). The cercaria develops from the germinal cells of the sporocyst or redia and emerges from the snail (intermediate host) to later become (in most species) the metacercaria.

Cercospora (Hyphomycetes) : Teleomorph: Mycosphaerellaceae Widespread plant pathogens that cause leaf spot on many plants.

Cerebellum : The wrinkled-looking,posterior part of the brain.

Cerebrum : The large,anterior portion of the brain.

Certified transplants : Plants that are produced under the guidelines of an organized program to ensure apparent freedom from injurious insects and diseases.

Cervical Vertebrate : Those in the neck region.

Cervix : The neck or a neck-like portion on another body organ,such as the bottom,narrowed end of the uterus.

Cess pools : This system is similar to a septic tank. In performance. Sewage water usually seeps through the open bottom and portholes in the sides of the walls. These can also clog up with overuse and the introduction of detergents and other material which slow up the bacterial action.

CF : Cystic fibrose. ***OR*** Compare, refer to.

CFNS : Chills, fever, night sweats.

CFU : Colony forming units. Viable micro-organisms (bacteria, yeasts & mould) capable of growth under the prescribed conditions (medium, atmosphere, time and temperature) develop into visible colonies (colony forming units) which are counted. The term colony forming unit (CFU) is used because a colony may result from a single microorganism or from a clump / cluster of microorganisms. ***OR*** Viable micro-organisms (bacteria, yeasts & mould) capable of growth under the prescribed conditions (medium, atmosphere, time and temperature) develop into visible colonies (colony forming units) which are counted. The term colony forming unit (CFU) is used because a colony may result from a single micro-organism or from a clump / cluster of micro-organisms

CH : Characteristic; chest; chief; chronic.

Chaetomium **(Ascomycetes) :** A common fungus in soils, dung, decaying organic matter, seeds, and wood or other cellulose-containing materials. Can be found indoors in water-damaged buildings on sheet rock, wallpaper, and other paper products. It is a common cause of food spoilage. Some species are allergenic but rarely cause human infections.

Chaetomium sp. **:** Large ascomycetous fungus producing perithecia. It is found on a variety of substrates containing cellulose including paper and plant compost It has been found on paper in sheetrock. It is reported to be allergenic. Can produce an Acremonium like state on fungal media.

Chancre (shang_ker) : The primary lesion of syphilis, occurring at the site of entry of the infection.

Chancroid : A sexually transmitted disease caused by the gram-negative bacterium *Haemophilus ducreyi*. Worldwide, chancroid is an important cofactor in the transmission of the AIDS virus. Also known as genital ulcer disease due to the painful circumscribed ulcers that form on the penis or entrance to the vagina.

Chaperonin : A protein that aids in the correct folding of other proteins and the assembly of multisubunit structures.

Charles Darwin : Person who published *The Origin of Species* in 1859.

Charles Lyell : Person who lived from 1797 to1875 and refined Hutton's ideas on geological. Processes and the age of the Earth.

CHD : Coronary heart disease.

ChE : Cholinesterase.

Chediak-Higashi syndrome : Chediak-Higashi syndrome is caused by a defect in a protein involved in intracellular vesicle fusion. Phagocytic cell function is affected as lysosomes fail to fuse properly with phagosomes and there is impaired killing of ingested bacteria.

Cheilocystidium : A cystidium on the edge of a gill.

Chelate (chelat) : Organic chemical that forms ring compound in which a metal is held between two or more atoms strongly enough to diminish the rate at which it becomes fixed by soil, thereby making it more available for plant and microbial uptake. ***OR*** A molecule that contains more than one binding site and frequently binds to another molecule through more than one binding site at the same time.

Chemical oxygen demand (COD) : The amount of chemical oxidation required to convert organic matter in water and wastewater to CO_2.

Chemical precipitation : Precipitation induced by addition of chemicals; the process of softening water by the addition of lime and soda ash as the precipitants.

Chemiosmotic coupling : The coupling of ATP synthesis to an electrochemical potential gradient across a membrane.

Chemiosmotic hypothesis : The hypothesis that a proton gradient and an electrochemical gradient are generated by electron transport and then used to drive ATP synthesis by oxidative phosphorylation.

Chemoautotroph : Organism that obtains energy from the oxidation of reduced inorganic compounds or elements and obtains carbon from carbon dioxide.

Chemokines : Chemokines are small chemoattractant proteins that stimulate the migration and activation of cells, especially phagocytic cells and lymphocytes. They have a central role in inflammatory responses.

Chemolithotroph : Organism that obtains energy from the oxidation of inorganic compounds and uses inorganic compounds as electron donors.

Chemolithotrophic autotrophs : Microorganisms that oxidize reduced inorganic compounds to derive both energy and electrons; CO_2 is their carbon source. Also called chemolithoautotrophs.

Chemoganotroph : Organism that obtains energy and electrons (reducing power) from the oxidation of organic compounds.

Chemoprophylaxis : Use of a chemotherapeutic agent to prevent a specific disease.

Chemecepts : Special protein receptors in the plasma membrane or periplasmic space that bind chemicals and trigger the appropriate chemotaxic response.

Chemostat : Continuous culture device usually controlled by the concentration of limiting nutrient and dilution rate. ***OR*** A continuous culture apparatus that feeds medium into the culture vessel at the same rate as medium containing microorganisms is removed; the medium in a chemostat contains one essential nutrient in a limiting quantity.

Chemotaxin : Chemotactic factor.

Chemotaxis : Movement along a chemical concentration gradient; migration of granulocytes and macrophages toward higher concentrations of chemotaxins during the inflammatory response ***OR*** Oriented movement of a motile organism with reference to a chemical agent. May be positive (toward) or negative (away) with respect to the chemical gradient. ***OR*** Oriented movement of a motile organism with reference to a chemical agent. May be positive (toward) or negative (away) with respect to the chemical gradient. ***OR*** The pattern of microbial behavior in which the microorganism moves toward chemical attractants and/or away from repellents.

Chemotherapeutic agents : Compounds used in the treatment of disease that destroy pathogens or inhibit their growth at concentrations low enough to avoid doing undesirable damage to the host.

Chemotherapy : The use of drugs to treat cancer ***OR*** the treatment of a disease with chemical agents or drugs; chemicals or drugs that affect a pathogenic microorganism negatively without harm to the host

Chemotroph : An organism that obtains its energy from the oxidation of chemical compounds.

Chernozem : Rich,black top soil with a lower layer of lime.

CHF : Congestive heart failure.

CHG : Change.

Chickenpox : A highly contagious skin disease, usually affecting 2- to 7-year-old children; it is caused by the varicella-zoster virus, which is acquired by droplet inhalation into the respiratory system.

Chimera : A recombinant plasmid containing foreign DNA, which is used as a cloning vector in genetic engineering.

Chimeric DNA : Recombinant DNA whose components originate from two or more different sources.

ChIP : Chromatin immunoprecipitation.

Chiral : Having handedness: consisting of one or another stereochemical form.

Chiral compound : A compound that can exist in two forms that are non-superimposable images of one another.

Chitin : A tough, resistant, nitrogen containing polysaccharide forming the walls of certain fungi, the exoskeleton of arthropods, and the epidermal cuticle of other surface structures of certain protists and animals. ***OR*** A carbohydrate polymer found in the cell walls of fungi and in the exoskeletons of arthropods, which provides strength for support and protection; chitinous- adj.

Chl : Chlorophyll.

Chlamydiae : Members of the genus *Chlamydia*: Gram-negative, coccoid cells that reproduce only within the cytoplasmic vesicles of host cells using a life cycle that alternates between elementary bodies and reticulate bodies.

Chlamydial pneumonia : A pneumonia caused by *Chlamydia pneumoniae*. Clinically, infections are mild and 50% of adults have antibodies to the chlamydiae.

***Chlamydomonas* :** A genus of single-celled organisms within Kingdom Protista; Division Chlorophyta.

Chlamydoconidium (pl. Chlamydoconidia) : A thick-walled, thallic conidium formed within the vegetative hyphae. Chlamydoconidia function as organs of perennation rather than dissemination.

Chlamydospe : Thickwalled resting structure that forms from the cell wall of a fungal hypha; usually formed under conditions where the hypha is no longer able to function optimally.

Chlamines : Compounds formed by the reaction of hypochlorous acid.

Chlamphenicol : A broad spectrum antibiotic that is produced by *Streptomyces venezuelae* or synthetically; it binds to the large ribosomal subunit and inhibits the peptidyl transferase reaction.

Chlination : The application of chlorine to water or wastewater, generally for the purpose of disinfection, but frequently for accom-plishing other biological or chemical results.

Chlophyll : Green pigment required for photosynthesis. ***OR*** The green photosynthetic pigment that consists of a large tetrapyrrole ring with a magnesium atom in the center. ***OR*** Several types of green,photosynthetic pigments,the chemical structure of which includes a porphyrin ring with magnesium in the center,which are found in chloroplasts.

Chlophyta : Division within Kingdom Prostist commonly known as the green algae.

Chloplast : Chlorophyll containing organelle of photosynthetic eukaryotes cells, that can harness light energy.

Chloplast : A photosynthetic organelle in plant and protista cells.

Chlotic (chlosis) : A yellow, or pale green, condition of normally green tissue due to low chlorophyll levels. Chlorosis may be caused by nutrient deficiencies, low light conditions and plant diseases. Chlorotic tissue is a common symptom of virus infection.

CHMP : Committee for Human Medicinal Products.

CHO : Chinese hamster ovary.

Choler : Yellow bile,one of the four humors people used to think were present in our bodies,and which was thought to be produced by the liver.

Cholera : An acute infectious enteritis, endemic and epidemic in Asia, which periodically spreading to the Middle East, Africa, Southern Europe, and South America; caused by Vibrio cholerae.

Choleragen : The cholera toxin; an extremely potent protein molecule elaborated by strains of *Vibrio cholerae* in the small intestine after ingestion of feces-contaminated water or food. It acts on epithelial cells to cause hypersecretion of chloride and bicarbonate and an outpouring of large quantities of fluid from the mucosal surface.

Chondrodystrophic dwarfism : A dominant semi lethal form of dwarfism.

Chionic villi sampling : An invasive means of sampling the placenta to do chromosomal testing during pregnancy.

CHR : Chronic; chronological.

Chromatin : The nucleoprotein fibers of eukaryotic chromosomes ***OR*** The DNA containing portion of the eucaryotic nucleus; the DNA is almost always complexed with histones. It can be very condensed (heterochromatin) or more loosely organized and genetically active (euchromatin). ***OR*** Organization of the DNA and associated proteins in a non-dividing nucleus of a cell.

Chromatography : Any technique used to separate different species of molecules (or ions) by subjecting them to two different carrier phases mobile and stationary phases. ***OR*** A chemical technique used to separate different molecules in a mixture. The mixture is passed over or through a material which retards the molecules in the mixture to different degrees. This means that the components of the mixture pass through the chromatography material at different rates, and can be separated.

Chromelosporium : Teleomorph: peziza (Ascomycetes).

Chromoblastomycosis : A chronic fungal infection of the skin, producing wartlike nodules that may ulcerate. It is caused by the black molds *Phialophora verrucosa* or *Fonsecaea pedrosoi*.

Chromogen : A colorless substrate that is acted on by an enzyme to produce a colored end product.

Chromogenic : Producing color; a chromogenic colony is a pigmented colony.

Chromophore group : A chemical group with double bonds that absorbs visible light and gives a dye its color.

Chromosomal aneuploidy : When there are more or less chromosomes than normal.

Chromosomal translocations : Most lymphoid tumors, and many other tumors, bear chromosomal translocations that mark points of breakage and rejoining of different chromosomes. These chromosomal breaks are particularly frequent in lymphomas and leukemias.

Chromosome : A structure of DNA and protein found in the cell nucleus. Each chromosome contains hundreds or thousands of the genes that form our hereditary blueprint. Humans have 23 pairs of chromosomes (each parent contributes one chromosome in each pair), containing a total of 50,000 to 100,000 genes from each parent. ***OR*** A distinguishable unit of genetic material in the nucleus of a dividing cell,made of DNA and protein and carries genes. ***OR*** A thread-like structure, visible in the cell nucleus during metaphase, that carries the hereditary information The bodies that have most or all of the cell's DNA and contain most of its genetic information (mitochondria and chloroplasts also contain DNA and genes).

Chromosome puff : A swollen region of a giant chromosome; the swelling reflects a high degree of transcription activity.

Chronic carrier : An individual who harbors a pathogen for a long time.

Chronic granulomatous disease : Chronic granulomatous disease is an immunodeficiency disease in which multiple granulomas form as a result of defective elimination of bacteria by phagocytic cells. It is caused by defects in the NADPH oxidase system of enzymes that generate the superoxide radical involved in bacterial killing.

Chronic infection : The converse of acute infections, i.e. Prolonged and stubborn. Caused by viruses which are able to persist in the body.

Chronic lymphocytic leukemias, CLLS : Chronic lymphocytic leukemias (CLLS) are B-cell tumors that are found in the blood. The great majority express CD5 and unmutated V genes and are therefore thought to arise from B-1 cells.

Chronic (see acute also) : Course of a disease persisting over a long period of time. ***OR*** The converse of acute infections, i.e. Prolonged and stubborn. Caused by viruses which are able to persist in the body.

Chrysolaminarin : The polysaccharide storage product of the chrysophytes and diatoms. ***OR*** A cystidium that stains yellow in KOH or ammonia ***OR*** The polysaccharide storage product of the chrysophytes and diatoms)

Chrysonilia (Hyphomycetes) : Teleomorph: Neurospora (Ascomycetes). This genus is widespread; being found in food and indoors. *Chrysonilia sitophila* is popularly referred to as the red bread mold that occurs on breads, baked goods, meat, and fruits.

Chrysosporium (Hyphomycetes) : Teleomorph: Various Ascomycetes. A common soil saprobe occasionally isolated from human or animal skin and nail. *Chrysosporium inops* is xerophilic and occurs in food.

Chytrid : Fungal organism in the phylum Chytridiomycota that consists of a spherical cell from which short thin filamentous branches (rhizoids) grow that resemble fine roots.

Chytrids : A group of chytridiomycetes, which are simple terrestrial and aquatic fungi that produce motile zoospores with single, posterior, whiplash flagella. Also considered protists.

CI : Cardiac insufficiency; cerebral infarction.

CIC : Crisis intervention clinic.

CICU : Cardiac intensive care unit.

Cilia : Threadlike appendages extending from the surface of some protozoa that beat rhythmically to propel them; cilia are membranebound cylinders with a complex internal array of microtubules, usually in a 9 + 2 pattern. ***OR*** Many, short "hair-like" structures (consisting of micro tubules arranged according to the "9+2formula")on the surface of some types of cells as a means of locomotion.

Ciliates : A class of protozoans distinguished by short hairs on all or part of their bodies.

Ciliophora : Division (Phylum) of organisms within Kingdom Protista which use the numerous cilia for locomotion.

Cilium : A hairlike structure on some cells. It beats rapidly to move the cell or produce feeding currents. (plural cilia). ***OR*** Short, threadlike appendages that extend from the surface of some protozoa and beat rhythmically to propel them.

Cinereous : Ash grey in colour.

CIRC : Circulation; circumference.

Circadian rhythm : Having an activity cycle of about 24 hours.

Cis bond : C=C double bond where the attached pieces of the carbon chain are on the same side of the double bond.

Cis dominance : Property of a sequence or a gene that exerts a dominant effect on a gene to which it is linked.

Cistron : A genetic unit that encodes a single polypeptide chain.

Citric acid : Derived from citrus fruit or by fermentation of crude sugar, also used as antioxidant, sequestrant, dispersing agent. Helps adjust pH No toxicity in diluted amounts.

Citric acid cycle : See tricarboxylic acid (TCA) cycle. ***OR*** Another name for the Krebs cycle

CK : Check.

c-Kit : The cell-surface receptor c-Kit, present on developing B cells and other developing white blood cells, binds to the stem cell factor on bone marrow stromal cells. Kit has protein tyrosine kinase activity.

CL : Clavicle; clean; closed; contact lens.

Cl complex, Clq, Clr, Cls, C4, C2 : The Cl complex of complement components comprises one molecule of Clq bound to two molecules each of the zymogens Clr and Cls. Clq initiates the classical pathway of complement activation by binding to a pathogen surface or to bound antibody. This binding activates the associated Cl r, which in turn cleaves and activates Cl s. The active form of Cls then cleaves the next two components in the pathway, C4 and C2.

Cl inhibit, ClINH : Cl inhibitor (clinh) is a protein that inhibits the activity of activated complement component Cl by binding to and inactivating its Clr: Cls enzymatic activity. It also inhibits other serine proteases including kallikrein. Deficiency in clinh is the cause of the disease hereditary angioneurotic edema, in which the production of vasoactive peptides, kinins, leads to subcutaneous and laryngeal swelling.

Cladogenesis : Splitting of one species into two or more species.

Cladosporium (Hyphomycetes) : Teleomorph: Mycosphaerella (Ascomycetes). Widely distributed as plant pathogens and saprobes. It is the most frequently found fungus in outdoor air. Indoors, it usually occurs at low concentrations in damp or humid areas, but may be found in high concentrations in water-damaged building materials. Its ability to sporulate heavily and to get airborne makes it an important fungal allergen. Frequently isolated as a contaminant in foods. Only occasionally associated with disease in humans; one species can cause chronic subcutaneous infection.

Cladosporium fulvum (Fulvia fulva) **:** Conidia dimensions 12-47 x 4-10 microns. It is found on the leaves of tomatoes.

Cladosporium herbarum **:** Aw 0.88. Conidia dimensions 5-23 x 3-8 microns. It is found on dead plants, woody plants, food, straw, soil, paint and textiles.

Cladosporium macrocarpum **:** Conidia dimensions 9-29 x 5-13 microns. It is found on dead plants, woody plants, food, straw, soil, paint and textiles.

Cladosporium sp. **:** Aw 0.88; Aw 0.84. Most commonly identified outdoor fungus. The outdoor numbers are reduced in the winter. The numbers are often high in the summer. Often found indoors in numbers less than outdoor numbers. It is a common allergen. Indoor *Cladosporium* sp. May be different than the species identified outdoors. It is commonly found on the surface of fiberglass duct liner in the interior of supply ducts. A wide variety of plants are food sources for this fungus. It is found on dead plants, woody plants, food, straw, soil, paint and textiles. It can cause mycosis. Produces greater than 10 antigens. Antigens in commercial extracts are of variable quality and may degrade within weeks of preparation. Common cause of extrinsic asthma (immeadiate-type hypersensitivity: type I). Acute symptoms include edema and bronchiospasms, chronic cases may develop pulmonary emphysema.

Cladosporium sphaerospermum **:** Conidia dimensions 3-4.5 microns. It is found as a secondary invader of plants, food, soil, paint and textiles.

Clamp connection : Small branch of a fungal hypha that connects two compartments separated by a septum and helps to maintain a dikaryon in each hyphal compartment; characteristic of fungi in the phylum Basidiomycota. ***OR*** semi-circular side branches that grow around the septa in many basidiomycetes.

Clarification : A process in which suspended material is removed from a wastewater. This may be accomplished by sedimentation, with or without chemicals, or filtration.

Clarifier : Settling tank, sedimentation basin. A tank or basin in which wastewater is held for a period of time, during which the heavier solids settle to the bottom and the lighter material will float to the water surface.

Class : Taxonomic classification below phylum (or subphylum) and above order.

Class II transactivator CIITA : See MHC class II transactivator.

Class II-associated invariant chain peptide, CLIP : The class II-associated invariant chain peptide (CLIP) is a peptide of variable length cleaved from the class II invariant chain by proteases. It remains associated with the MHC class II molecule in an unstable form until it is removed by the HLA-DM protein.

Class switching : See isotype switching.

Classes : See isotypes.

Classical pathway : The classical pathway of complement activation is the pathway activated by Cl binding either directly to bacterial surfaces or to antibody, that serves as a means of flagging the bacteria as foreign. See also alternative pathway; mannan-binding lectin.

Classification : The arrangement of organisms into groups based on mutual similarity or evolutionary relatedness. ***OR*** (i) Arrangement of organisms into groups based on mutual similarity or evolutionary relatedness. (ii) Systematic arrangement of soils into groups or categories on the basis of their characteristics.

Clavate : Club-shaped. ***OR*** The collar bone.

Clay : Soil particle < 0.002 mm in diameter.

Clean : Defined as giving careful consideration and effort to avoid crosscontamination. May involve ethanol washing, flaming instruments, etc., and may approach sterile conditions.

Cleistothecium : Closed ascocarp of fungi in the phylum Ascomycota.

CLIA : Chemiluminescent immunoassay.

Cliate : Protozoan that moves by means of cilia on the surface of the cell.

Climax : Most advanced successional community of plants capable of development under, and in dynamic equilibrium with, the prevailing environment.

Cline : Gradual geographical variation in some phenotypic character.

Clinical trial : A carefully planned study that evaluates the benefits and risks of treatments and screening tests on humans. Also called research studies or medical research, clinical trials help establish the safety and effectiveness of new drugs or treatments.

Clitis : Female erectile structure that is homologous to the penis.

Clonal deletion : Clonal deletion is the elimination of immature lymphocytes on binding to self antigens to produce tolerance to self, as required by the clonal selection theory. Clonal deletion is the main mechanism of central tolerance and can also occur in peripheral tolerance.

Clonal expansion : Clonal expansion is the proliferation of antigen-specific lymphocytes in response to antigenic stimulation and precedes their differentiation into effector cells. It is an essential step in adaptive immunity, allowing rare antigen-specific cells to increase in number so that they can effectively combat the pathogen that elicited the response.

Clonal selection theory : The clonal selection theory is a central paradigm of adaptive immunity. It states that adaptive immune responses derive from individual antigen-specific lymphocytes that are self-tolerant. These specific lymphocytes proliferate in response to antigen and differentiate into antigen-specific effector cells that eliminate the eliciting pathogen, and memory cells to sustain immunity. The theory was formulated by Sir Macfarlane Burnet and in earlier forms by Niels Jerne and David Talmage.

Clone : (i) Population of cells all descended from a single cell. (ii) Number of copies of a DNA fragment to be replicated by a phage or plasmid. ***OR*** (Noun) A genetically identical copy of a living organism (Verb) To make a clone, or a genetically identical copy of a living organism. Often refers to a process that can only occur in a laboratory, although when organisms like bacteria divide, they naturally produce clones.

Cloned T-cell line : A cloned T-cell line is a continuously growing line of T cells derived from a single progenitor cell. Cloned T-cell lines must be stimulated with antigen periodically to maintain growth. They are useful for studying T-cell specificity, growth, and effector functions.

Cloning vector : DNA molecule that is able to bring about the replication of foreign DNA fragments. ***OR*** A self-replicating entity to which foreign DNA can be covalently attached for purposes of amplification in host cells.

Clonotypic : A feature unique to individual cells or members of a clone is said to be clonotypic. Thus, a monoclonal antibody that reacts with the receptor on a cloned T-cell line is said to be a clonotypic antibody and to recognize its clonotype or the clonotypic receptor of that cell. See also idiotype; idiotypic.

Closterium **:** Genus of Desmidsin Kingdom Protista; Division Chlorophyta which bear a gypsum-containing vacuole at each tip of their cells.

Clostridial myonecrosis : Death of individual muscle cells caused by *Clostridia*. Also called gas gangrene.

Clusters of differentiation, CD : Clusters of differentiation (CD) are groups of monoclonal antibodies that identify the same cell-surface molecule. The cell-surface molecule is designated CD followed by a number (e.g. CD1, CD2, etc.).

CM : Carboxymethyl.

cM : Centimorgan.

CML : Chemical markup language.

CMO : Contract manufacturing organization.

CMP : Camp receptor protein (catabolite activator protein).

CMP : Cytidine monophosphate.

CMS : Circulation, motion, sensation.

CMV : Cytomegalovirus.

CN : Cranial nerve.

CNR : Cell nuclear replacement

CNS : Central nervous system.

CNV : Copy number variation

CNV : Cranial nerve number 5.

CNX : Cranial nerve number 10.

CO : Carbon monoxide; cardiac output; check out.

CoA : Coenzyme A.

COA : Condition on admission.

Coactivator : A molecule that functions together with a protein apoactivator. For example, camp is a coactivator of the CAP protein.

Coaggregation : The collection of a variety of bacteria on a surface such as a tooth surface because of cell-to-cell recognition of genetically distinct bacterial types. Many of these interactions appear to be mediated by a lectin on one bacterium that interacts with a complementary carbohydrate receptor on another bacterium.

Coagulants : Chemicals which cause very fine particles to clump (floc) together into larger particles. This makes it easier to separate the solids from the water by settling, skimming, draining, or filtering.

Coagulase : An enzyme that induces blood clotting; it is characteristically produced by pathogenic *Staphylococci*.

Coagulation system : The coagulation system is a proteolytic cascade of plasma enzymes that triggers blood clotting when blood vessels are damaged.

CoA-SH : Reduced coenzyme A.

Cobalamin : Vitamin B_{12} which contains a cobalt atom.

Coccidioidomycosis : A fungal disease caused by *Coccidioides immitis* that exists in dry, highly alkaline soils. Also known as valley fever, San Joaquin fever, or desert rheumatism.

Coccus : Spherical bacterial cells. ***OR*** (singular) sphericalshaped cell; cocci (plural) (see also bacillus, spirillum).OR A roughly spherical bacterial cell.

Coccyx : Four fused vertebrae which comprise the tail bone.

Cochlea : Coiled organ in inner ear in which receptor nerve endings are located

COD : Chemical oxygen demand the amount of oxygen in mg/l required to oxidize both organic & oxidizable inorganic compounds.

COD : Cause of death; condition on discharge.

Code degeneracy : The genetic code is organized in such a way that often there is more than one codon for each amino acid.

Coding joint : A coding joint is formed by the imprecise joining of a V gene segment to a (D)J gene segment in immunoglobulin or T-cell receptor genes.

Co-dominant : The expression of both alleles in the case of a heterozygous genotype. ***OR*** The expression of a gene is said to be codominant when both alleles at one locus are expressed in roughly equal amounts in heterozygotes. Most genes show this property, including the highly polymorphic MHC genes.

Codon : A sequence of three nucleotides in mRNA that directs the incorporation of an amino acid during protein synthesis or signals the start or stop of translation. ***OR*** In a messenger RNA molecule, a sequence of three bases that represents a particular amino acid.

Coelomycetes : An artificial class of fungi characterized by asexual spores that are produced within a cavity lined by fungal tissue or fungal and host tissues. Most are saprobes or pathogens on plants, fungi, and lichens.

Coenocytic : Fungal hypha without crosswalls (septa), so that the nuclei present in the cytoplasm are freefloating and mobile. ***OR*** Infrequently septate, multinucleate hyphae as in the Zygomycetes. ***OR*** Refers to a multinucleate cell or hypha formed by repeated nuclear divisions not accompanied by cell divisions.

Coenurus : Larva of *Taenia multiceps* which is a fluid filled cyst containing many invaginated scolecies.

Coenzyme : Lowmolecularweight chemical which participates in an enzymatic reaction by accepting and donating electrons or functional groups. ***OR*** An organic molecule that associates with enzymes and affects their activity. ***OR*** A loosely bound cofactor that often dissociates from the enzyme active site after product has been formed.

Coenzyme : An organic,non-protein enzyme helper.

Coevolution : Evolution of two species totally dependent on each other

Cofactor : A non-protein enzyme helper. ***OR*** The nonprotein component of an enzyme; it is required for catalytic activity. ***OR*** A small molecule required for enzyme activity. It could be organic in nature, like a coenzyme, or inorganic in nature, like a metallic cation.

Cohesion : Force holding a solid or liquid together, owing to attraction between like molecules.

Co-immunoprecipitation : See immunoprecipitation analysis.

Co-isogenic : See congenic.

Cold sore : A lesion caused by the herpes simplex virus; usually occurs on the border of the lips or nares. Also known as a fever blister or herpes labialis.

Colicin : A plasmidencoded protein that is produced by enteric bacteria and binds to specific receptors on the cell envelope of sensitive target bacteria, where it may cause lysis or attack specific intracellular sites such as ribosomes.

Coliform : Gramnegative, nonsporeforming facultative rod that ferments lactose with gas formation with 48 hours at 35ºC. Often an indicator organism for fecal contamination of water supplies. *Escherichia coli* and *Enterobacter* are important members. ***OR*** A gram-negative, nonsporing, facultative rod that ferments lactose with gas formation within 48 hours at 35°C.

Colitis : Inflammation of the colon.

Collagen : Long proteins whose structure is wound into a triple helix. The resulting fibers have a high tensile strength. Collagen is a primary component of mammalian hair.

Collarette : A small collar. Usually, a remnant of a cell wall present at the tip of a phialide, or around a sporangiophore.

Collectins : Collectins are a structurally related family of calcium-dependent sugar-binding proteins or lectins containing collagen-like sequences. An example is mannan-binding lectin.

Colloid fraction : Organic and inorganic matter with very small particle size and a correspondingly large surface area per unit of mass.

Colloids : Very small, finely divided solids (particles that do not dissolve) that remain dispersed in a liquid for a long time due to their small size and electrical charge.

Colon : The major portion of the large intestine extending from the cecum to the rectum; ascending, transverse, descending, and sigmoid portions. ***OR*** The main part of the large intestine

Colonization : Establishment of a community of microorganisms at a specific site or ecosystem.

Colonscopy : A procedure where a long and flexible tubular instrument called a colonoscope is inserted into the patient"s anus in order to view the lining of the colon and rectum. It is performed to test for colorectal cancer and other bowel diseases, and enables the doctor to collect tissue samples for laboratory analysis.

Colony : Clone of bacterial cells on a solid medium that is visible to the naked eye. ***OR*** A cluster or assemblage of microorganisms growing on a solid surface such as the surface of an agar culture medium; the assemblage often is directly visible, but also may be seen only microscopically. ***OR*** Clone of bacterial cells on a solid medium that is visible to the naked eye. ***OR*** A population of cells growing on solid medium (plates) arising from a single colony forming unit.

Colony forming unit (CFU) : The minimum number of separable cells on the surface of or in semisolid agar medium which gives rise to a visible colony of progeny on the order of tens of millions of cells in number. Cfus may consist of pairs, chains and clusters as well as single cells. ***OR*** The number of microorganisms that can form colonies when cultured using spread plates or pour plates, an indication of the number of viable microorganisms in a sample.

Color break : A symptom in which there are streaks of white or other colors in flower parts, usually petals. (Also called flower break).

Colado tick fever : A disease that occurs in the mountainous regions of the western United States. It is caused by an RNA virus of the genus Coltivirus that is spread from ground squirrels, rabbits, and deer to humans by the tick, *Dermacentor andersoni*. Complications are rare.

Colless sulfur bacteria : A diverse group of nonphotosynthetic proteobacteria that can oxidize reduced sulfur compounds such as hydrogen sulfide. Many are lithotrophs and derive energy from sulfur oxidation. Some are unicellular, whereas others are filamentous gliding bacteria.

Colostrum : A fluid that is rich in antibodies. It is a baby's first food and is secreted from the breasts before milk is produced.

Columella (pl. columellae) : A sterile dome-like structure at the tip of a sporangiophore or within a sporangium.

Columnar : Forming a column.

Combinatial biology : Introduction of genes from one micro-organism into another microorganism to synthesize a new product or a modified product, especially in relation to antibiotic synthesis.

Combinatial diversity : Antigen receptors manifest two distinct types of combinatorial diversity generated by the combination of separate units of genetic information. Receptor gene segments are joined in many different combinations to generate diverse receptor chains, and then two different receptor chains (heavy and light in immunoglobulins; 0, and P or y and 6 in T-cell receptors) are combined to make the antigenrecognition site.

Combined available chlorine : The concentration of chlorine which is combined with ammonia (NH_3) as chloramine or as other chloro derivatives, yet is still available to oxidize organic matter.

Combined sewer : A sewer designed to carry both sanitary wastewaters and storm or surface-water runoff.

Comedo : A plug of dried sebum in an excretory duct of the skin.

Comet : Computerized molecular evaluation of toxicity. ***OR*** Heavenly body with long, luminous tail and elliptical orbit.

Cometabolism : The modification of a compound not used for growth by a microorganism, which occurs in the presence of another organic material that serves as a carbon and energy source. ***OR*** Transformation of a substrate by a microorganism without deriving energy, carbon, or nutrients from the substrate. The organism can transform the substrate into intermediate degradation products but fails to multiply at its expense.

Commensal : Micro-organisms which live in close contact with the body without causing disease. Commensal micro-organisms may be good for one organism, but cause disease in another organism. ***OR*** Living on or within another organism without injuring or benefiting the other organism.

Commensalism : Interaction between organisms where one organism benefits from the association while the second organism remains unaffected or a relationship between two species (commensals) in which one is benefited and the other is not affected, neither negatively nor positively.

Commercialisation : To take an idea and turn it into a business in order to make money from it.

Comminution : Shredding. A mechanical treatment process which cuts large pieces of waste into smaller pieces so that they won't plug pipes or damage equipment. ***OR*** Reduction in the size of organic materials as a result of feeding by soil organisms; shredding is one form of comminution.

Common cold : An acute, self-limiting, & highly contagious virus infection of the upper respiratory tract that produces inflammation, profuse discharge, and other symptoms.

Common lymphoid progenits : Common lymphoid progenitors are stem cells that give rise to all lymphocytes. They are derived from pluripotent hematopoietic stem cells.

Common variable immunodeficiency : Common variable immunodeficiency is a relatively common deficiency in antibody production whose pathogenesis is not yet understood. There is a strong association with genes mapping within the MHC.

Common vehicle transmission : The transmission of a pathogen to a host by means of an inanimate medium or vehicle.

Common vehicle transmission : The transmission of a pathogen to a host by means of an inanimate medium or vehicle.

Common g chain : The common g chain (gc) is a trans-membrane polypeptide chain (CD132) that is common to a subgroup of class I cytokine receptors. It plays a key role in the intracellular signaling mediated by these receptors as shown by gene knockout.

Commonsource epidemic : An epidemic that is characterized by a sharp rise to a peak and then a rapid, but not as pronounced, decline in the number of individuals infected; it usually involves a single contaminated source from which individuals are infected.

Communicable disease : A disease associated with a pathogen that can be transmitted from one host to another.

Community : All organisms that occupy a common habitat and interact with one another.

Compensaty growth : Accelerated population growth in response to grazing or predation.

Competent : A bacterial cell that can take up free DNA fragments and incorporate them into its genome during transformation.

Competition : An interaction between two organisms attempting to use the same resource (nutrients, space, etc.).

Competitive binding assays : Competitive binding assays are serological assays in which unknowns are detected and quantitated by their ability to inhibit the binding of a labeled known ligand to its specific antibody. This is also referred to as a competitive inhibition assay.

Competitive exclusion principle : Two competing organisms overlap in resource use, which leads to the exclusion of one of the organisms.

Competitive inhibition assay : When known sources of antibody or antigen are used as competitive inhibitors of antigen-antibody interactions, this assay is referred to as a competitive inhibition assay.

Complement fixation (CF) : An assay for detecting the presence of antibodies reactive against a particular antigen, e.g. A virus.

Complement receptors, CRs : Complement receptors (crs) are cell-surface proteins on various cells that recognize and bind complement proteins that have bound an antigen such as a pathogen. Comple-ment receptors on phagocytes allow them to identify pathogens coated with complement proteins for uptake and destruction. Complement receptors include CR1, CR2, CR3, CR4, and the receptor for Clq.

Complement system : A group of plasma proteins that plays a major role in an animal's defensive immune response. ***OR*** A group of plasma proteins that plays a major role in an animal's defensive immune response complementary. In reference to base pairing, the ability of two polynucleotide sequences to form a doublestranded helix by hydrogen bonding between bases in the two sequences.

Complement, Complement activation : The complement system is a set of plasma proteins that act together to attack extracellular forms of pathogens. Complement activation can occur spontaneously on certain pathogens or by antibody binding to the pathogen. The pathogen becomes coated with complement proteins that facilitate pathogen removal by phagocytes and can also kill certain pathogens directly.

Complementarity determining regions, CDRs : The complementarity determining regions (cdrs) of immunoglobulins and T-cell receptors are the parts of these molecules that determine their specificity and make contact with specific ligand. The cdrs are the most variable part of the molecule, and contribute to the diversity of these molecules. There are three such regions (CDR1, CDR2, and CDR3) in each V domain.

Complementary : In reference to base pairing, the ability of two polynucleotide sequences to form a doublestranded helix by hydrogen bonding between bases in the two sequences.

Complementary base pairing : The pairing of adenine (A) with thymine (T) and cytosine (C) with guanine (G) in a DNA double helix. ***OR*** the pairing of adenine (A) with uracil (U) and cytosine (C) with guanine (G) in a DNA-RNA complex (e.g., during transcription). ***OR*** For a given sequence of nucleic acids, the nucleic acids that are related to them by the rules of base pairing.

Complementary DNA : (cDNA) A DNA copy of an RNA molecule (e.g., a DNA copy of an mRNA). ***OR*** A DNA molecule made *in vitro* using mRNA as the template and reverse transcriptase as the enzyme.

Complete flower : A term referring to a flower which has all four layers of flower parts.

Complex medium : Culture medium that contains some ingredients of unknown chemical composition. ***OR*** Medium whose precise chemical composition is unknown. Also called undefined medium.

Complex viruses : Viruses with capsids having a complex symmetry that is neither icosahedral nor helical.

Compositae : Plant family in which flowers are arranged as a "head".

Compost : Organic residues which have been mixed, piled, and moistened, with or without addition of fertilizer and lime, and generally allowed to undergo thermophilic decomposition until the original organic materials are substantially altered or decomposed.

Composting : The microbial processing of fresh organic matter under moist, aerobic conditions, resulting in the accumulation of a stable humified product, which is suitable for soil improvement and stimulation of plant growth.

Compromised host : A host with lowered resistance to infection and disease for any of several reasons. The host may be seriously debilitated (due to malnutrition, cancer, diabetes, leukemia, or another infectious disease), traumatized (from surgery or injury), immuno-suppressed, or have an altered microbiota due to prolonged use of antibiotics.

Concatemer : A long DNA molecule consisting of several genomes linked together in a row.

Concentration factor : The ratio of initial aliquot volume divided by final volume of a solution ($V_i \div V_f$)

Concentric ringspot : A symptom typical of some plant viruses in which a yellow or necrotic spot is surrounded by rings of tissue that are also yellow or necrotic. Also sometimes described as a thumbprint or bull's eye pattern.

Condensation : The joining of two organic compounds.

Conditional mutations : Mutations that are expressed only under certain environmental conditions.

Configuration : The spatial arrangement in which atoms are covalently linked in a molecule.

Confocal fluoresent microscopy : In confocal fluorescent microscopy, it is possible to use optics to produce images at very high resolution by having two origins of fluorescent light that come together only at one plane of a thicker section.

Confocal microscopy : A type of microscopy where light from a very thin section or plane is seen, whilst interfering light from other sections is removed.

Confocal scanning laser microscope (CSLM) : A light microscope in which mono-chromatic laser-derived light scans across the specimen at a specific level and illuminates one spot at a time to form an image. Stray light from other parts of the specimen is blocked out to give an image with excellent contrast and resolution.

Conformation : The three-dimensional arrangement adopted by a molecule, usually a complex macromolecule. Molecules with the same configuration can have more than one conformation.

Conformational epitopes : Conformational epitopes, or discontinuous spitopes, on a protein antigen are formed from several separate regions in the primary sequence of a protein brought together by protein folding. Antibodies that bind conformational epitopes bind only native folded proteins.

Congenic : Congenic strains of mice are genetically identical at all loci except one. Each strain is generated by the repetitive back-crossing of mice carrying the desired trait onto a strain that provides the genetic background for the set of congenic strains. The most important congenic strains in immunology are the congenic strains, developed by George Snell, that differ from each other at the MHC.

Congenital (neonatal) herpes : An infection of a newborn caused by transmission of the herpesvirus during vaginal delivery.

Congenital disorder of glycosylation : The disorder can result in several diseases including seizures and psychomotor retardation and in severe cases, death before adulthood.

Congenital rubella syndrome : A wide array of congenital defects affecting the heart, eyes, and ears of a fetus during the first trimester of pregnancy, and caused by the rubella virus.

Congenital syphilis : Syphilis that is acquired in utero from the mother.

Conidia (conidium, singular) : Asexual non-motile spores.

Conidiogenous cell : A cell that forms conidia.

Conidiophore : A specialized hypha upon which conidia develop. ***OR*** An asexual, thinwalled spore borne on hyphae and not contained within a sporangium; it may be produced singly or in chains. ***OR*** An asexual, thin-walled spore borne on hyphae and not contained within a sporangium; it may be produced singly or in chains.

Conidium (plural, conidia) : Nonmotile, asexual spore resulting from mitotic nuclear division and formed from the ends or sides of a hypha; produced in abundant numbers by the asexual phase of soil fungi in the phyla Ascomycota and Basidiomycota.

Conidobolus sp. **:** Can cause a chronic inflammatory disease of the nasal mucosa (entomophthoro-mycosis).

Conjugants : Complementary mating types that participate in a form of protozoan sexual reproduction called conjugation.

Conjugate vaccines : Conjugate vaccines are vaccines made from capsular polysaccharides bound to proteins of known immunogenicity, such as tetanus toxoid.

Conjugated linoleic acids : Natural components of animal foods which have been shown to have health benefits including inhibiting growth of cancers, stimulating the immune system and improving the health of the gut.

Conjugation : In prokaryotes, transfer of genetic information from a donor cell to a recipient cell by celltocell contact. ***OR*** "bacterial sex;" the exchange of genetic material from one bacterial cell (donor) to another (recipient) by the close association of the two cells and the passage of the bacterial chromosome or plasmid through the F pilus or a temporary joining of two bacteria to transfer genetic material. ***OR*** 1. The form of gene transfer and recombination in bacteria that requires direct cell-to-cell contact. 2. A complex form of sexual reproduction commonly employed by protozoa. ***OR*** A type of sexual reproduction in which the contents of the male gamete cell go over into and unite with the female cell

Conjugative plasmid : A plasmid that carries the genes for sex pili and can transfer copies of itself to other bacteria during conjugation. ***OR*** Selftransmissible plasmid; a plasmid that encodes all the functions needed for its own intercellular transmission by conjugation.

Conoi : A hollow cone of spirally coiled filaments in the anterior tip of certain apicomplexan protozoa.

Consensus sequence : In nucleic acids, the "average" sequence that signals a certain type of action by a specific protein. The sequences actually observed usually vary around this average.

Constium : A physical association of two different organisms, usually beneficial to both organisms. ***OR*** Two or more members of a natural assemblage in which each organism benefits from the other. The group may collectively carryout some process that no single member can accomplish on its own.

Constant region (CL and CH) : The part of an antibody molecule that does not vary greatly in amino acid sequence among molecules of the same class, subclass, or type.

Constant region, C region : The constant region (C region) of an immunoglobulin or T-cell receptor is that part of the molecule that is relatively constant in amino acid sequence between different molecules. In an antibody molecule the constant regions of each chain are composed of one or more C domains. The constant region of an antibody determines its particular effector function. Cf. Variable region.

Constipation : Diagnosed when bowel movements occur less than is normal or less than three times per week. In constipation, bowel motions may consist of small amounts of hard stool that are difficult to pass. Constipation may also be accompanied by feelings of bloatedness, abdominal pain and a swollen abdomen.

Constitutive enzyme : Enzyme always synthesized by the cell regardless of environmental conditions.

Constitutive mutant : A strain that produces an inducible enzyme continually, regardless of need, because of a mutation in either the operator or regulator gene.

Constructed wetlands : Intentional creation of marshland plant communities and their associated microorganisms for environmental restoration or to purify water by the removal of bacteria, organic matter, and chemicals as the water passes through the aquatic plant communities.

Consumer : An organism that feeds directly on living or dead animals, by ingestion or by phagocytosis. ***OR*** Research into the likes and dislikes of people buying a particular product.

Contact hypersensitivity reaction : A contact hypersensitivity reaction is a form of delayed-type hypersensitivity in which T cells respond to antigens that are introduced by contact with the skin. Poison ivy hypersensitivity is a contact hyper-sensitivity reaction due to T-cell responses to the chemical antigen pentadecacatechol in poison ivy leaves. ***OR*** Contact stabilization is a modification of the conventional activated sludge process. In contact stabilization, two aeration tanks are used. One tank is for separate reaeration of the return sludge for at least four hours before it is permitted to flow into the other aeration tank to be mixed with the primary effluent requiring treatment.

Contact transmission : Transmission of the pathogen by contact of the source or reservoir of the pathogen with the host.

Contagious disease : A disease capable of being transmitted from one person to another.

Contaminant : A substance that pollutes or infects something. ***OR*** When a substance is polluted or infected with something, such as a chemical or microorganism.

Contamination : The introduction of contaminants into a substance.

Context : The flesh of the fungus.

Continuous culture system : A culture system with constant environmental conditions maintained through continual provision of nutrients and removal of wastes.

Continuous epitopes : Continuous epitopes, or linear epitopes, are antigenic determinants on proteins that are contiguous in the amino acid sequence and therefore do not require the protein to be folded into its native conformation for antibody to bind. The epitopes detected by T cells are continuous.

Contractile vacuole : In protists and some animals, a clear fluid-filled cell vacuole that takes up water from within the cell and then contracts, releasing it to the outside through a pore in a cyclical manner. Contractile vacuoles function primarily in osmoregulation and excretion.

Control : Reduction of disease incidence, prevalence, morbidity or mortality to a locally acceptable level as a result of deliberate efforts. Continued intervention measures are required to maintain the reduction.

Convalescent carrier : An individual who has recovered from an infectious disease but continues to harbor large numbers of the pathogen.

Conventional treatment : The preliminary treatment, sedimentation, flotation, trickling filter, rotating biological contactor, activated sludge and chlorination of wastewater.

Convergent evolution : When sympatric species evolve similar characteristics due to similar environmental pressures.

Conversion : Changing from one substance to another. As food matter is changed to cell growth or to carbon dioxide. ***OR*** used to describe the central tissue of the gill when it curves towards the mid-line.

Convertase : A convertase is an enzymatic activity that converts a complement protein into its reactive form by cleaving it. Generation of the C3 convertase is the pivotal event in complement activation.

Convex : Used to describe a cap that is regularly rounded or broadly obtuse.

Coombs test : The Coombs test is a test for antibody binding to red blood cells. Red blood cells that are coated with antibody are agglutinated if they are exposed to an anti-immunoglobulin antibody. The Coombs test is important in detecting the nonaggluti-nating antibodies against red blood cells produced by Rh incompatibility in pregnancy.

Cooperative binding : A situation in which the binding of one ligand to a macromolecule favors the binding of another. For example, DNA cooperatively binds histone molecules, and hemoglobin cooperatively binds oxygen molecules.

Cooperativity : Two binding sites are said to demonstrate cooperativity in binding to their ligand when the binding of ligand to one site enhances the binding of ligand to the second site.

Codinate induction : The simultaneous expression of two or more genes.

COPD : Chronic obstructive pulmonary disease.

COPE : Chronic obstructive pulmonary emphy-sema.

Copiotrophic : Having a high nutrient level. ***OR*** inhabiting dung.

Coprinus (Basidiomycetes) : These mushrooms are popularly referred to as the inky caps because their gills dissolve into a black inky fluid at maturity. Found on wood, dung, humus, and soil. Some species are edible.

CoQ : Coenzyme Q (ubiquinone).

Co-receptor : A co-receptor is a cell-surface protein that increases the sensitivity of the antigen receptor to antigen by binding to associated ligands and participating in signaling for activation. CD4 and CD8 are MHC-binding co-receptors on T cells, whereas CD19 is part of a complex that makes up a co-receptor on B cells.

Corepress : A small molecule that inhibits the synthesis of a repressible enzyme.

Ciaceous : Leathery.

Corm : A swollen underground stem that serves as a propagative structure in Crocus, Gladiolus, Cyclamen, and other plants.

Cornea : The clear covering over the front of the eye.

Corona : See B-cell corona.

Conary artery : The artery supplying blood to the heart muscle.

Corpus luteum : A temporary secretory structure which from remains of an ovarian follicle after ovulation.

Cortex : The layer of a bacterial endospore that is thought to be particularly important in conferring heat resistance on the endospore.

Corticosteroids : Corticosteroids are a family of drugs related to steroids that are naturally produced in the adrenal cortex, such as cortisone. Corticosteroids can kill lymphocytes, especially developing thymocytes, inducing apoptotic cell death. They are useful anti-inflammatory, anti-lymphoid tumor, and immunosuppressive agents.

Cortina : A cob-web like partial veil of silky fibres.

Cyza : See common cold.

Cosmid : A plasmid vector with lambda phage cos sites that can be packaged in a phage capsid; it is useful for cloning large DNA fragments.

Co-stimulaty signal, co-stimulaty molecules : The proliferation of lymphocytes requires both antigen binding and the receipt of a co-stimulatory signal. Co-stimulatory signals are delivered to T cells by the co-stimulatory molecules, B7.1 and B7.2, related molecules that are expressed on the surface of the cell presenting antigen, and which bind the T-cell surface molecule CD28. B cells may receive co-stimulatory signals from common pathogen components such as LPS, from complement fragments, or from CD40 ligand expressed on the surface of an activated anti-genspecific helper T cell.

Cot curve : A curve that indicates the rate of DNA-DNA annealing as a function of DNA concentration and time.

Cottony : Having a loose and coarse texture.

Covalent : Nonionic chemical bond formed by a sharing of electrons between two atoms.

Cowpox : Cowpox is the common name of the disease produced by vaccinia virus, used by Edward Jenner in the successful vaccination against smallpox, which is caused by the related variola virus.

COX : Cyclooxygenase.

cpDNA : Chloroplast DNA.

CPE : Chronic pulmonary emphysema; complete physical examination.

CPG : Controlled porosity glass.

cpm : Counts per minute.

CPX : Complete physical examination.

CR : See complement receptors.

CR : Clinical records; closed reduction.

CR&C : Closed reduction and cast.

CR1, CD35 : CR1 (CD35) is one of several receptors on cells for various components of complement. It is used to remove immune complexes from the plasma.

CR2, CD21 : CR2 (CD21) is part of the B-cell co-receptor complex along with CD19 and CD81. It binds to antigens that have various breakdown products of C3, especially C3d, bound to them and, by cross-linking to the B-cell receptor, enhances sensitivity to antigen by at least a hundredfold. It is also used by the Epstein-Barr virus to invade B cells and produce the symptoms of infectious mononucleosis.

CR3, CD11b:CD18 : CR3 (CD11b:CD18) is a P2 integrin that functions both as an adhesion molecule and as a complement receptor. It binds iC3b, and stimulates phagocytosis.

CR4, CDllc:CD18 : CR4 (CDllc:CD18) is a P2 integrin that binds iC3b and stimulates phagocytosis.

CRAMS : Contract research and manufacturing services.

Cranium : Collectively,bones of the skull.

Crassulacean acid metabolism (CAM) : Process in some plants,named after the plant family Crassulaceae,in which CO_2 is stored in various organic acids for subsequent use in photosynthesis; Crassulaceae is the family to which *Sedum* belongs.

CRC : Custom research company.

C-reactive protein : C-reactive protein is an acute-phase protein that binds to phosphorylcholine, which is a constituent of the C-polysaccharide of the bacterium *Streptococcus pneumoniae*, hence its name. Many other bacteria also have surface phosphorylcholine that is accessible to C-reactive protein, so the protein can bind many different bacteria and opsonize them for uptake by phagocytes. C-reactive protein does not bind to mammalian tissues.

Credibility : Something that can be believed and trusted.

Crenate : Scalloped.

Crepuscular : When an organism is active at dawn and / or dusk.

CRF : Chronic renal failure.

Crista : Inner membrane in a mitochondrion, site of respiration.

Cristae : Infoldings of the inner mitochondrial membrane. ***OR*** The convolutions of the inner mitochondrial membrane.

CRIT : Critical hematocrit.

CRNA : Certified registered nurse anesthetist.

CRO : Clinical research organization or contract research organization.

Crohn's disease : A chronic inflammatory disease of the digestive tract, particularly the small intestine and colon.

CROM : Cervical range of motion.

Crossfeeding : (i) Specific type of syntrophy where two populations cooperate to metabolize a compound. (ii) One organism consuming products excreted by another organism.

Crossing-over : A process in which segments of two adjacent DNA strands are exchanged; breaks occur in both strands, and the exposed ends of each strand join to those of the opposite segment on the other strand.

Cross-matching : Cross-matching is used in blood typing and histocompatibility testing to determine whether donor and recipient have antibodies against each other's cells that might interfere with successful transfusion or grafting.

Cross-reaction : A cross-reaction is the binding of antibody to an antigen not used to elicit that antibody. Thus, if antibody raised against antigen A also binds antigen B, it is said to cross-react with antigen B. The term is used generically to describe the reactivity of antibodies or T cells with antigens other than the eliciting antigen.

Crown Research Institutes : Science research businesses owned by the New Zealand Crown (i.e. The New Zealand Government). They include Agresearch, Crop & Food Research, ESR, Scion, GNS Science, Hortresearch, Industrial Research, Manaaki Whenua Landcare Research, and NIWA.

CRP : Catabolite regulator protein.

CRT : Cell residence time - the amount of time in days that an average "bug" remains in the process. Also termed "sludge age".

Cruciferae : Plant family in which the flowers have four petals arranged opposite each other in a cross shape and six stamens.

Cryopreservation : The cooling of cells, tissues or whole organisms to sub-zero temperatures in order to stop all biological activity. If done in the right way, biological activity can be resumed when the samples are warmed.

Cryptins : Peptides produced by Paneth cells in the intestines. Cryptins are toxic for some bacteria, although their mode of action is not known.

Cryptococcosis : An infection caused by the basidiomycete, Cryptococcus neoformans, which may involve the skin, lungs, brain, or meninges.

***Cryptostroma corticale* :** Conidia dimensions 4-6.5 x 3.5-4 microns. Found on the bark of maple and sycamore trees and on stored logs.

Crystallizable fragment (Fc) : The stem of the Y portion of an antibody molecule. Cells such as macrophages bind to the Fc region, and it also is involved in complement activation.

CS : Cervical spine; cesarean section.

CSF : Colony stimulating factor.

CSF : Cerebrospinal fluid.

Csk, C-terminal Src kinase : The protein known as Csk or C-terminal Src kinase is constitutively active in lymphocytes and has the function of phosphorylating the C-terminal tyrosine of Src-family tyrosine kinases, thus inactivating them.

CSP : Chiral stationary phase.

CSP : Cervical spine.

CT : Chromatin transfer. ***OR*** Carpal tunnel; cerebral thrombosis; cervical-thoracic; cervical traction.

CTB : Confined to bed.

CTC : A compound (5cyano2,3ditolyl tetrazolium chloride) that is fluorogenic, producing intracellular crystals, in response to bacterial respiration.

CTL : Cytotoxic T lymphocyte. ***OR*** Cervical, thoracic, lumbar.

CTLA-4 : CTLA-4 is the high-affinity receptor for B7 molecules on T cells.

CTP : Cytidine triphosphate

CTS : Carpal tunnel syndrome.

CU : Cause unknown; clinical unit.

CUD : Cause undetermined.

Cull : Select for killing, often because of poor quality or surplus numbers.

Cultivar : A variety of plant created by selective breeding, usually under carefully controlled conditions.

Culture : Population of microorganisms cultivated in an artificial growth medium. A pure culture is grown from a single cell; a mixed culture consists of two or more microbial species or strains growing together.

Cunninghamella sp. **:** Can cause disseminated and pulmonary infections in immune compromised hosts.

Curvularia (Hyphomycetes) : Teleomorph: Cochliobolus (Ascomycetes) A common saprobe found in soil, plants, cereals, and cellulosic materials such as paper and archives. Some species are plant pathogens but can also occur indoors. It is allergenic and may cause infections in immunocompromised people.

Curvularia sp. **:** Reported to be allergenic. It may cause corneal infections, mycetoma and infections in immune compromised hosts.

Cutaneous : A skin disease caused by *Corynebacterium diphtheriae* that infects wound or skin lesions, causing a slow-healing ulceration.

Cutaneous anthrax : A form of anthrax involving the skin.

Cutaneous lymphocyte antigen, CLA : Cutaneous lymphocyte antigen (CLA) is a cell-surface molecule that is involved in lymphocyte horning to the skin in humans.

Cutaneous T-cell lymphoma : Cutaneous T-cell lymphoma is a malignant growth of T cells that home to the skin.

Cuticle : The surface layer of the cap or stem, usually distinct from the flesh.

Cuvette : A small, transparent, often tubular laboratory vessel.

CV : Cardiac volume; cervical vertebra; color vision.

CVA : Cerebrovascular accident.

CVD : Cardiovascular disease; cerebrovascular disease.

CVS : Cardiovascular status.

CW : Chest wall; crutch walking.

CX : Cervix; chest x-ray; consciousness.

Candidal vaginitis : Vaginitis caused by Candida species.

CXR : Chest x-ray.

Cyanobacteria : A major grouping within bacteria; produce carbohydrates and oxygen through photosynthesis. Found in freshwater and marine environments and may be solitary or colonial. Used to be referred to as the blue green algae. ***OR*** A large group of bacteria that carry out oxygenic photosynthesis using a system like that present in photosynthetic eucaryotes.

Cyanobacterium : Prokaryotic, oxygenic phototrophic bacterium containing chlorophyll a and phycobilins, formerly the "bluegreen algae."

Cyanophage : Virus that infects cyanobacteria.

Cyanophycin : Nonprotein nitrogen storage polymer in cyanobacteria.

Cyclic photophosphylation : Formation of ATP when light energy is used to move electrons cyclically through an electron transport chain during photosynthesis.

Cyclophosphamide : Cyclophosphamide is a DNA alkylating agent that is used as an immunosuppressive drug. It acts by killing rapidly dividing cells, including lymphocytes proliferating in response to antigen.

***Cyclospa cayetanensis* :** *Cyclospora cayetanensis* is an acidfast, coccidianlike, parasitic protozoa with the size of 810 micrometer in diameter. *Cyclospora cayetanesis* was recently recognized as a new (intestinal) protozoan pathogen of human (1993, species name was proposed in 1994) and was identified as the cause of the prolonged diarrhoea of travelers as well as immunocompetent and immunocompromised patients.

Cyclospin A : Cyclosporin A is a powerful immuno-suppressive drug that inhibits signaling from the T-cell receptor, preventing T-cell activation and effector function. It binds to cyclophilin, and this complex binds to and inactivates the serine / threonine phosphatase calcineurin.

Cylindrical : Cylindric, having parallel walls and circular cross-section.

Cyperaceae : Monocot plant family commonly known as sedges in which the stems are triangular in cross-section.

Cys : Cysteine.

Cyst : Resting stage formed by some bacteria, nematodes, and protozoa in which the whole cell is surrounded by a protective layer; not the same as endospore.

Cystic fibrosis : An inherited disease which affect the lungs and pancreas impairing breathing and digestion.

Cysticercoid : A single, evaginated scolex that is embedded in a small solid cyst that are typically found in small intermediate hosts such as arthropods (Cyclophyllidian tapeworms).

Cysticercus : Larval stage in the cestode life cycle which is a fluid filled cyst containing an attached single invaginated scolex typically found in mammal intermediate hosts (Cyclophyllidian tapeworms).

Cystidium : Special large sterile cells amongst the basidia on the gills of many fungi – often of distinctive shape and used in classification. Also found on surfaces of cap and stem.

Cytidine : A pyrimidine nucleoside found in DNA and RNA.

Cytochrome : Ironcontaining porphyrin ring (e.g., heme) complexed with proteins which act as electron carriers in an electrontransport chain. ***OR*** Heme proteins that carry electrons, usually as members of electron transport chains. ***OR*** An electron-carrying molecule with a porphyrin ring in its center.

Cytochrome c oxidase subunit : Part of the enzyme cytochrome c enzyme that is involved in cellular respiration. The gene encoding this subunit is used for DNA barcoding.

Cytokine : A general term for nonantibody proteins, released by a cell in response to inducing stimuli, which are mediators that influence other cells. Are produced by lymphocytes, monocytes, macrophages, and other cells.

Cytokine capture : See antigen capture.

Cytokine receptors : Cytokine receptors are cellular receptors for cytokines. Binding of the cytokine to the cytokine receptor induces new activities in the cell, such as growth, differentiation, or death.

Cytokines : Soluble molecules produced by cells that serve as mediators of intracellular reactions and biological response modifiers (e.g., lymphokinesor cytokines are proteins made by cells that affect the behavior of other cells. Cytokines made by lymphocytes are often called lymphokines or interleukins (abbreviated il), but the generic term cytokine is used in this book and most of the literature. Cytokines act on specific cytokine receptors on the cells that they affect. See also chemokines.

Cytokinesis : The process of division of the cytoplasm to form two cells.

Cytology : The study of cells.

Cytomegalovirus inclusion disease : An infection caused by the cytomegalovirus and marked by nuclear inclusion bodies in enlarged infected cells.

Cytopathic effect : The observable change that occurs in cells as a result of viral replication. Examples include ballooning, binding together, clustering, or even death of the cultured cells.

Cytoplasm : The fluid and organelles within a cell. ***OR*** Cellular contents inside the cell membrane, excluding the nucleus. ***OR*** The contents enclosed by the plasma (or cytoplasmic) membrane, excluding the nucleus.

Cytoplasmic matrix : The protoplasm of a cell that lies within the plasma membrane and outside any other organelles. In bacteria it is the substance between the cell membrane and the nucleoid.

Cytoplasmic membrane : Selectively permeable membrane surroun-ding the cell's cytoplasm.

Cytoproct : Site on a protozoan where undigestible matter is expelled.

Cytosine : A pyrimidine 2-oxy-4-aminopyrimidine found in nucleosides, nucleotides, and nucleic acids. ***OR*** A pyrimidine base found in DNA and RNA.

Cytoskeleton : Network of proteins making up the internal skeleton of a cell. Gives the cell its characteristic shape and enables it to move materials within itself and out from it. ***OR*** A network of microfilaments, microtubules, intermediate filaments & other components in the cytoplasm of eucaryotic cells that helps. ***OR*** The filamentous skeleton, formed in the eukaryotic cytoplasm, that is largely responsible for controlling cell shape.

Cytosol : The liquid portion of the cytoplasm, including the macromolecules but not including the larger structures like subcellular organelles or cytoskeleton.

Cytostome : A permanent site in the protozoan ciliate body through which food is ingested. ***OR*** Cell opening through which protozoa ingest food (mouth) or secrete wastes (anus).

Cytotoxic T (TC) cell : A cell that is capable of recognizing virus-infected cells through the major histo-compatability molecules and developing into an activated cell that destroys the infected cells. T cells that can kill other cells are called cytotoxic T cells. Most cytotoxic T cells are MHC class-I restricted CD8 T cells, but CD4 T cells can also kill in some cases. Cytotoxic T cells are important in host defense against cytosolic pathogens.

Cytotoxic T lymphocyte (CTL) : The activated T cell that can attack and destroy virus-infected cells, tumor cells, and foreign cells.

Cytotoxin : A toxin, or poison, that has a negative effect on living cells. ***OR*** A toxin or antibody that has a specific toxic action upon cells; cytotoxins are named according to the cell for which they are specific (e.g., nephrotoxin). ***OR*** Cytotoxins are proteins made by cytotoxic T cells that participate in the destruction of target cells. Perforins and granzymes are the major defined cytotoxins.

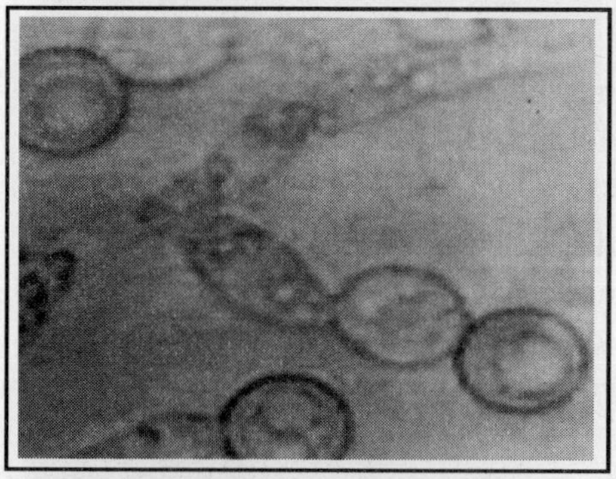

Candida

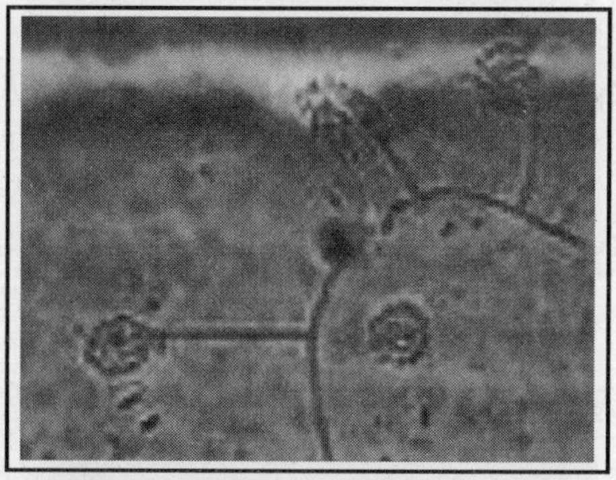

Cephalosporium

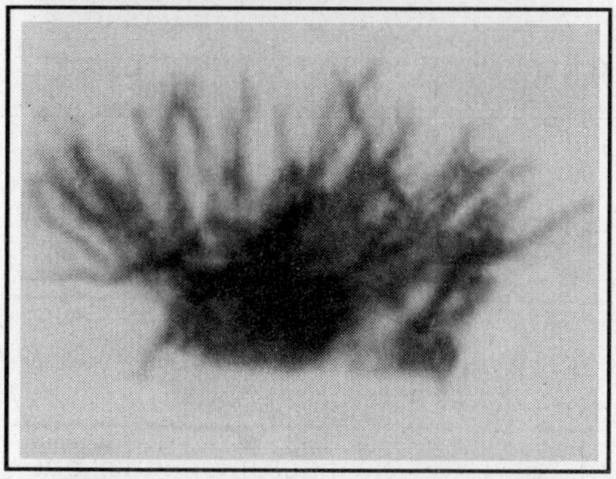

Cercospora

Chaetomium

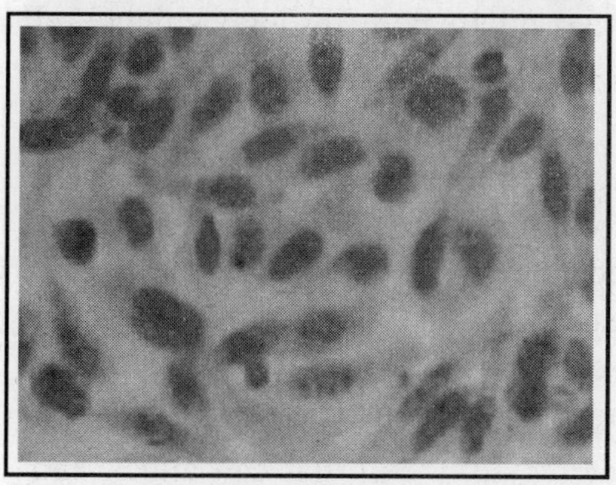

Chlamydiae

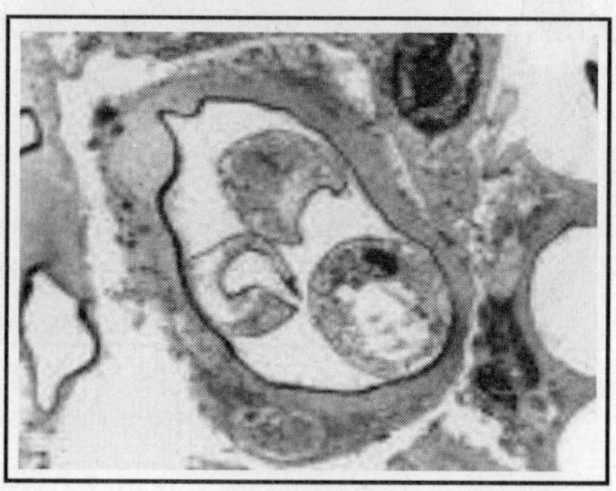

Chytrid

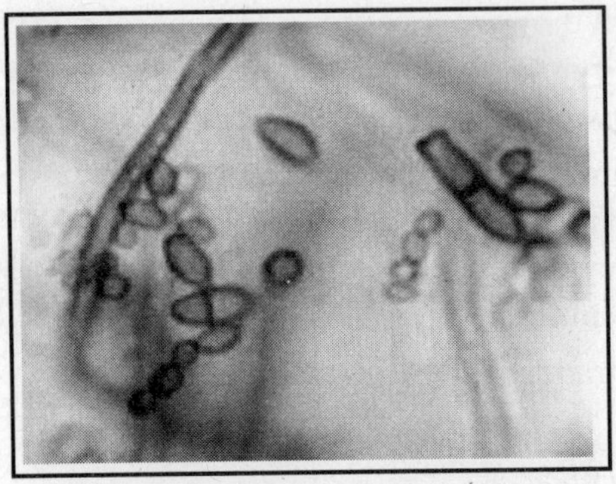

Cladosporium

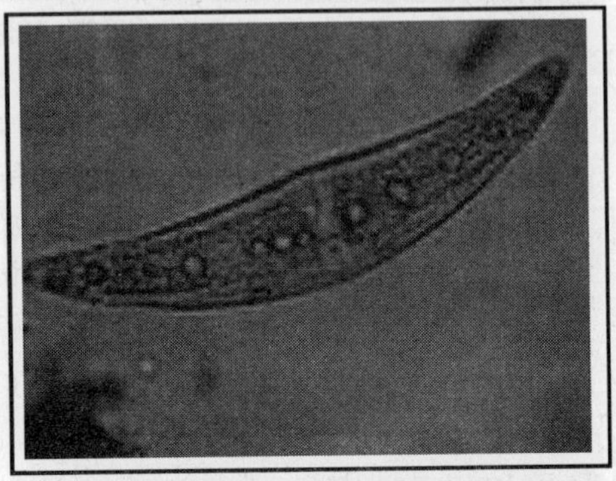

Closterium

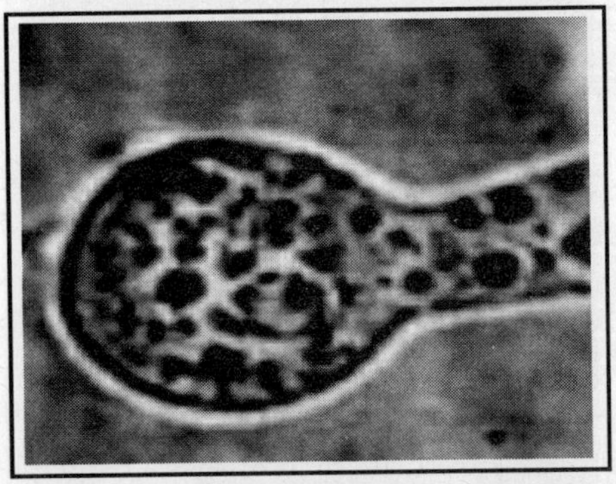

Conidiobolus

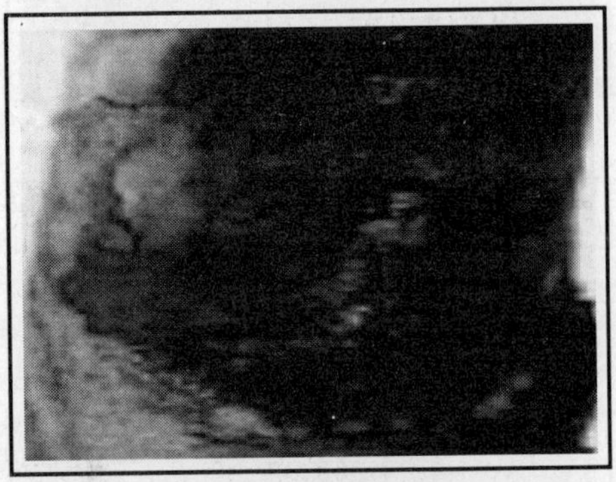

Cryptostroma

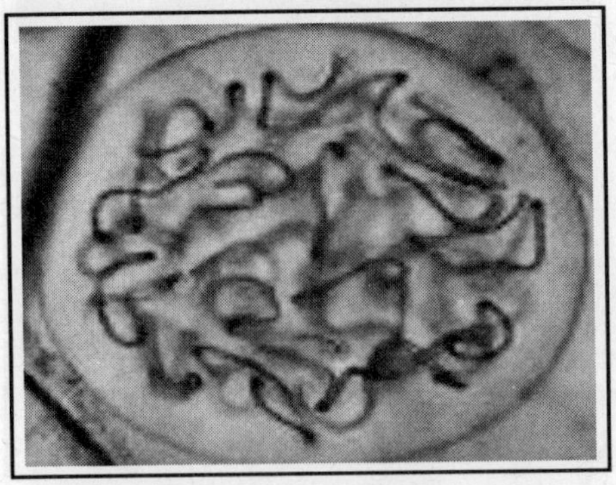

Cynobacteria

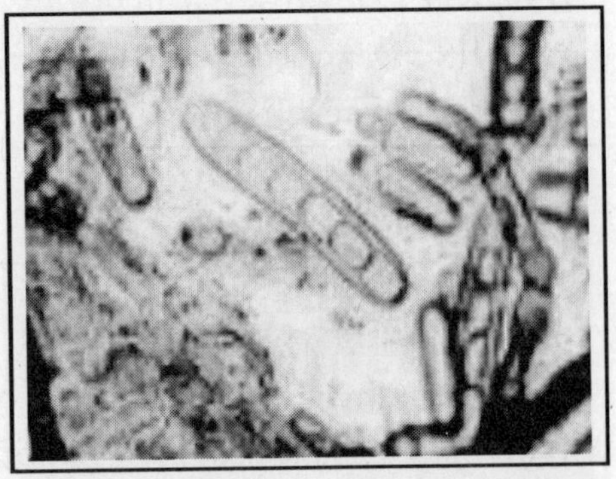

Dreschlera

D

D gene segments, diversity gene segments : D gene segments, or diversity gene segments, are short DNA sequences that join the V and J gene segments in rearranged immunoglobulin heavy-chain genes and in T-cell receptor β and δ chain genes. See gene segments.

D loop : An extended loop of single-stranded DNA displaced from a duplex structure by an oligonucleotide.

D value : See decimal reduction time.

D. Phil : An advanced research degree that is usually completed after doing a first degree. D.Phil. is the abbreviation given by Oxford University for Doctor of Philosophy, whereas other universities normally use the abbreviation Ph.D.

D/A : Date of accident; date of admission.

D/DW : Dextrose in distilled water.

D/I unit : Deionizing unit, frequntly used to maintain water quality in aquariums. Advantages does not waste water like the R/O unit, is designed to be hooked up to either a faucet or household piping system, the anion & cation resins can be regenerated (with another expensive unit) indefinitely, and these systems allow a larger water flow (up to 2,000 gallons a day), than an R/O system, but cost dramatically more too.

D/NS : Dextrose in normal saline.

D5%DW : Dextrose 5% in distilled water.

D5%NS : Dextrose 5% in normal saline.

D5LR : Dextrose 5% with lactated ringers.

D5W : Dextrose 5% in water.

Da : Dalton.

DA : Degenerative arthritis.

dAb : Single domain antibody.

DAF : Dissolved air flotation-one of many designs for waste treatment.

DAF : DNA amplification fingerprinting.

DAG : Directed acyclic graph.

Dalton : A unit of mass equivalent to the mass of a hydrogen atom (1.66 x 10^{-24} g).

DAMD : Directed amplification of minisatellite DNA.

Dane particle : A 42 nm spherical particle that is one of three that are seen in hepatitis B virus infections. The Dane particle is the complete virion.

Dark reactions : Reactions that can occur in the dark, in a process that is usually associated with light, such as the dark reactions of photosynthesis.

Dark reactivation : The excision and replacement of thymine dimers in DNA that occurs in the absence of light.

Dark zone : See germinal centers.

Dark-field microscopy : Microscopy in which the specimen is brightly illuminated while the background is dark.

DAT : Diet as tolerated.

dATP : Deoxy-adenosine triphosphate. One of the four types of nucleotides used to make a DNA molecule. The other nucleotides are dTTP (deoxy-thymine triphosphate), dGTP (deoxy-guanine triphosphate) and dCTP (deoxy-cytosine triphosphate)**.** ***OR*** deoxyadenosine 5'-triphoshate.

DB : Disability; date of birth.

DBP : Diastolic blood pressure.

DC : Discharge; discontinue; Doctor of Chiropractic.

DC&B : Dilation, curettage and biopsy.

dCTP : Deoxycytidine 5'-triphosphate.

dd : Dideoxy.

DD : Discharge diagnosis; dry dressing.

DDD : Degenerative disc disease.

ddNTP : Dideoxynucleotide (triphoshate).

ddRNAi : DNA-directed RNAi.

DDS : Drug delivery system.

DDT : Dichlorodiphenyltrichloroethane. An insecticide used by farmers in the 1950's to kill pests like grass grubs. It is toxic to humans and other mammals when it is swallowed or absorbed through the skin.

De novo pathway : A biochemical pathway that starts from elementary substrates and ends in the synthesis of a biochemical.

DEAE : Diethylaminoethyl.

Deamination : The removal of amino groups from amino acids. ***OR*** The enzymatic removal of an amine group, as in the deamination of an amino acid to an alpha keto acid.

Death domains : Death domains were originally defined in proteins encoded by genes involved in programmed cell death or apoptosis, and are now known to be involved in protein-protein interactions.

Death phase : The decrease in viable microorganisms that occurs after the completion of growth in a batch culture.

DEC : Deceased; decrease.

Decant : To gently pour a liquid into a new container so that any sediment or solid material in the liquid is left behind.

Decay-accelerating factor DAF, CD55 : The decay-accelerating factor (DAF or CD55) is a cell-surface molecule that protects cells from lysis by complement. Its absence causes the disease paroxysmal nocturnal hemoglobinuria.

Decimal reduction time : The time required to kill 90% of the microorganisms or spores in a sample at a specified temperature.

Declaration : A written statement of facts.

Declining growth : A growth phase in which the availability of food begins to limit cell growth.

Decomposer : Heterotrophic organism that breaks down organic compounds. ***OR*** Chemical breakdown of a compound into simpler compounds, often accomplished by microbial metabolism OR An organism that breaks down complex materials into simpler ones, including the release of simple inorganic products. Often a decomposer such as an insect or earthworm physically reduces the size of substrate particles.

Decurrent : Gills running down the stem (like chanterelles).

DEF : Defecation; deferred; deficient.

Defective endogenous retroviruses : Defective endogenous retroviruses are partial retroviral genomes integrated into host cell DNA and carried as host genes. There are a great many defective endogenous retroviruses in the mouse genome.

Defensin : Small antimicrobial cationic peptides that are widely distributed and disrupt membrane integrity; Abundant in neutrophilic granules. ***OR*** Specific peptides produced by neutrophils that permeabilize the outer and inner membranes of certain microorganisms, thus killing them.

Defined medium : Culture medium made with components of known composition. ***OR*** Medium whose exact chemical composition is quantitatively known.

Definitive host : Found in all parasitic life cycles. The definitive host is the host in which the sexual stages of the parasite ("adults") mature. If there are no sexual stages in the life cycle, then the definitive host is the host in the life cycle considered most important to humans.

DEG : Degeneration.

Degradation : A growth phase in which the availability of food begins to limit cell growth. ***OR*** Process whereby a compound is usually transformed into simpler compounds.

Degradative enzyme class : Of protein capable of catalytic catabolic reactions.

Dehalespiration : The utilization of haloorganic compounds by anaerobes as terminal electron acceptors.

Dehydration synthesis : Removal of water from two smaller molecules to bond them together in to one, larger molecule.

Dehydrogenase : An enzyme that catalyzes the removal of a pair of electrons (and usually one or two protons) from a substrate molecule.

Delayed-type hypersensitivity : Delayed-type hypersensitivity, or type IV hypersensitivity, is a form of cell-mediated immunity elicited by antigen in the skin and is mediated by CD4 TH1 cells. It is called delayed-type hypersensitivity because the reaction appears hours to days after antigen is injected. Cf. Immediate hypersensitivity.

Delta agent : A defective RNA virus that is transmitted as an infectious agent, but cannot cause disease unless the individual is also infected with the hepatitis B virus. See hepatitis D.

Delta-proteobacteria : One of the five subgroups of proteobacteria, each with distinctive 16S rRNA sequences. Chemoorganotrophic bacteria that usually are either predators on other bacteria or anaerobes that generate sulfide from sulfate and sulfite.

Deltoid : The shoulder muscle.

Dematiaceous : A dark brown, greenish gray or black colour.

Demographics : The study of vital statistics (birth and death rates) of populations.

Denaturation : Process where doublestranded DNA unwinds and dissociates into two single strands. The reverse of DNADNA hybridization. ***OR*** A change in the shape of an enzyme that destroys its activity; the term is also applied to changes in nucleic-acid shape. ***OR*** The disruption of the native folded structure of a nucleic acid or protein molecule; may be due to heat, chemical treatment, or change in pH. ***OR*** Loss of a protein's normal three-dimensional shape,its native conformation.

Denature : The process where by a protein looses its native conformation.

Dendrite : The "incoming" end of a neuron.

Dendritic cell : A key part of the immune system; these cells present foreign substances to other immune system cells. ***OR*** Dendritic cells, also known as interdigitating reticular cells, are found in T-cell areas of lymphoid tissues. They have a branched or dendritic morphology and are the most potent stimulators of T-cell responses. Nonlymphoid tissues also contain dendritic cells, but these are not able to stimulate T-cell responses until they are activated and migrate to lymphoid tissues. The dendritic cell derives from bone marrow precursors. It is distinct from the follicular dendritic cell that presents antigen to B cells.

Dendritic epidermal T cells, (dETCs) : Dendritic epidermal T cells (detcs), are a specialized class of $\gamma : \delta$ T cells found in the skin of mice and some other species, but not humans. All detcs have the same $\gamma : \delta$ T-cell receptor; their function is unknown.

Dendrogram : A treelike diagram that is used to graphically summarize mutual similarities and relationships between organisms.

Denitrification : Reduction of nitrate or nitrite to molecular nitrogen or nitrogen oxides by microbial activity (dissimilatory nitrate reduction) or by chemical reactions involving nitrite (chemical denitrification) ***OR*** An anaerobic biological reduction of nitrate nitrogen to nitrogen gas, the removal of total nitrogen from a system, and/or an anaerobic process that occurs when nitrite ions are reduced to nitrogen gas and bubbles are formed as a result of this process. The bubbles attach to the biological floc in the activated sludge process and float the floc to the surface of the secondary clarifiers. This condition is often the cause of rising sludge observed in secondary clarifiers or gravity thickeners. (See Nitrification).

Density : Mass per volume of a substance or solution (g/ml).

Density-gradient centrifugation : The separation, by centrifugation, of molecules according to their density, in a gradient varying in solute concentration.

Dental plaque (plak) : A thin film on the surface of teeth consisting of bacteria embedded in a matrix of bacterial polysaccharides, salivary glycoproteins, and other substances.

Denticle : A small projection or peg on which conidia are produced.

Deoxyribo nucleicacid (DNA) : The genetic code material, consisting of pairs of nitrogenous bases bonded to a "ladder"of phosphate and ribose molecules.

Deoxyribonucleic acid : (DNA) Polymer of nucleotides connected via a phosphatedeoxyribose sugar backbone; the genetic material of the cell. ***OR***

The nucleic acid that constitutes the genetic material of all cellular organisms. It is a polynucleotide composed of deoxyribo-nucleotides connected by phosphodiester bonds.

Depressed : Used to describe a cap where the central portion lower than the margin.

Derepressible enzyme : Enzyme that is produced in the absence of a specific inhibitory compound acting at the transcriptional level.

Dermatitis : An inflammation of the skin making it red and sometimes dry, flaky or itchy.

Dermatomycosis : A fungal infection of the skin; the term is a general term that comprises the various forms of tinea, and it is sometimes used to specifically refer to athlete's foot (tinea pedis).

Dermatophyte : A fungus parasitic on the skin or a fungus belonging to the genera Epidermophyton, Microsporum, or Trichophyton with the ability to utilize keratin to infect hair, nail and skin.

Desensitization : Desensitization is a procedure in which an allergic individual is exposed to increasing doses of allergen in hopes of inhibiting their allergic reactions. It probably involves shifting the balance between CD4 TH1 and TH2 cells and thus changing the antibody produced from IgE to IgG. ***OR*** To make a sensitized or hypersensitive individual insensitive or nonreactive to a sensitizing agent.

Desert crust : A crust formed by microbial binding of sand grains in the surface zone of desert soil; crust formation primarily involves cyanobacteria.

Designer baby : A term used to describe children whose genetic make-up is selected by their parents before birth.

Detachment : Transfer of biofilm particulate constituents (cells, polymers, precipitates) from the biofilm to the fluid bathing the biofilm.

Detergent : An organic molecule, other than a soap, that serves as a wetting agent and emulsifier; it is normally used as cleanser, but some may be used as antimicrobial agents.

Determinant spreading : See epitope spreading.

Determinate conidiophore : The conidiophore does not alter in length after the formation of conidia.

Detrimental : Harmful.

Detritus : Dead plant and animal matter, usually consumed by bacteria, but some remains.

Detritus : Partially decomposed leaf litter.

Deuteromycetes : An artificial subdivision to accommodate those fungi where only the asexual state is known. ***OR*** In some classification systems, the deuteromycetes or Fungi Imperfecti are a class of fungi. These organisms either lack a sexual stage or it has not yet been discovered.

DEV : Deviation.

Dew Point : The temperature to which air with a given quantity of water vapor must be cooled to cause condensation of the vapor in the air.

Dextrinoid : Staining brick red with Meltzers iodine.

Dextrose : Another name for glucose because it rotates polarized light to the right.

DF : Dorsiflexion.

DG : Sn-1,2-diacylglycerol.

DGGE : Denaturing gradient gel electrophoresis.

dGTP : Deoxyguanosine 5'-triphoshate.

DHAP : Dihydroxyacetone phosphate.

DHF : Dihydrofolate.

DHF : Dorsi-hyperflexion.

DHFR : Dihydrofolate reductase.

DI : Diabetes insipidus; date of injury.

Diabetes : A disease where there is too much sugar in the blood. This disease is caused by either too little insulin in the blood (type I diabetes) or insensitivity to insulin (type II diabetes).

Diabetes : A high blood sugar level often due to insufficient insulin secretion.

Diacylglycerol, DAG : Diacylglycerol (DAG) is most commonly released from inositol phospholipids by the action of phospholipase C-b. Diacylglycerol production is stimulated by the ligation of many receptors & it acts as an intracellular signaling molecule, activating cytosolic protein kinase C, which further propagates the signal.

Diadelphous : In a flower, having the stamens arranged in two bundles or sets by the fusion of the filaments.

Diagnostics : The science and practice of diagnosis of disease.

Dialysis : Removal of small molecules from a macromolecule preparation by allowing them to pass across a semipermeable membrane. ***OR*** Diapedesis is the movement of blood cells, particularly leukocytes, from the blood across blood vessel walls into tissues.OR passage of leukocytes and other plasma proteins out through blood vessel walls and into surrounding tissues.

Diaphragm : A thin membrane that separates the gleba from the stem in a puffball. ***OR*** The muscle which draws air into the lungs and separates the chest cavity from the abdominal cavity.

Diarrhoea : Frequent and watery bowel movements. ***OR*** Very loose,watery feces ; this can be dangerous if too much water is lost, causing dehydration.

Diastole : Residual blood pressure in the arteries when the heart is relaxed.

Diatom : Alga with siliceous cell walls that persist as a skeleton after death. Any of the microscopic unicellular or colonial alga constituting the class Bacillariophyceae.

Diatomaceous earth : Geologic deposit of fine, grayish siliceous material composed chiefly or wholly of the remains of diatoms. It may occur as a powder or as a porous, rigid materiaor l. A fine, siliceous (made of silica) "earth" composed mainly of the skeletal remains of diatoms (single cell microscopic algae with rigid internal structure consisting mainly of silica). Tests prove that DE leaches unacceptable amounts of silicate into the water for fish health. If used as a filter substance, a silicone removing resin should be employed afterwards.

Diauxic growth : A biphasic growth pattern or response in which a microorganism, when exposed to two nutrients, initially uses one of them for growth and then alters its metabolism to make use of the second. ***OR*** Biphasic growth on a mixture of two carbon sources in which one carbon source is used up before the other one. For example, in the presence of glucose.

Diazotroph : Organism that can use dinitrogen as its sole nitrogen source, i.e. Capable of N_2 fixation.

Dichotomous : A type of hyphal branching into two equal forks.

Dicotyledones : The Class (in Division Spermatophyta; Subdivision Angiospermae) of plants that Have two nutrient storage are as in their seeds.

Dictyoconidium (pl. dictyoconidia) : A conidium with both longitudinal and transverse septa; a muriform conidium.

Dicyma (Hyphomycetes) : Teleomorph: ascotricha (Ascomycetes)

Dideoxynucleotide : A two celled conidium. ***OR*** A modified nucleotide that is similar to the nucleotides used to make DNA, but that contains one less oxygen atom. DNA is made by adding nucleotides together. When a dideoxynucleotide is added, however, no other nucleotides can be added in. Dideoxynucleotides are used in DNA sequencing.

didN : Dideoxynucleotide.

Difference spectra : Plots comparing the absorption spectra of a molecule or an assembly of molecules in different states, for example, those of mitochondria under oxidizing or reducing conditions.

Differential centrifugation : Separation of molecules and/or organelles by sedimentation rate.

Differential interference contrast (DIC) microscope : A light microscope that employs two beams of plane polarized light. The beams are combined after passing through the specimen and their intereference is used to create the image.

Differential media : Culture media that distinguish between groups of microorganisms based on differences in their growth and metabolic products.

Differential signaling hypothesis : The differential signaling hypothesis proposes that qualitatively different antigens might mediate the positive and negative selection of T cells in the thymus. Cf. Avidity hypothesis.

Differential staining procedures : Staining procedures that divide bacteria into separate groups based on staining properties.

Differentiate : To change into a cell type with a specialised function, such as a heart, muscle, or brain cell.

Differentiated cell : A cell in a multicellular organism that has been allocated a specific 'job' and has made changes in the way it works in order to do that job. In plants, some differentiated cells can switch back to being undifferentiated, or unspecialised. In animals, differentiated cells can not change back to being undifferentiated.

Differentiation antigens : Differentiation antigens are proteins detected on some cells by means of specific antibodies. Many differentiation antigens have important functional roles characteristic of the differentiated phenotypes of the cell on which they are expressed, such as cell-surface immunoglobulin on B cells.

Differentiation : A change in the form and pattern of a cell and the genes it expresses as a result of growth and replication, usually during development of a multicellular organism. Also occurs in microorganisms (e.g. In sporulation).

Diffused air aeration : A diffused air activated sludge plant takes air, compresses it, and then discharges the air below the water surface of the aerator through some type of air diffusion device.

Diffusely adhering *E. coli* (DAEC) : DAEC strains of E. Coli adhere over the entire surface of epithelial cells and usually cause diarrheal disease in immunologically naive and malnourished children.

Diffusion (nutrient) : Movement of nutrients in soil that results from a concentration gradient.

DiGeorge's syndrome : Digeorge's syndrome is a recessive genetic immunodeficiency disease in which there is a failure to develop thymic epithelium, and is associated with absent parathyroid glands and large vessel anomalies. It seems to be due to a developmental defect in neural crest cells.

Digester : A tank in which sludge is placed to allow decomposition by microorganisms. Digestion may occur under anaerobic (most common) or aerobic conditions.

Digestion : The process of breaking food down into molecules small enough to be absorbed by the body.

Digestive system : The group of organs, including the stomach and intestines, that are involved in the breakdown of food so that the nutrients can be absorbed by the body, are referred to as the digestive system.

Digestive tract (alimentary canal) : The "tube" that starts at the mouth and continues to the anus and is technically "outside" of the body proper consisting of the mouth, pharynx, esophagus, stomach, small intestine and large intestine.

Dikaryon : Two nuclei present in the same hyphal compartment; they constitute a homokaryon when both nuclei are genetically the same or a heterokaryon when each nucleus is genetically different from the other.

Dikaryotic stage : In fungi, having pairs of nuclei within cells or compartments. Each cell contains two separate haploid nuclei, one from each parent.

DIL : Dilated.

Diluent : A substance,such as water,used to dilute or dissolve another substance,there by forming a solution.

Dilution factor : The ratio of final volume (diluent + aliquot) divided by the initial aliquot volume ($V_f \div V_i$).

Dilution plate count method : Method for estimating the viable numbers of microorganisms in a sample. The sample is diluted serially and then transferred to agar plates to permit growth and quantification of colonyforming units.

DIM : Diminished.

Dimer : Structure resulting from the association of two subunits.

Dimitic : When two types of hyphae are present in tissues of members of Polyporaceae.

Dimorphic : Having two different morphological forms.

Dinitrogen fixation : Conversion of molecular dinitrogen (N_2) to ammonia and subsequently to organic combinations or to forms useful in biological processes.

Dinoflagellata : A Division (Phylum) of organisms within Kingdom Protista.

Dinoflagellate : An algal protist characterized by two flagella used in swimming in a spinning pattern. Many are bioluminescent and an important part of marine phytoplankton, some also are important marine pathogens.

Dinosteranes/Dinosteroids : Chemicals found in dinoflagellates, which have been useful in documenting their existence early in the fossil record.

Dioecious : Referring to a plant which has separate male and female flowers. On separate male and female plants.

Dipeptide, Tripeptide : Two or three aminoacids bonded together.

Diphtheria : An acute, highly contagious childhood disease that generally affects the membranes of the throat and less frequently the nose. It is caused by *Corynebacterium diphtheriae*.

Dipicolinic acid : A substance present at high concentrations in the bacterial endospore. It is thought to contribute to the endospore's heat resistance.

DIPJ : Distil interphalangeal joint.

Diplococcus : A pair of cocci.

Diploid : In eukaryotes, an organism or cell with two chromosome complements, one derived from each haploid gamete. ***OR*** Having two sets of chromosomes.

Diploid cell : A cell that contains two chromosomes (2N) of each type.

Dipole : A separation of charge within a single molecule.

Direct Coombs test : The direct Coombs test uses anti-immunoglobulin to agglutinate red blood cells as a way of detecting whether they are coated with antibody *in vivo* due to autoimmunity or maternal anti-fetal immune responses (see Coombs test; indirect Coombs test).

Direct count : Method of estimating the total number of microorganisms in a given mass of soil by direct microscopic examination.

Directed mutagenesis : In a DNA sequence, an intentional alteration that can be genetically inherited.

Directed- adaptive mutation : A mutation that seems to be chosen so the organism can better adapt to its surroundings.

DIS : Disabled; disease.

Disaccharide : A slightly more complex sugar consisting of two monosaccharide bonded together by dehydration synthesis.

Disclimax : Shortened from "disturbance climax," used to describe an ecosystem in which the climax Community is held at a "lower" level due to repeated, unpredictable events (for example,maintenance of a prairie area by periodic fires which kill invading trees).

Discontinuous epitopes : See conformational epitopes.

Disease : A deviation or interruption of the normal structure or function of any part of the body that is manifested by a characteristic set of symptoms and signs. ***OR*** An abnormal condition of body function(s) or structure that is considered to be harmful to the affected individual (host); any deviation from or interruption of the normal structure or function of any part, organ, or system of the body; dependent upon the dosage and virulence of the organism/agent and varies inversely with the resistance of the host

Disease syndrome : A set of signs and symptoms that are characteristic of the disease.

DISI : Dorsal intercalary segment instability.

Disinfection : Agent that kills microorganisms. ***OR*** An agent, usually chemical, that disinfects; normally, it is employed only with inanimate objects. ***OR*** The process designed to kill most microorganisms in wastewater, including essentially all pathogenic (disease-causing) bacteria. There are several ways to disinfect, with chlorine being the most frequently used in water and wastewater treatment plants.

Disjunct cell : An empty cell that fragments and/or undergoes lysis to release a conidium.

DISL : Dislocate; dislocation.

DISP : Disposition.

Disseminated intravascular coagulation (DIC) : Disorder characterized by a reduction in the elements involved in blood coagulation due to their utilization in widespread blood clotting within the vessels; late stages marked by profuse hemorrhaging.

Dissimilaty nitrate reduction : The process in which some bacteria use nitrate as the electron acceptor at the end of their electron transport chain to produce ATP. The nitrate is reduced to nitrite or nitrogen gas. ***OR*** (DNRA) Use of nitrate by organisms as an alternate electron acceptor in the absence of oxygen resulting in the reduction of nitrate to ammonium. ***OR*** Use of nitrate by organisms as an alternate electron acceptor in the absence of oxygen resulting in the reduction of nitrate to ammonium.

Dissociation constant : An equilibrium constant for the dissociation of a molecule into two parts (e.g., dissociation of acetic acid into acetate anion and a proton); Kd.

Dissolved solids : Chemical substances either organic or inorganic that are dissolved in a waste stream and constitute the residue when a sample is evaporated to dryness.

DIST : Distal; distended.

Distant : Of gills that are widely spaced.

Distribution box : Serves to distribute the flow from the septic tank evenly to the absorption field or seepage pits. It is important that each trench or pit receive an equal amount of flow. This prevents overloading of one part of the system.

Distribut : The rotating mechanism that distributes the wastewater evenly over the surface of a trickling filter or other process unit.

Disulfide bridge : A covalent linkage formed by oxidation between two cysteine SH groups either in the same polypeptide chain or in different polypeptide chains. Reversible by adding reducing agents.

Diuretic : A substance which stimulates the kidneys to excrete more water, which promotes urination.

Diurnal : Active during daytime.

Diurnal oxygen shifts : The changes in oxygen levels that occur in waters when algae produce and use oxygen on a cyclic basis during day and night.

Diverge : Separate or become different.

Divergent : Used to describe the central flesh of gills where hyphae turn outwards from the mid-line.

DJD : Degenerative joint disease.

DKA : Did not keep appointment.

DKB : Deep knee bends.

DLS : Dynamic light scattering.

DM : Diabetes mellitus; diastolic murmur.

DMD : Disease-modifying drug.

DMF : N,N-dimethylformamide.

DMS : Dimethyl sulfate.

DMSO : Dimethyl sulfoxide.

DNA : A molecule shaped like a twisted ladder that carries genetic information in the (deoxyribonucleic acid) nucleus of a cell. ***OR*** The genetic material found in cells is called deoxyribonucleic acid (DNA).

DNA barcode : The unique nucleotide sequence of a specific part of the genome, usually the CO1 gene. Within-species variations in this gene are much smaller than the variations that occur between species.

DNA cloning : The propagation of individual segments of DNA as clones.

DNA fingerprint : The unique sequence of DNA bases in the genome of an individual organism.

DNA fingerprinting : Molecular genetic techniques to assess possible differences among DNA in a samples.

DNA library : Collection of cloned DNA fragments which in total contain genes from the entire genome of an organism; also called a gene library. ***OR*** A mixture of clones, each containing a cloning vector and a segment of DNA from a source of interest.

DNA ligase : An enzyme that joins two DNA fragments together through the formation of a new phosphodiester bond.

DNA microarrays (DNA chips) : DNA microarrays are created by placing a different DNA on a small part of a microchip, and using them to assess RNA expression in normal or malignant cells. ***OR*** Solid supports that have DNA attached in highly organized arrays and are normally used to evaluate gene expression.

DNA polymerase : A DNA sequence that is highly variable between individuals of a species. ***OR*** An enzyme that synthesizes new DNA using a parental DNA strand as a template. ***OR*** An enzyme that catalyzes the formation of 3'-5' phosphodiester bonds from deoxy-ribonucleotide triphosphates.

DNA profile : A pattern obtained by a scientific process that reflects the unique DNA sequence found in a living organism. See also DNA.

DNA replication : The copying of DNA prior to cell division. This involves an unwinding of the DNA double helix into two separate strands which are then both copied.

DNA sequencing : DNA sequencing is used to determine the exact sequence of nucleotides in a sample of DNA. To find out more, get in-the-lab sheet: DNA sequencing.

DNA vaccination : When vaccinating with plasmid DNA, it was seen that an adaptive immune response to the encoded protein occurred, leading to the term DNA vaccination. This lead to the realization that bacterial DNA, which is loaded with unmethylated cpg dinucleotides, was adjuvant for this type of vaccination.

DNA vaccine : A vaccine that contains DNA which encodes antigenic proteins. It is injected directly into the muscle; the DNA is taken up by the muscle cells and encoded protein antigens are synthesized. This produces both humoral and cell-mediated responses.

DNA : Deoxyribonucleic acid. A polydeoxy-ribonucleotide in which the sugar is deoxyribose; the main repository of genetic information in all cells and most viruses. ***OR*** "Deoxyribonucleic acid". The nucleic acid which carries the genetic code of an organism. It is the primary component of chromosomes.

DNA-dependent kinase : The genetic defect in scid mice, which cannot rearrange their T- or B-cell receptor genes and have a severe combined immunodeficiency phenotype, is in the enzyme DNA-dependent kinase. This enzyme is part of a complex of proteins that bind to the hairpin ends of doublestranded breaks in DNA, and its catalytic subunit is critical for VDJ recombinationd.

DNAPL : See NAPL.

DND : Died a natural death.

DNKA : Did not keep appointment.

DNP : 2,4-dinitrophenol.

DNR : Dorsal nerve root.

DNS : Did not show for appointment.

DO : Dissolved Oxygen - a measure of the oxygen dissolved in water expressed in milligrams per liter.

DOA : Day of admission; dead on arrival.

DOB : Date of birth.

DOC : Died of other causes.

DOD : Date of death.

DoE : Design of experiment.

DOE : Date of examination.

DOI : Date of injury.

Dol : Dolichol.

Dolipe septum : Specialized crosswall separating compartments of a hypha of fungi in the phylum Basidiomycota; consisting of a central pore covered with perforated membranes on both sides (called a parenthosome).

Dolipore septum : A characteristic septum found in the Basidiomycetes that flares out near the pore to form an elongate channel.

Domain : Highest level of biological classification, superseding kingdoms. The three domains of biological organisms are the Bacteria, the Archaea, and the Eukarya. ***OR*** A segment of a folded protein structure showing conformational integrity. A domain could include the entire protein or just a fraction of the protein. Some proteins, such as antibodies, contain many structural domains.

Domains : 1. Compact, self-folding, structurally independent regions of proteins (usually around 100-300 amino acids in length); large proteins may have two or more domains connected by less structured stretches of polypeptide. In the antibody molecule, they are the loops, along with about 25 amino acids on each side, that form compact, globular sections. 2. The primary taxonomic groups above the kingdom level; all living organisms may be placed in one of three domains.

Dominant : Refers to a trait or characteristic that appears even if only one copy of the gene is present.

Dominant : Describing an allele whose phenotype is expressed regardless of whether the organism is homozygous or heterozygous for that allele.

Donor : In tissue grafting experiments, the grafted tissues come from a donor and are placed in a recipient or host. ***OR*** A person that gives something, for example, an organ or blood.

DOPA : L-3,4-dihydroxyphenylalanine.

Doratomyces : A saprobe commonly found on decaying plant materials, straw, dung, wood, and in soil. It produces dark, sooty colonies. It has the ability to penetrate cellophane and to decompose cellulose. Doratomyces stemonitis is suspected to be the causal agent of "speck rot" on potatoes.

DOS : Date of surgery.

Dosage (inoculum size) : The number of pathogenic microorganisms entering the host.

DOT : Date of transfer.

DoTS : Database of transcribed sequences.

Double diffusion agar assay (Ouchterlony technique) : An immunodiffusion reaction in which both antibody and antigen diffuse through agar to form stable immune complexes, which can be observed visually.

Double helix : The structure of DNA, made up of two spiraling strands of nucleotides wrapping around each other.

Double helix : A structure in which two helically-twisted polynucleotide strands are held together by hydrogen bonding and base stacking.

Double septum : A two layered septum that may undergo centripetal separation (schizolysis) to release a conidium.

Double-blind : A type of clinical trial where neither the researcher or patient knows whether they are receiving the new drug or treatment. This is used to prevent the results from being influenced by peoples' opinions.

Double-negative thymocytes : Double-negative thymocytes are immature T cells within the thymus that lack expression of the two co-receptors, CD4 and CD8. In a normal thymus, these represent about 5% of thymocytes.

Double-positive thymocytes : Double-positive thymocytes are an intermediate stage in T-cell development in the thymus and are characterized by expression of both the CD4 and the CD8 co-receptor proteins. They represent the majority (80%) of thymocytes.

Doubling time : Time needed for a population to double in number or biomass.

Dour : Dissolved Oxygen Uptake Ratio, reported as mg DO/L/hour, a test that is used to measure how much oxygen is being consumed by the microbes in a wastewater treatment plant, to assure that the biomass is receiving sufficient dissolved oxygen.

Downy : Having a short and dense mycelial texture.

DP : Deep pulse; diastolic pressure.

DPD : Diffuse pulmonary disease.

DPP IV : Dipeptidyl peptidase IV.

DPT : Diphtheria, pertussis, tetanus immuni-zation.

DPT (diphtheria-pertussis-tetanus) vaccine : A vaccine containing three antigens that is used to immunize people against diphtheria, pertussis or whooping cough, and tetanus.

DR : Diabetic retinopathy; diagnostic radiology; dorsal root; dressing.

Draining lymph node : The term draining lymph node is used for any lymph node that is downstream of a site of infection and thus receives antigens and microbes from the site via the lymphatic system. Draining lymph nodes often enlarge enormously during an immune response and can be palpated; they were originally called swollen glands.

Drench : A technique used to deliver anti-parasite drugs to livestock, usually by injecting the drug into the animals' mouths.

Dreschlera (Hyphomycetes) : Mostly plant pathogens that cause leaf spot, seedling blight, leaf stripe, or net blotch.

Dreschlera sp. **:** Conidia dimensions 40-120 x 17-28 microns. Found on grasses, grains and decaying food. It can occasionally cause a corneal infection of the eye.

Droplet nuclei : Small particles (1 to 4 mm in diameter) that represent what is left from the evaporation of larger particles (10 mm or more in diameter) called droplets.

DSC : Differential scanning calorimetry.

dscDNA : Double-stranded complementary DNA.

DSD : Discharge summary dictated; dry sterile dressing.

dsDNA : Double-stranded DNA.

DSG : Dressing; dry sterile gauze.

DSP : Decreased sensory perception.

dsRNA : Double-stranded RNA.

DT : Delerium tremens; date and time; date of treatment.

dTDP : Deoxythymidine diphosphate.

dTMP : Deoxythymidine monophosphate.

DTP : Distal tingling on percussion (Tinel's sign).

DTR : Deep tendon reflex.

dTTP : Deoxythymidine triphosphate.

dTTP : Deoxythymidine 5'-triphosphate.

DU : Diagnosis undetermined; duodenal ulcer.

Duodenum : The short (approximately twelve fingerbreadths in length = about 10 inches) Ushaped upper (proximal) portion of the small intestine extending from the pyloric sphincter of the stomach to the jejunum, encompassing the superior, descending, transverse and ascending portions, in that order; hepatic ducts (from liver), pancreatic duct, and cystic duct (from gallbladder) join and enter into the intestine at the descending duodenum.

Duodenum : The first section of the small intestine.

Duplex : Same as double helix.

Duties : Things each of us ought to do. For example, if you have a baby, you have the duty to look after it.

DVA : Distance visual acuity.

DVT : Deep venous thrombosis.

DW : Distilled water; doing well.

DX : Diagnosis; discharged; disease.

DXD : Discontinued.

Dyad symmetry : Property of a structure that can be rotated by 180° to produce the same structure.

Dysentery : Inflammation of intestines (especially the colon (colitis) of the large intestine) with accompanying severe abdominal cramps, tenesmus, and frequent, lowvolume stools containing blood, mucus, and fecal leukocytes (PMN's).

Dysgonic : A slow growing variant.

Dyspepsia : (Or indigestion). A general term covering a group of symptoms in the digestive tract including a feeling of fullness, bloating, nausea, heartburn, or gassy discomfort in the chest or abdomen. The symptoms develop during meals or shortly afterward. In most cases, indigestion is a minor problem that often clears up without professional treatment.

Dysuria : Painful or difficult urination.

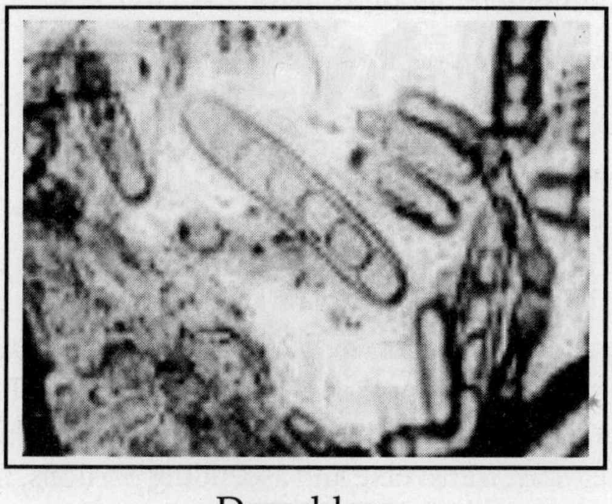

Dreschlera

E

E : Reduction potential.

Early induced responses : The early induced responses or early nonadaptive responses are a series of host defense responses that are triggered by infectious agents early in infection. They are distinct from innate immunity because there is an inductive phase, and from adaptive immunity in that they do not operate by clonal selection of rare antigen-specific lymphocytes.

Early mRNA : Messenger RNA produced early in a virus infection that codes for proteins needed to take over the host cell and manufacture viral nucleic acids.

Early pro-B cell : Early pro-B cell: see pro-B cell.

Earthworm : Soildwelling worm with a segmented body structure. Through their activities, earthworms can stimulate microbial activity, mix soils and aide in the formation of soil structure, and translocate plant material from the surface to lower soil strata.

Eating disorders : These appear as disturbed eating patterns. Some common eating disorders are anorexia nervosa, bulimia nervosa and binge eating disorder. Eating disorders are not just about food but can be an outward sign of deep psychological and emotional turmoil.

EB : Embryonic body.

EBL : Estimated blood loss.

Ebola virus hemorrhagic fever : An acute infection caused by a virus that produces varying degrees of hemorrhage, shock, and sometimes death.

EBV : Epstein-Barr virus or estimated breeding value.

EC : Enzyme commission.

Eccentric : Off set to one side. Used to describe how the stem attaches to the cap.

Ecdysone : A hormone that stimulates the molting process in insects.

ECF : Extended care facility.

ECG : Electrocardiogram.

Echinacea : A herb that is commonly used to reduce cold and flu symptoms.

Echiniulate : Covered in spines.

Echinulate : Covered with delicate spines.

ECHO : Echocardiogram.

Eclipse period : The initial part of the latent period in which infected host bacteria do not contain any complete virions.

ECM : Extracellular matrix.

E-coli **:** *Escherichia coli* - one of the non-pathogenic coliform organisms used to indicate the presence of pathogenic bacteria in water.

Ecological niche : See niche.

Ecological relationship (ecological community; ecosystem) : The relationship between organisms and their environment (including other organisms).

Ecology : The study of the relationship between organisms and their environment. ***OR*** Science which studies the interrelations among organisms and between organisms and their environment. ***OR*** The science of interrelations between living organisms and their environment,Including both physical and biotic factors,as well as intra-and inter specific interactions.

Ecosystem : Community of organisms and the environment in which they live. ***OR*** Groupings of various organisms interacting with each other and their environment. ***OR*** A self-regulating biological community and its associated physical and chemical environment. ***OR*** The biological, physical, and chemical relationships between a community of organisms and the environment they live in.

Ecotype : A genetic strain of a population that is adapted to the unique local environmental conditions.

Ectomycorrhiza (EM) : Mycorrhizal type in which the fungal mycelia extend inward, between root cortical cells, to form a network (Hartig net) and outward into the surrounding soil. Usually the fungal hyphae also form a mantle on the surface of the root. ***OR*** Referring to a mutualistic association between fungi and plant roots in which the fungus surrounds the root tip with a sheath.

Ectoparasite : A parasite that lives on the surface of its host.

Ectoplasm : The outer stiffer portion or region of the cytoplasm in a protozoan, which may be differentiated in texture from the inner portion or endoplasm.

Ectosymbiosis : A type of symbiosis in which one organism remains outside of the other organism.

Ectothrix : Natural hair invasion by a dermatophyte characterized by arthroconidia on the outside of the hair shaft.

Eczema : The common skin disease eczema is seen mainly in children; its etiology is poorly understood. ***OR*** An inflammation of the skin making it red and sometimes dry, flaky or itchy.

ED : Edema.

Edaphoric : (i) Of or pertaining to the soil. (ii) Resulting from or influenced by factors inherent in the soil or other substrate, rather than by climatic factors.

Edaphoric climax : Term used to describe a successional climax at a "lower" level due to existing environmental conditions.

EDC : Endocrine disrupting compound.

Edema : Abnormal volume of fluid in intercellular tissue. ***OR*** In immunology, edema is the swelling caused by the entry of fluid and cells from the blood into the tissues, which is one of the cardinal features of the process of inflammation.

Edema : Accumulation of unreturned blood plasma in various body tissues/ organs.

Edman degradation : A systematic method of sequencing proteins, proceeding by stepwise removal of single amino acids from the amino terminus of a polypeptide chain.

EDRF : Endothelium-derived relaxing factor.

EDTA : Ethylenediamine tetraacetic acid.

EDV : Essential derivation of varities.

EEG : Electroencephalogram.

EENT : Eyes, ears, nose, throat.

EF : Elongation factor.

Effacing lesion : The type of lesion caused by enteropathogenic strains of *E. coli* (EPEC) when the bacteria attach to and destroy the brush border of intestinal epithelial cells. The term AE (attaching-effacing) *E. coli* is now used to designate true EPEC strains that are an important cause of diarrhea in children from developing countries and in traveler's diarrhea.

Effect lymphocytes : Effector lymphocytes can mediate the removal of pathogens from the body without the need for further differentiation, as distinct from naive lymphocytes, which must proliferate and differentiate before they can mediate effector functions, and memory cells, which must differentiate and often proliferate before they become effector cells. They are also called armed effector cells in this book, to indicate that they can be triggered to effector function by antigen binding alone.

Effect mechanisms : Effector mechanisms are those processes by which pathogens are destroyed and cleared from the body. Innate and adaptive immune responses use most of the same effector mechanisms to eliminate pathogens.

Efferent lymphatic vessel : Lymphocytes leave a lymph node through the efferent lymphatic vessel.

Efficacy : The ability of a drug or substance to produce a desired effect, such as treating a disease.

Effluent : Wastewater or other liquid raw (untreated), partially or completely treated flowing from a reservoir, basin, treatment process, or treatment plant. ***OR*** Wastewater or other liquid - raw (untreated), partially or completely treated - flowing from a reservoir, basin, treatment process, or treatment plant. ***OR*** Liquid discharged as waste, as from an industrial plant or sewage works.

Effuse : Spread out, radiate.

EGF : Epidermal growth factor.

EGFR : Epidermal growth factor receptor.

EGL : Energy grade line a line that represents the elevation of energy head in feet of water flowing in a pipe, conduit, or channel.

EGS : External guide sequence.

E_h : Potential generated between an oxidation or reduction halfreaction and the H electrode in the standard state.

Ehrlichiosis : A tick-borne (*Dermacentor andersoni*, *Amblyomma americanum*) rickettsial disease caused by *Ehrlichia chaffeensis*. Once inside leukocytes, a nonspecific illness develops that resembles Rocky Mountain spotted fever.

EIA : Enzyme immunoassay.

Eicosanoid : Any fatty acid with 20 carbons.

EJ : Elbow jerk.

Ejaculation : Expulsion of semen.

EKG : Electrocardiogram (also ECG).

EL : Elbow.

Electrolytic process : A process that causes the decomposition of a chemical compound by the use of electricity.

Electron acceptor : Substance that accepts electrons during an oxidation reduction reaction. An electron acceptor is an oxidant. ***OR*** A substance that accepts electrons during an oxidation-reduction reaction. Examples for bacteria are oxygen (aerobes), nitrate (nitrate reducing bacteria), and sulfate (sulfate reducers).

Electron donor : Substance that donates electrons in an oxidationreduction reaction. An electron donor is a reductant. ***OR*** Compound that donates electrons in an oxidationreduction reaction. Examples are organic carbon and hydrogen.

Electron microscope : A microscope that bounces atom particles (electrons) off objects to create an image on film. ***OR*** An extremely powerful microscope that uses beams of electrons instead of light to magnify objects. Traditional light microscopes magnify images 1,000-2,000 times; electron microscopes can magnify 300,000 times or more.

Electron transport chain : A system of electron carriers embedded into the inner membrane of a mitochondrion to transfer energy from electrons to O_2 to form water. ***OR*** A series of electron carriers that operate together to transfer electrons from donors such as NADH and $FADH_2$ to acceptors such as oxygen OR Final sequence of reactions in biological oxidations composed of a series of oxidizing agents arranged in order of increasing strength and terminating in oxygen.

Electrontransport chain phosphylation : See oxidative phosphorylation.

Electrophilic compounds : Chemicals that attack or are drawn to regions in other chemicals in which electrons are readily available; oxidizing agents act as electrophilic compounds.

Electrophesis : Separation of charged molecules, such as nucleic acids, in an electrical field. ***OR*** Electrophoresis is the movement of molecules in a charged field. In immunology, many forms of electrophoresis are used to separate molecules, especially protein molecules, to determine their charge, size, and subunit composition. ***OR*** A process by which enzymes are separated based on differences in electrical charge. A technique that separates substances through differences in their migration rate in an electrical field due to variations in the number and kinds of charged groups they have. ***OR*** The movement of particles in an electrical field. A commonly-used technique for analysis of mixtures of molecules in solution according to their electrophoretic mobilities. ***OR*** The application of an electric field to create temporary pores in the plasma membrane in order to insert DNA into the cell and transform it.

Electrostatic : Electric charge that is stationary.

Elementary body : A small, dormant body that serves as the agent of transmission between host cells in the chlamydial life cycle.

ELISA : Enzymelinked immunosorbent assay. An immunoassay that uses specific antibodies to detect antigens or antibodies in body fluids. The antibodycontaining complexes are visualized through enzyme coupled to the antibody. Addition of substrate to the enzymeantibodyantigen complex results in a colored product. ***OR*** Enzyme Linked Immunosorbent Assay is a type of serological test that can be used to test for plant viruses and other plant pathogens.

ELISCO : Entrepreneurial life science company.

ELISPOT assay : ELISPOT assay is an adaptation of ELISA in which cells are placed over antibodies or antigens attached to a plastic surface. The antigen or antibody traps the cells' secreted products, which can then be detected using an enzyme-coupled antibody that cleaves a colorless substrate to make a localized colored spot.

Ellipsoid : Rugby football shaped.

Elliptical : Oval, with a symmetric curve.

Elongate : Lengthened.

Elongation cycle : The cycle in protein synthesis that results in the addition of an amino acid to the growing end of a peptide chain.

Elongation factors : Protein factors uniquely required during the elongation phase of protein synthesis. Elongation factor G (EF-G) brings about the movement of the peptidyl tRNA from the A site to the P site of the ribosome.

Eluate : The fluid that has passed through (eluted from) a chromatographic column.

Eluviation : Removal of soil material from a layer of soil as a suspension.

Emarginate : Of gills notched near the stem, similar to sinuate but with a much more conspicuous notch.

Embden-Meyerhof pathway : A pathway that degrades glucose to pyruvate; the six-carbon stage converts glucose to fructose 1,6-bisphosphate, and the three-carbon stage produces ATP while changing glyceraldehyde 3-phosphate to pyruvate.

Embden Meyerhof Parnas pathway : A biochemical pathway that degrades glucose to pyruvate; the sixcarbon stage converts glucose to fructose1, 6bisphosphate, and the threecarbon stage produces ATP while changing glyceraldehyde3phosphate to pyruvate.

Embolus : A moving blood clot which can "get stuck" somewhere and cause problems.

Embolus (singular) : A clot in a vessel obstructing blood flow; emboli (plural)

Embryo : Plant or animal at an early stage of developmentor A mass of living cells between the point of conception and the time when it starts to look like something recognisable. In humans, this is at about 8 weeks from conception.

Embryonated egg (Also called a "Larvated egg") : A nematode egg with a developed larva inside it. Most nematode eggs leave the host in the morula stage and develop in the environment to the embryonated stage (the stage just before hatching). A few nematode eggs (ex. *Strongyloides* spp., *Dictyocaulus arnfieldi*) are embryonated at the time they leave the host.

Embryonic stem cells (ES cells) : A stem cell found in embryonic tissue or cord blood. ***OR*** Embryonic stem cells (ES cells) are mouse embryonic cells that will grow continuously in culture and that retain the ability to contribute to all cell lineages. ES cells can be genetically manipulated in tissue culture and then inserted into mouse blastocysts to generate mutant lines of mice; most often, genes are deleted in ES cells by homologous recombination and the mutant ES cells are then used to generate gene knockout mice. They have also been used to clone sheep, and could soon be used to replace body parts in humans.

Emericella (Ascomycetes) : Anamorph: Aspergillus (Hyphomycetes) Usually found in soil, potatoes, grain, citrus, and stored seeds. Can be found in food and the indoor environment. *Emericella nidulans* can produce a sterigmatocystin mycotoxin and can be pathogenic to man and animals.

EMG : Electromyogram.

Emigration : Moving out of a location.

Empedocles : Person who lived around 490 to 444 BC and thought there were four original elements: Earth, Air, Fire, and Water.

Emphysema : A lung disease which makes it difficult to breath and causes shortness of breath. ***OR*** Rupture and / or coalescing of alveoli,with scar tissue formation causing loss of elasticity in the lungs.

Empyema : An infection in the chest cavity,outside of the lungs. ***OR*** Pus accumulation in the lung(s) or other body cavity.

EMR : Electron magnetic resonance.

EMS : Electrical muscle stimulation.

EMT : Embryo multification and transfer.

Emulsifier : A substance that can help one liquid mix with another when they normally wouldn't do so, such as oil and water.

Emulsion : A liquid mixture of two or more liquid substances not normally dissolved in one another, one liquid held in suspension in the other.

Enantiomphs : Isomers that are mirror images of one another.

Encapsulated bacteria : Encapsulated bacteria have thick carbohydrate coats that protect them from phagocytosis. Encapsulated bacteria can cause extracellular infections and are effectively engulfed and destroyed by phagocytes only if they are first coated with antibody and complement produced in an adaptive immune response.

Encapsulation : A method of getting a substance, usually an enzyme or bacterium, into a small package or capsule while it is still working or alive.

Encode : Encyclopedia of DNA elements.

Encystation : The formation of a cyst.

End product inhibition : See feedback inhibition.

Endemic : A disease or pathogen present or usually prevalent in a given population or geographic region at all times (see also epidemic).

Endemic (murine) typhus : A form of typhus fever caused by the rickettsia *Rickettsia typhi* that occurs sporadically in individuals who come into contact with rats and their fleas.

Endemic disease : A disease that is commonly or constantly present in a population, usually at a relatively steady low frequency.

Endergonic reaction : Chemical reaction that proceeds with the consumption of energy. ***OR*** A reaction with a positive standard free energy change.

Endocarditis : Inflammation of the endocardium (see also myocarditis).

Endocardium : Membrane that lines interior of heart; innermost of two membranes that surround the heart

Endocrine gland : One of the ductless glands which secretes hormones. ***OR*** Specialized tissues whose function is to synthesize and secrete hormones.

Endocrine tissues : Tissues that produce and secrete hormones into the blood so that they can then have a widespread effect throughout the body.

Endocytosis : Cellular uptake of material from the environment into an intracellular vesicle by invagination of the cell membrane; includes both phagocytosis and pinocytosisor. The process in which a cell takes up solutes or particles by enclosing them in vesicles pinched off from its plasma membrane.

Endoenzyme : Enzyme that operates along the internal portions of a polymer.

Endogenote : The recipient bacterial cell's own genetic material into which the donor DNA can integrate.

Endogenous : Originating from inside of host (see endogenous flora).

Endogenous flora : Microbial flora occupying niche(s) that are in or on the body of the host.

Endogenous infection : An infection by a member of an individual's own normal body microbiota.

Endogenous pyrogen : A substance such as the lymphokine interleukin-1, which is produced by host cells and induces a fever response in the host. It also is called simply a pyrogen.

Endogenous pyrogens : Cytokines that can induce a rise in body temperature are called endogenous pyrogens, as distinct from exogenous substances such as endotoxin from gram-negative bacteria that induce fever by triggering endogenous pyrogen synthesis and release.

Endogenous respiration : A reduced level of respiration (breathing) in which organisms break down compounds within their own cells to produce the oxygen they need.

Endolithic : Rock dwelling.

Endometrium : The richly-vascularized lining of the uterus which brings food to a developing baby.

Endomycorrhiza : Mycorrhizal association with intracellular penetration of the host root cortical cells by the fungus as well as outward extension into the surrounding soil. ***OR*** Referring to a mutualistic association of fungi and plant roots in which the fungus penetrates into the root cells and arbuscules and vesicles are formed.

Endonuclease : Endoenzyme that cleaves phosphodiester bonds within a nucleic acid molecule. ***OR*** An enzyme that breaks a phosphodiester linkage at some point within a polynucleotide chain.

Endopeptidase : A parasite that lives inside the body of its host OR An enzyme that breaks a polypeptide chain at an internal peptide linkage.

Endoperidium : The innermost wall of the fruit body of Gasteromycetes.

Endophyte : Organism growing within a plant. The association may be symbiotic or parasitic. ***OR*** A microorganism living within a plant, but not necessarily parasitic on it.

Endoplasm : The central portion of the cytoplasm in a protozoan.

Endoplasmic reticulum : The convoluted network of membranes through out a cell. ***OR*** A system of membranous tubules and flattened sacs (cisternae) in the cytoplasmic matrix of eucaryotic cells. Rough or granular endoplasmic reticulum (RER or GER) bears ribosomes on its surface; smooth or agranular endoplasmic reticulum (SER or AER) lacks them. ***OR*** A system of double membranes in the cytoplasm that is involved in the synthesis of transported proteins. The rough endoplasmic reticulum has ribosomes associated with it.

END : Electron-nuclear double resonance.

Endoscopy : An endoscope used for gastroenterology is a hollow, thin, flexible tube that uses a lens or miniature camera to view various areas of the gastrointestinal tract. The performance of an exam using an endoscope is referred to as endoscopy. Biopsies and therapeutic procedures can also be done using endoscopes.

Endosomes : A membranous vesicle formed by endocytosis. ***OR*** Antigen taken up by phagocytosis generally enters the endosomes, the acidified vesicles present in cells. Protein antigens entering by this route are presented by MHC class II molecules.

Endospore : Differentiated cell formed within the cells of certain Grampositive bacteria and extremely resistant to heat and other harmful agents. ***OR*** seedlike structure; formed to preserve life when conditions turn bad. ***OR*** An extremely heat- and chemical-resistant, dormant, thick-walled spore that develops within bacteria.

Endosymbiont : An organism that lives within the body of another organism in a symbiotic association.

Endosymbiotic theory : A type of symbiosis in which one organism is found within another organism. ***OR*** The theory that eucaryotic organelles such as mitochondria and chloroplasts arose when bacteria established an endosymbiotic relationship with the eucaryotic ancestor and then evolved into eucaryotic organelles.

Endothermic hypothesis : Term used to describe an animal capable of controlling its body temperature Internally.

Endothrix : Natural hair invasion by a dermatophyte characterized by the development of arthroconidia within the hair shaft only.

Endotoxin : Complex bacterial toxin; Lipid A portion of lipopolysaccharide from Gramnegative cell walls; LPS is composed of Lipid A + Core Polysaccharide + O Antigen (a.k.a., O polysaccharide side chain) and is released upon lysis of the cell during infection ; Lipid A component is responsible for endotoxicity effects on the host; O side chain is the antigenic portion of the LPS molecule. ***OR*** Endotoxins are bacterial toxins that are released only when the bacterial cell is damaged, as opposed to exotoxins, which are secreted bacterial toxins. The most important endotoxin is the lipopoly-saccharide of Gram-negative bacteria, which is a potent inducer of cytokine synthesis and the proximal cause of endotoxic shock.

End-product (feedback) inhibition : The inhibition of the first enzyme in a pathway by the end product of that pathway.

Energy : The capacity to do work or cause particular changes.

Energy charge : The fractional degree to which the AMP-ADP-ATP system is filled with high-energy phosphates (phosphoryl groups).

ENGL : European Network of GMO Laboratories.

Enhanced rhizosphere degradation : Enhanced biodegradation of contaminants near plant roots where compounds exuded by the roots increase microbial biodegra-dation activity. Other plant processes such as water uptake by the plant roots can enhance biodegradation by drawing contaminants to the root zone.

Enhancer : A DNA sequence that can stimulate transcription at an appreciable distance from the site where it is located. It acts in either orientation and either upstream or downstream from the promoter. ***OR*** Within genomic DNA, there are specific sequences that act as cellspecific enhancers of RNA transcription.

Enology : The science of wine making.

Enrichment culture : Technique in which environmental (including nutritional) conditions are controlled to favor the development of a specific organism or group of organisms.

Ensheathed : A larval nematode that is within the retained cuticle of the previous stage.

ENT : Ear, nose, throat.

Entamoeba histolytica **:** A parasitic member of Kingdom Protista; Division Rhizopoda which causes amoebic dysentery.

Enteric : Of intestinal origin, especially applied to wastes or bacteria.

Enteric (entero) : Relating to the intestine (e.g., enteric infection).

Enteric bacteria : General term for a group of bacteria that inhabit the intestinal tract of humans and other animals. Among this group are pathogenic bacteria such as Salmonella and Shigella.

Enterohemorrhagic *E. coli* (EHEC) : EHEC strains of *E. coli* (0157:H7) produce several cytotoxins that provoke fluid secretion in traveler's diarrhea; however, their mode of action is unknown.

Enteroinvasive *E. coli* (EIEC) : EIEC strains of *E. coli* cause traveler's diarrhea by penetrating and binding to the intestinal epithelial cells. EIEC may also produce a cytotoxin and enterotoxin.

Enteropathogenic *E. coli* (EPEC) : EPEC strains of *E. coli* attach to the brush border of intestinal epithelial cells and cause a specific type of cell damage called effacing lesions that lead to traveler's diarrhea.

Enterotoxigenic *E. coli* (ETEC) : ETEC strains of *E. coli* produce two plasmid-encoded enterotoxins (which are responsible for traveler's diarrhea) and are distinguished by their heat stability: heat-stable enterotoxin (ST) and heat-labile enterotoxin (LT).

Enterotoxin : An exotoxin with enteric activity, i.e., affects the intestinal tract or a toxin specifically affecting the cells of the intestinal mucosa, causing vomiting and diarrhea.

Enterovirus : Small virus that infects the gastrointestinal tract and can spread to other areas, especially the nervous system. Examples include coxsackie, hepatitis A, and polio.

Entner Doudoff pathway (ED pathway) : A pathway that converts glucose to pyruvate and glyceraldehyde3phosphate by producing 6phosphogluconate and then dehydrating it.

Entomology : The study of insects.

Entropy : The randomness of a system. ***OR*** A measure of the randomness or disorder of a system; a measure of that part of the total energy in a system that is unavailable for useful work.

Envelope : A lipid membrane enveloping a virus particle. ***OR*** (i) All the structures outside the plasma membrane in bacterial cells. (ii) In virology it is an outer membranous layer that surrounds the nucleocapsid in some viruses.

Environmental resistance : All biotic and abiotic factors combining to limit explosion.

Environmental sustainability : Working and behaving in a way that protects the sources of raw materials to ensure that they are available in an ongoing way to future generations.

Enzootic : The moderate prevalence of a disease in a given animal population.

Enzyme : Protein catalyst which drives chemical reactions within a cell. Enzymes display a high degree of specificity with regard to both the substrate on which they act and the type of reaction they catalyze. Enzymes may also be introduced from outside the cell to cause biochemical reactionsor Organic substances (proteins) produced by living organisms and act as catalysts to speed up chemical changes. ***OR*** A protein catalyst with specificity for both the reaction catalyzed and its substrates.

Enzymelinked immunosbent assay (ELISA) : Immunoassay that uses specific antibodies to detect antigens or antibodies. The antibodycontaining complexes are visualized through an enzyme coupled to the antibody. Addition of substrate to the enzymeantibodyantigen complex results in a colored product. ***OR*** A technique used for detecting and quantifying specific antibodies and antigens. ***OR*** The enzyme-linked immunosorbent assay (ELISA) is a serological assay in which bound antigen or antibody is detected by a linked enzyme that converts a colorless substrate into a colored product. The ELISA assay is widely used in biology and medicine as well as in immunology.

Enzymes : Proteins that act as catalysts (cause other chemical store act without being part of that chemical reaction).

Eosinophils : 25% of total leukocytes; stain with eosin dye; antiparasitic (e.g., intestinal worms)or eosinophils are white blood cells thought to be important chiefly in defense against parasitic infections; they are activated by the lymphocytes of the adaptive immune response. The level of eosinophils in the blood is normally quite low. It can increase markedly in several situations, such as atopy, resulting in eosinophilia, an abnormally large number of eosinophils in the blood. ***OR*** a polymorphonuclear leukocyte that has a two-lobed nucleus and cytoplasmic granules that stain yellow-red. A mobile phagocyte that is highly antiparasitic.

Eotaxin-1, eotaxin-2 : Eotaxin-1 and eotaxin-2 are CC chemokines that act specifically on eosinophils.

EP : Ectopic pregnancy.

EPG : Expected progeny difference.

EPI : Epinephrine.

Epicoccum (Hyphomycetes) : A cosmopolitan saprobe isolated from air, soil, grain, seeds, textiles, paper products, and food materials. Can be a plant pathogen, and is a common cause of leaf spots of various plants. Can be found in indoor environments, where it can grow under conditions of low humidity. It is a known allergen, and is occasionally isolated from human skin and sputum.

Epicoccum sp. **:** Conidia dimensions 15-25 microns. A common allergen. It is found in plants, soil, grains, textiles and paper products.

Epidemic : The rapid spreading of an infectious disease to many individuals in an area. ***OR*** A disease that suddenly increases in occurrence above the normal level in a given population OR A disease caused by *Rickettsia prowazekii* that is transmitted from person to person by the body louse.

Epidemiologist : A person who specializes in epidemiology.

Epidemiology : The study of factors influencing the occurence, transmission, distribution, prevention and control of disease in a defined population.

Epidermis : The outer layer of cells surrounding the body of an organism,such as a plant leaf or our skin.

Epidermophyton sp. **:** Can cause infections of skin and nails.

Epididymis : Coiled tubules,found with in the scrotum,that store sperm and are the site of their final maturation.

Epiglottis : A flap above the larynx which closes to block off the trachea during swallowing.

Epigynous : Term used to describe plants with flower parts (petals,sepals,etc.) Attached above the ovary (so the ovary is inferior).

Epilimnion : The surface water in a lake or ocean into which light penetrates,thus photosynthesis can occur.

Epimers : Two stereoisomers with more than one chiral center that differ in configuration at one of their chiral centers.

Epinephrine : Another name for adrenaline,a hormone produced by the adrenal glands which triggers the "fight-or-flight" response.

Epipetalous : In a flower,when the stamens arise from the corolla tissue.

Episome : Plasmid that replicates by inserting itself into the bacterial chromosome. ***OR*** A plasmid that can exist either independently of the host cell's chromosome or be integrated into it.

Epitheca : The larger of two halves of a diatom frustule (shell).

Epithelial tissue : In an animal,tissue which coats or lines various body parts.

Epithelium : Tissue or cells that cover external and internal body surfaces, for example skin cells and cells that line the stomach.

Epitope : The region of an antigen to which the variable region of an antibody binds.OR An epitope is a site on an antigen recognized by an antibody or an antigen receptor; epitopes are also called antigenic determinants. A T-cell epitope is a short peptide derived from a protein antigen. It binds to an MHC molecule and is recognized by a particular T cell. B-cell epitopes are antigenic determinants recognized by B cells and are typically discontinuous in the primary structure.

Epitope spreading : Epitope spreading describes the fact that responses to autoantigens tend to become more diverse as the response persists. This is also called determinant spreading or antigen spreading.

Epitopes : The chemical group that is recognised by a specific antibody, resulting in an immune response.

Epizootic : A sudden outbreak of a disease in an animal population.

Epizootiology : The field of science that deals with factors determining the frequency and distribution of a disease within an animal population.

EPO : Erythropoietin.

EPR : Electron paramagnetic resonance.

EPS : Generic term referring to polymers of varying chemical composition that are excreted outside the cell. Synonymous with exopolysaccharide, exocellular polymer, extracellular polymer, exopolymer. Term does not distinguish between bound polymer (capsule) and polymer excreted into the surrounding environment. In biofilms, EPS may refer to the polymers that extend between cells, forming a matrix and anchoring the cells to each other and to the substratum. (See glycocalyx).

Epsilon-proteobacteria : One of the five subgroups of proteobacteria, each with distinctive 16S rRNA sequences. Slender Gram-negative rods, some of which are medically important (Campylobacter and Helicobacter).

Epstein-Barr virus, EBV : The Epstein-Barr virus (EBV) is a herpesvirus that selectively infects human B cells by binding to complement receptor 2 (CR2, also known as CD21). It causes infectious mononucleosis and establishes a lifelong latent infection in B cells that is controlled by T cells. Some B cells latently infected with EBV will proliferate *in vitro* to form lymphoblastoid cell lines.

Equalizing basin : A holding basin in which variations in flow and composition of liquid are averaged. Such basins are used to provide a flow of reasonably uniform volume and composition to a treatment unit. Also called a balancing reservoir.

Equilibrium : The state of a system in which no net change is occurring and free energy is at a minimum; in a chemical reaction at equilibrium, the rates in the forward and reverse directions exactly balance each other out. ***OR*** The point at which the concentrations of two compounds are such that the interconversion of one compound into the other compound does not result in any change in free energy.

Equilibrium dialysis : The affinity of an antibody for its antigen can be determined by equilibrium dialysis, a technique in which antibody in a dialysis bag is exposed to varying amounts of a small antigen able to diffuse across the dialysis membrane. The amount of antigen inside and outside the bag at the equilibrium diffusion state is determined by the amount and affinity of the antibody in the bag.

Equinox : The two days of the year—one in spring and one in autumn—when day length and night length are equal.

ER : Endoplasmic reticulum.

ER : Emergency room; external rotation.

Eradication : Permanent reduction to zero of the worldwide incidence of infection caused by a specific agent as a result of deliberate efforts. Intervention measures are no longer needed.

Eratosthenes : Person who lived in the third century BC and calculated the diameter of the earth from the difference in sun angle in Greece vs Egypt

ERE : External rotation in extension.

Erect : Upright.

ERF : External rotation in flexion.

Ergot : The dried sclerotium of *Claviceps purpurea*. Also, an ascomycete that parasitizes rye and other higher plants causing the disease called ergotism.

Ergotism : The disease or toxic condition caused by eating grain infected with ergot; it is often accompanied by gangrene, psychotic delusions, nervous spasms, abortion, and convulsions in humans and in animals.

Ericoid mycorrhiza : Type of mycorrhiza found on plants in the Ericales. The hyphae in the root are able to penetrate cortical cells (endomycorrhizal habit); however, no arbuscules are formed. Major forms are ericoid, arbutoid, and monotropoid.

E-rosettes : E-rosettes are human T cells that will bind to treated red blood cells from sheep; the many red blood cells bound to each T cell give it the appearance of a rosette and increase its buoyant density so that the T cells can be isolated by gradient centrifugation. E-rosetting is often used for isolating human T cells.

Erp57 : Erp57 is a chaperone molecule involved in loading peptide onto MHC class I molecules in the endoplasmic reticulum.

Erysipelas : An acute inflammation of the dermal layer of the skin, occurring primarily in infants and persons over 30 years of age with a history of streptococcal sore throat.

Erythema infectiosum : A disease in children caused by the parvovirus B19. This disease is common in children between 4 and 11 years of age and is sometimes called fifth disease, since it was the fifth of six erythematous rash diseases in children in an older classification.

Erythro-blastotic : A disorder in which immature RBCs, erythroblasts, are present in circulating blood.

Erythroblastosis fetalis : Erythroblastosis fetalis is a severe form of Rh hemolytic disease in which maternal anti-Rh antibody enters the fetus and produces a hemolytic anemia so severe that the fetus has mainly immature erythroblasts in the peripheral blood.

Erythrocyte : Red blood cell

Erythromycin : An intermediate spectrum macrolide antibiotic produced by *Streptomyces erythreus*.

ES : Electrical stimulation.

ESC : Embryonic stem cell

Eschar : A slough produced on the skin by a thermal burn, gangrene, or the *Anthrax bacillus*.

Escherichia coli **:** A species of fecal bacteria,a normal part of our intestinal flora which makes vitamins K and B-12 for us,but some strains of which may cause illness if ingested in to the upper GI tract.

Escherichia coli (E. coli)**. :** A Gram negative bacterium commonly found in the vertebrate intestine. It is the bacterium most frequently used in the study of biochemistry and genetics.

E-selectin : See selectins.

Esophagus : The tube down which food travels from the pharynx to the stomach.

ESR : Electron spin resonance.

ESRD : End stage renal disease.

ESRF : End stage renal failure.

EST : Expressed sequence tag.

Established cell line : A group of cultured cells derived from a single origin and capable of stable growth for many generations.

Estivation : A short-term (overnight) decrease in body functions.

Estrous cycle : Periodic sexual receptiveness in the female, followed by absorption of the unused endometrium.

Estuaries : Bodies of water which are located at the lower end of a river and are subject to tidal fluctuations.

Estuary : An area where fresh and salt water mix,such as at the mouth of a river where it enters the ocean—salt concentration of the water is variable.

ET : Embryo transfer.

Ethanol : An alcohol made (when yeast ferments sugar) from the 2-carbon hydrocarbon, ethane, CH_3CH_2OH, present in beverages such as beer and wine and added to some brands of automobile fuel.

Ether : A molecule containing two carbons linked by an oxygen atom.

Ethics : Reasoned views about why certain things are morally wrong and others right. There is no single ethical framework on which everyone agrees. However, three widely used ethical frameworks use the principles of rights and duties, utilitarianism and virtue.

Ethiopian realm : The biogeographical realm which includes the continent of Africa and nearby areas.

Ethology : The study of animal behavior.

Etiology : The cause or origen of a disease; the study of the cause(s) of disease.

ETL : Economic trait locus.

ETOH : Ethyl alcohol.

ETS : External transcribed spacer.

ETT : Embryo transfer technology.

ETT : Exercise tolerance test.

EUA : Examination under anesthetic.

Eubacteria : Old term for the Bacteria.

Eucarya : The domain that contains organisms composed of eucaryotic cells with primarily glycerol fatty acyl diesters in their membranes and eucaryotic rRNA.

Eucaryotes : The kingdom that describes allorganisms (plants, fungi, animals, fish, birds, etc.). That have an organized nucleus surrounded by a nuclear membrane. The other major kingdom is the Procaryotes (containing only the primitive bacteria and Archae).

Eucaryotic cells : Cells that have a membrane-delimited nucleus and differ in many other ways from procaryotic cells; protists, algae, fungi, plants, and animals are all eucaryotic.

***Euglena* :** Genus name for organisms in Kingdom Protista; Division Euglenophyta which possess a flagellum, a reddish stigma ("eyespot"), and a chloroplast.

Euglenoids : A group of algae (the division Euglenophyta) or protozoa (order Euglenida) that normally have chloroplasts containing chlorophyll a and b. They usually have a stigma and one or two flagella emerging from an anterior gullet.

Euglenophyta : Division within Kingdom Protista whose members have a flagellum and a light-sensitive stigma.

Eukarya : Phylogenetic domain containing all eukaryotic organisms.

Eukaryote : A cell or organism that has a membrane-bound nucleus. ***OR*** Organism having a unit membranebound nucleus and usually other organelles. ***OR*** Single or multicellular organism whose cell or cells have their genetic material contained within a nuclear membrane. Includes all fungi, plants, animals, and many single-celled organisms. Excludes bacteria, which are prokaryotes. ***OR*** Organisms whose cells have a true nucleus.

Eukaryotic : Cell has nucleus; divides through mitosis. The cytoplasmic membrane of prokaryotes is fundamentally similar to that of a eukaryotic cell. The variety of proteins in the cytoplasmic membrane in prokaryotes is greater because of the significantly greater number of functions performed by it. As with eukaryotes, the membrane controls the permeability of the cell and is the major site of energy metabolism. In bacteria, either the electron transport system or the site of photosynthetic energy metabolism is located in the cytoplasmic membrane. The ribosomes of the prokaryotes are the site of protein synthesis. Many prokaryotes are immotile, but some have either a single flagellum or multipole flagella as their means of locomotion.

Eumycota : A division of fungi in some classification systems. These are the true fungi consisting of the Zygomycetes, Ascomycetes, Basidiomycetes, and Chytridiomycetes.

Eumycotic mycetoma **:** See maduromycosis.

Eurotium (Ascomycetes) : Anamorph: Aspergillus (Hyphomycetes) Can be found in stored food, fruit juices, grains, nuts, milled rice, spices, meat products, and peas. Also commonly occurs in indoor environments. *Eurotium herbariorum* may cause keratitis and indigestion in man.

Euryhaline : Having a wide range of salt concentration tolerance.

Eurythermal : Bodies of water which are located at the lower end of a river and are subject to tidal fluctuations.

Eurythermal : Having a wide range of temperature tolerance.

Eutrophic : Lakes have high nutrient status and high productivity ***OR*** Having high concentrations of nutrients optimal, or nearly so, for plant or animal growth. Can be applied to nutrient or soil solutions and bodies of water.

Eutrophic : A nutrient-enriched environment. ***OR*** Having high concentrations of nutrients optimal, or nearly so, for plant or animal growth. Can be applied to nutrient or soil solutions and bodies of water.

Eutrophication : The enrichment of an aquatic environment with nutrients. ***OR*** Enrichment of natural waters with excess nutrients that leads to algae blooms and subsequent oxygen deficiency when the algae die and bacteria degrade them.

Evanescent : Ephemeral, soon disappearing.

Evolution : Changes,overtime,in the percent ages or frequencies of the various alleles of the genes in a population.

Evolutionary distance : A quantitative indication of the number of positions that differ between two aligned macromolecules, and presumably a measure of evolutionary similarity between molecules and organisms.

Evolve : A gradual change in traits of an organism that occurs over time.

EX : Exacerbate; examination; excision.

EXC : Excellent; except; exchange.

Excision repair : DNA repair in which a damaged region is replaced.

Excited state : An energy-rich state of an atom or a molecule, produced by the absorption of radiant energy.

Excystation : The escape of one or more cells or organisms from a cyst.

Exdoperidium : The outermost wall of the fruit body of Gasteromycetes

Exergonic reaction : Chemical reaction that proceeds with the liberation of energy. ***OR*** A reaction that spontaneously goes to completion as written; the standard free energy change is negative, and the equilibrium constant is greater than one.

Exergonic reaction : A chemical reaction that takes place with a negative change in standard free energy.

Exfoliative toxin : An exotoxin produced by *Staphylococcus aureus* that causes the separation of epidermal layers and the loss of skin surface layers. It produces the symptoms of the scaled skin syndrome.

Exit site (E site) : The location on a ribosome to which an empty tRNA moves from the P site before it finally leaves the ribosome during protein synthesis.

Exobiology : Branch of biology concerned with the effects of extraterrestrial environments on living organisms.

Exoenzyme : Enzymes that are secreted by cells. ***OR*** Enzyme that acts at the end of a polymer cleaving off monomers and dimers and sometimes larger chain fragments

Exogenote : The piece of donor DNA that enters a bacterial cell during gene exchange and recombination.

Exogenous : Originating external to host (see exogenous flora).

Exogenous flora : Microbial flora normally existing externally to the body of the host.

Exon : A DNA sequence which has a corresponding mRNA equivalent. Exons alternate with introns in many eukaryotic genes. ***OR*** The region in a split or interrupted gene that codes for RNA which ends up in the final product (e.g., mRNA). ***OR*** A segment within a gene that carries part of the coding information for a protein.

Exonuclease : An enzyme that breaks a phosphodiester linkage at one or the other end of a polynucleotide chain so as to release single or small nucleotide residues.

Exophiala (Hyphomycetes) : Widely distributed as a saprobe in soil, water, on plants and decaying wood. It is an occasional contaminant of feet and nails. Exophiala infections have also been reported in animals, including fish.

Exothermic, Ectothermic : Term used to describe an animal which controls its body temperature externally(by basking in the sun to warm up or moving to a shady place to cool off).

Exotoxin : Potent toxic substance formed and released extracellularly by species of certain bacteria; genetic control can be encoded either chromosomally, on a plasmid, or by a lysogenic bacteriophage (prophage) incorporated in either the chromosome or a plasmid. ***OR*** A heat-labile, toxic protein produced by a bacterium as a result of its normal metabolism or because of the acquisition of a plasmid or prophage that redirects its metabolism. It is usually released into the bacterium's surroundings.

EXP : Expand; explain.

Experimental allergic encephalomyelitis, (EAE) : (EAE) is an inflammatory disease of the central nervous system that develops after mice are immunized with neural antigens in a strong adjuvant.

Expertise : Having excellent knowledge or skills in a particular area. Being an expert in something.

Exponential growth : Period of sustained growth of a microorganism in which the cell number constantly doubles within a fixed time period.

Exponential phase : Period during the growth cycle of a population in which growth increases at an exponential rate. As referred to as logarithmic phase. ***OR*** The phase of the growth curve during which the microbial population is growing at a constant and maximum rate, dividing and doubling at regular intervals.

Expressed sequence tag (EST) : A partial gene sequence unique to a gene that can be used to identify and position the gene during genomic analysis.

Expression vector : A special cloning vector used to express recombinant genes in host cells; the recombinant gene is transcribed and its protein synthesized.

Exsheathment : The final step in molting of the infective stage when the old cuticle (sheath) is lost and the new nematode stage emerges inside the definitive host. It is a requirement for the establishment for infection in the host.

EXT : Extend; external; extremity.

Exteins : Polypeptide sequences of precursor self-splicing proteins that are joined together during formation of the final, functional protein. They are separated from one another by intein sequences, which they flank.

Extens : A muscle which straightens a joint.

Extinction : The specific organism no longer exists in nature or the laboratory.

Extracellular : Outside the cell.

Extracutaneous spotrichosis : An infection by the fungus *Sporothrix schenckii* that spreads throughout the body.

Extravasation : The movement of cells or fluid from within blood vessels to the surrounding tissues is called extravasation.

Extreme barophilic bacteria : Bacteria that require a high-pressure environment to function.

Extreme environment : An environment in which physical factors such as temperature, pH, salinity, and pressure are outside of the normal range for growth of most microorganisms; these conditions allow unique organisms to survive and function.

Extremophiles : Microorganisms that grow under harsh or extreme environmental conditions such as very high temperatures or low pHs.

Extrinsic factor : An environmental factor such as temperature that influences microbial growth in food.

Exudate : Low molecular weight metabolites that leak from plant roots into soil. ***OR*** fluid, cells, or cellular debris that has escaped from blood vessels and has been deposited on or in tissues. ***OR*** Droplets of fluid formed on the surface of a colony.

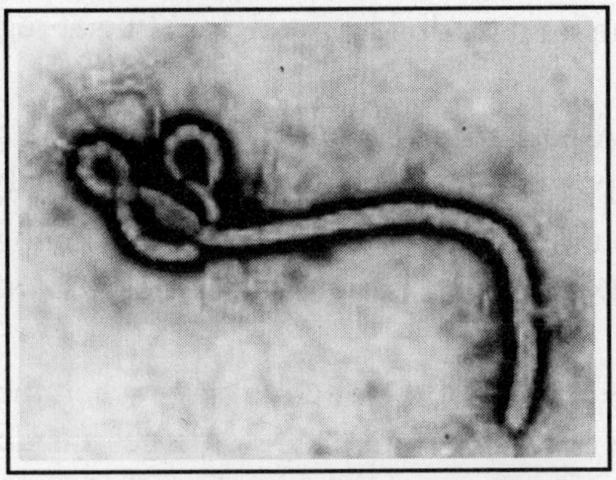

Ebola virus

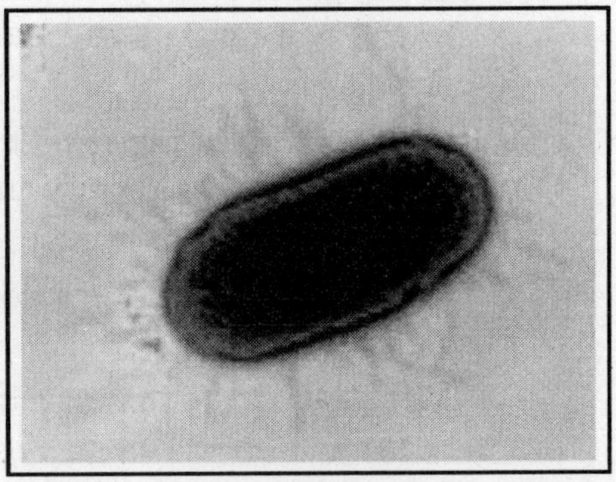

Escherichia coli

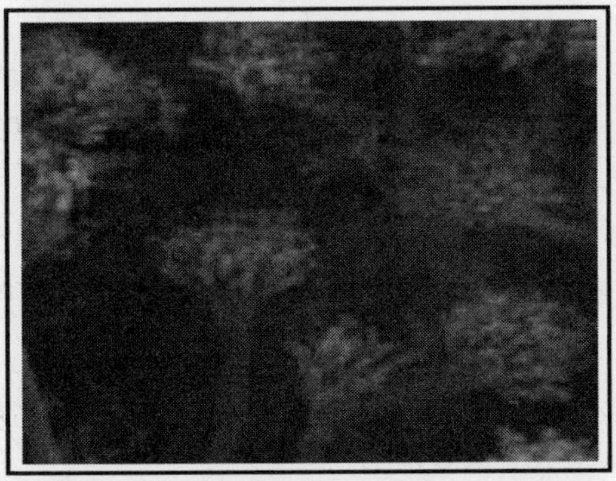

Emericella

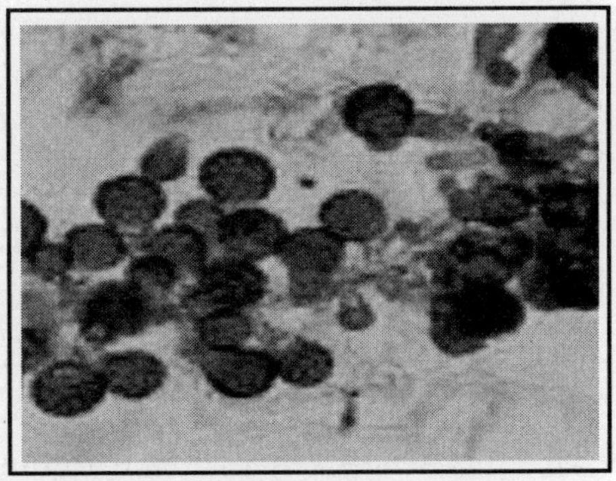

Epicoccum

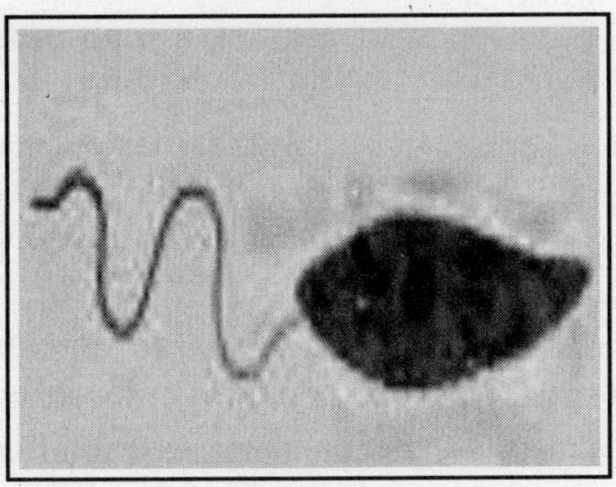

Euglena

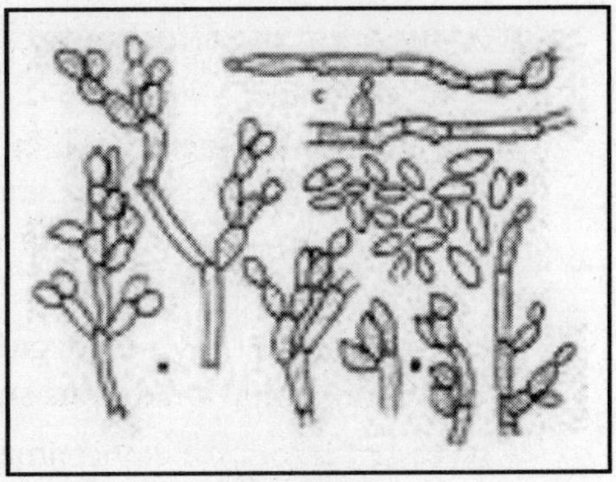

Exophilia

F

F : Faraday constant. ***OR*** Food-represents BOD in the F/M ratio. Expressed in pounds.

F factor : The fertility factor, a plasmid that carries the genes for bacterial conjugation and makes its *E. coli* host cell the gene donor during conjugation. ***OR*** A large bacterial plasmid, known as the sex-factor plasmid because it permits mating between F+ and F- bacteria.

F value : The time in minutes at a specific temperature (usually 250°F) needed to kill a population of cells or spores.

F/I : Fever due to infection.

F/M : A ratio of the amount of food to the amount of organisms. Used to control an activated sludge process.

F/U : Follow up.

F_ plasmid : An F plasmid that carries some bacterial genes and transmits them to recipient cells when the F_ cell carries out conjugation; the transfer of bacterial genes in this way is often called sexduction.

F1 : Filial generation 1 (initial hybrid generation)

F1 particle : Particle on the inner mitochondrial membrane, which is the site of ATP synthesis by oxidative phosphorylation.

F1P : Fructose-1-phosphate.

F2 : Filial generation 2 (second hybrid generation).

F6P : Fructose-6-phosphate.

FA : Failed appointment; forearm.

F_{AB} : Antibody molecule fragment that binds antigen.

Fab fragment : IgG antibody molecules can be cleaved into three fragments by the enzyme papain. Two of these are identical Fab fragments, so called because they are the fragment with specific antigen binding. The Fab fragment consists of the light chain and the N-terminal half of the heavy chain held together by an interchain disulfide bond. Another protease, pepsin, cuts in the same general region of the antibody molecule as papain but on the carboxy-terminal side of the disulfide bonds. This produces the F(ab')2 fragment, in which the two arms of the antibody molecule remain linked. See also Fc fragment.

FABERE : Flexion, abduction, external rotation and extension.

Facilitated diffusion : Diffusion across the plasma membrane that is aided by a carrier.

FACOG : Fellow of the American College of Obstetrics and Gynecology.

FACP : Fellow of the American College of Physicians.

FACS : Fluorescence-activated cell sorting.

FACS : Fellow of the American College of Surgeons.

FACS® : See fluorescence-activated cell sorter.

Factors B, D, H, I and P : Factors B, D, H, I, and P are all components of the alternative pathway of complement activation. Factor B plays a role very similar to that of C2b in the classical pathway. Factor D is a serine protease that cleaves factor B. Factor H is an inhibitory protein with a role similar to decay-accelerating factor. Factor I is a protease that breaks down various components of the alternative pathway. Factor P, or properdin, is a positive regulatory component of the alternative pathway. It stabilizes the C3 convertase of the alternative pathway on the surface of bacterial cells.

Facultative : Ability to adapt and live under various conditions (see facultative anaerobe) or prefers but not absolute, e.g., facultative anaerobe or Indicates that a bacterium is able to grow in the presence or absence of an environmental condition (e.g., facultative aerobe).

Facultative aerobe : An anaerobe that can survive with or without oxygenor An organism that can use molecular oxygen in its metabolism but that also can live anaerobically. ***OR*** A bacterium capable of growing under aerobic conditions or anaerobic conditions in the presence of an inorganic ion ie. SO_4, NO_3. ***OR*** Microorganisms that do not require oxygen for growth, but do grow better in its presence.

Facultative organism : Organism that can carry out both options of a mutually exclusive process (e.g., aerobic and anaerobic metabolism).

Facultative pond : The most common type of pond in current use. The upper portion (supernatant) is aerobic, while the bottom layer is anaerobic. Algae supply most of the oxygen to the supernatant.

Facultative psychrophile : See psychrotroph.

FAD : Flavin adenine dinucleotide.

$FADH_2$: Reduced flavin adenine dinucleotide.

FADIR : Flexion, adduction and internal rotation.

FADIRE : Flexion, adduction, internal rotation and extension.

Faecal contamination : Traces of faeces (stools) coming into contact with areas of the body where they could potentially cause harm. For example, faecal contamination of an open wound, such as a pressure sore, could cause infection.

Faeces : Solid waste excreted by an animal. Also called stool.

Falcate : Curved like a sickle.

FAME : Fatty Acid Methyl Ester, a means of identifying bacteria by analysis of the fatty acids in their cells. This is often done as an initial screening until a company, such as Alken-Murray Corporation, is sure that a bacterial strain is one they wish to use commercially, and then 16S rRNA identification may be completed for a more definite identification and fingerprint of the strain, to keep competitors from copying that strain.

Family : Taxonomic level below order and above genus.

Farinaceous : Smells and tastes of flour

Farmer's lung : Farmer's lung is a hypersensitivity disease caused by the interaction of IgG antibodies with large amounts of an inhaled allergen in the alveolar wall of the lung, causing alveolar wall inflammation & compromising gas exchange.

FART tax : Farmers Against Ridiculous Taxes was a group set up to protest taking farmers for the methane emissions of their stock.

Fas gene : The gene that is active in target cells which are susceptible to killing by cells expressing the Fas ligand, a member of the TNF family of cytokines and cell surface molecules.

Fas, Fas ligand : Fas is a member of the TNF receptor family; it is expressed on certain cells and makes them susceptible to killing by cells expressing Fas ligand, a cell-surface member of the TNF family of proteins. Binding of Fas ligand to Fas triggers apoptosis in the Fas-bearing cell.

Fascia : Sheet of fibrous tissue below the skin or surrounding muscles or groups of muscle fibers (see fasciitis).

Fasciitis : Inflammation of the fascia.

Fastidious : Complex nutritional or cultural requirements, making isolation and culture of a fastidious organism more difficult.

Fatty acid synthetase : The multienzyme complex that makes fatty acids; the product usually is palmitic acid.

Fatty acid : A long-chain hydrocarbon containing a carboxyl group at one end. Saturated fatty acids have completely saturated hydrocarbon chains. Unsaturated fatty acids have one or more carbon-carbon double bonds in their hydrocarbon chains.

Fauna : Ecological term that generically refers to animals; flora generically refers to plants or animals.

FB : Foreign body.

FBP : Fructose-1,6-bisphosphate.

FBPase : Fructose bisphosphatase.

FBS : Fetal bovine serum.

Fc fragment : IgG antibody molecules can be cleaved into three fragments by the enzyme papain. One of these is the Fc fragment, so-called for Fragment crystallizable. The Fc fragment consists of the C-terminal halves of the two heavy chains disulfide-bonded to each other by the residual hinge region. See also Fab fragments.

Fc receptors : Fc receptors are receptors for the Fc portion of immunoglobulin isotypes. They include the Fcã and Fcå receptors.

FCA : Fracture, complete, angulated.

FCC : Fracture, complete, comminuted; fracture, compound, comminuted.

FCCC : Fracture, complete, comminuted, compound.

FCD : Fracture, complete, deviated.

Fcã receptors, FcãRI, RII, RIII : Fcã receptors, including fcãRI, RII, and RIII, are cell-surface receptors that bind the Fc portion of IgG molecules. Most Fcã receptors bind only aggregated IgG, allowing them to discriminate bound antibody from free IgG. They are expressed on phagocytes, B lymphocytes, NK cells, and follicular dendritic cells. They have a key role in humoral immunity, linking antibody binding to effector cell functions.

Fcå-receptor, FcåRI : The high-affinity Fcå-receptor (fcåRi) on the surface of mast cells and basophils binds free IgE. When antigen binds this IgE and crosslinks FcåRI, it causes mast-cell activation.

Fd : Ferredoxin.

FEB : Fever.

Fecal coliform : Coliforms whose normal habitat is the intestinal tract and that can grow at 44.5°C.

Fecal enterococci : Enterococci found in the intestine of humans and other warm-blooded animals. They are used as indicators of the fecal pollution of water.

Feces : Waste matter eliminated from the bowels; excrement. (The word "feces" is plural but can be used with either a singular or plural verb.).

Fecundity : The reproductive capacity of an individual, measured by the number of offspring it produces.

Feedback inhibition : Inhibition by an end product of the biosynthetic pathway involved in its synthesis. ***OR*** A negative feedback mechanism in which a pathway end product inhibits the activity of an enzyme in the sequence leading to its formation; when the end product accumulates in excess, it inhibits its own synthesis.

Femur : The thigh bone.

FER : Flexion, extension, rotation.

Fermentation : The enzymatic breakdown (catabolism) of carbohydrates generally in the absence of oxygen. ***OR*** Metabolic process in which organic compounds serve as both electron donors and electron acceptors. ***OR*** A type of heterotrophic metabolism in which an organic compound rather than oxygen is the terminal electron (or hydrogen) acceptor. Less energy is generated from this incomplete form of glucose oxidation than is generated by respiration, but the process supports anaerobic growth. ***OR*** The anaerobic breakdown of a nutrient molecule such as sugar to produce energy. Other byproducts may also be produced, for example, the production of alcohol when yeast breaks down fruit sugars in grapes. ***OR*** An energy-yielding process in which an energy substrate is oxidized without an exogenous electron acceptor. Usually organic molecules serve as both electron donors and acceptors.

Fermentation : Anaerobic conversion of glucose to some waste product,including glycolysis plus step store generate NAD^+.

Fertiliser : A substance that provides nutrients to plants need in order to grow. ***OR*** Any organic or inorganic material of natural or synthetic origin (other than liming materials) added to a soil to supply one or more elements essential to plant growth.

Fever : A response to disease characterized by a rise in core body temperature and activation of the immune system. ***OR*** A complex physiological response to disease mediated by pyrogenic cytokines and characterized by a rise in core body temperature and activation of the immune system. ***OR*** any elevation of the body temperature above the normal; functions to speed up immune reactions and to limit/slow bacterial growth and multiplication

Fever blister : See cold sore.

FF : Forward flexion.

FFD : Free from disability.

FFF : Field-flow fractionation.

FFROM : Full, free range of motion.

FFT : Forward flexion: fingertips to toes.

FGF : Fibroblast growth factor.

FH : Familial hypercholesterolemia

FIA : Fluorescence immoassay.

Fibrillose : Used of cap or stem surface with thin thread-like filaments which are more or less parallel.

Fibrin : Insoluble protein that is main constiuent in blood clotting; formed by enzymatic cleavage of fibrinogen by thrombin.

Fibrinolysin (staphylokinase; streptokinase; plasmin) : Enzyme capable of breaking down fibrin and thus dissolving blood clots.

Fibroblast : Connective tissue cell that that produces collagenous fibers and forms fibrous tissues.

Fibronectin : Two forms of adhesive glycoprotein found in host, important in crosslinking collagen in connective tissue (celltocell binding), tissue repair and platelet aggregation.

Fibrose : Composed of tough, stringy material.

Fibrosis : Formation of fibrous connective tissue; usually occurring with scar and granuloma formation.

Fibula : The thin, outer (little-toe side) shinbone.

Field capacity : Content of water, on a mass or volume basis, remaining in a soil after being saturated with water and after free drainage is negligible.

Filament : The stalk-like portion of the stamen ("male" flower part) which supports the anther.

Filamentous : In the form of very long rods, many times longer than wide (for bacteria), in the form of long branching strands (for fungi). ***OR*** In the form of very long rods, many times longer than wide (for bacteria), in the form of long branching strands (for fungi). ***OR*** of caps with a cuticle composed of elongate hyphae as in the mycelium (not globose cells).

Filamentous organisms : Organisms that grow in a thread or filamentous form. Common types are Thiothrix, Actinomycetes & Cyanobacteria (aka blue-green algae). This is a common cause of sludge bulking in the activated sludge process. Variously known as "pond scum", "blue-green algae", or "moss", when it appears in a pond/lake, and confused with algae because it looks a lot like algae. Cyanobacteria forms a symbiotic relationship with some varieties of algae, making the combination very difficult to combat in lakes and ponds. Filamentous organisms and Actinomycetes will naturally stick to solid surfaces. Common types of Cyanobacteria are Oscillatoria, Anabaena, and Synechococcus. Other filament formers include Spirogyra, Cladophora, Rhizoclonium, Mougeotia, Zygnema and Hydrodictyon. Nocardia is another filament former, which causes foaming and interferes with flocculation in a waste treatment plant.

Filter aid : A chemical (usually a polymer) added to water to help remove fine colloidal suspended solids.

Fimbria : A fine, hairlike protein appendage on some gram-negative bacteria that helps attach them to surfaces. ***OR*** Short filamentous structure on a bacterial cell; although flagellalike in structure, generally present in many copies and not involved in motility. Plays a role in adherence to surfaces and in the formation of pellicles. ***OR*** modern term for short, hairlike projections or appendages (organelles) on the outer surface of certain bacteria composed of protein subunits (pilin) extending outward from the surface that act as a virulence factor by promoting adherence; formerly known as pili; fimbria (singular).

Fimbriate : Fringed with cystidia or finely torn.

Final host : The host on/in which a parasitic organism either attains sexual maturity or reproduces.

Fingerprinting : The characteristic two-dimensional paper chromatogram obtained from the partial hydrolysis of a protein or a nucleic acid.

First filial generation : The"children" in a genetic cross.

First law of thermodynamics : Energy can be neither created nor destroyed (even though it can be changed in form or redistributed).

First set rejection : When tissue or organ grafts are placed in an unmatched recipient, they are rejected by a first set rejection, which is an immune response by the host against foreign antigens in the graft. Cf. Second set rejection.

FISH : Fluorescent in situ hybrdization.

Fission : Type of cell division in which overall cell growth is followed by formation of a crosswall which typically divides the fully grown cell into two similar or identical cells.

Fixation : The process in which the internal and external structures of cells and organisms are preserved and fixed in position.

FJRM : Full joint range of motion.

FK506 : See tacrolimus.

FL : Flank; flexion; fluid.

Flagellate : Protozoan that moves by means of one to several flagella.

Flagellin : The protein used to construct the filament of a bacterial flagellum.

Flagellum : Bacterial cell appendage used for motility. ***OR*** whiplike bacterial locomotory (provide motility) organelles anchored in the cell membranes that are composed of helicallycoiled protein subunits (flagellin); flagella (plural). ***OR*** One or two, longer, "hair-like"or" tail-like" structures on the surface of some types of cells as a means of locomotion,contain microtubules organized according to the "9+2 formula".

Flat plane warts : Small, smooth, slightly raised warts.

Flatulence : Caused by excessive production of gas in the intestines. ***OR*** Excessive gas in the stomach and intestines, which can cause farting.

Flavin adenine dinucleotide : An electron carrying cofactor often involved in energy production (for example, in the tricarboxylic acid cycle and the b-oxidation pathway).

Flavin adenine dinucleotide : (FAD, FADH)another electron carrier, similar to NAD^+ (plays a role in the Krebs cycle).

Flavonoid : A group of compounds that give plants their colour and are thought to offer some health benefits.

Flex : A muscle which bends a joint.

Flexuous : Wavy.

Floating matter : Matter which passes through a 2000 micron sieve and separates by flotation for an hour.

Floc : Clumps of bacteria and particulate impurities or coagulants that have come together and formed a cluster. Found in aeration tanks and secondary clarifiers.

Floccose : Fluffy or cottony. ***OR*** Covered in loose cottony scales.

Flocculation : The process of forming floc particles when a chemical coagulant or flocculent such as alum or ferric chloride is added to the wastewater.

Flora : (Microbiology Definition); Microbiota microorganisms present in or characteristic of a special location (niche) (Note flora generically refers to plants; fauna generically refers to animals).

Flow cytometer : Instrument with a tiny orifice through which particles (such as bacteria) flow through one at a time. As they pass through a laser beam, biochemicals may be determined on a per cell basis. ***OR*** a laboratory technique in which cell numbers and characteristics, such as size, shape or density, are measured using fluorescent dyes.

Flow equalization system : A device or tank designed to hold back or store a portion of peak flows for release during low-flow periods.

Flower break : See color break.

Fluid mosaic model : The currently accepted model of cell membranes in which the membrane is a lipid bilayer with integral proteins buried in the lipid, and peripheral proteins more loosely attached to the membrane surface.

Fluoresce : Produce light when exposed to radiation of a particular wavelength. For example, some substances fluoresce when exposed to ultraviolet light.

Fluescence *in situ* hybridisation : A technique used to look at chromosomes.

Fluorescence microscope : A microscope that exposes a specimen to light of a specific wavelength and then forms an image from the fluorescent light produced. Usually the specimen is stained with a fluorescent dye or fluorochrome.

Fluescence : The emission of light by an excited molecule in the process of making the transition from the excited state to the ground state.

Fluorescence-activated cell sorter, FACS® : Individual cells can be characterized and separated in a machine called a fluorescence-activated cell sorter (FACS®) that measures cell size, granularity, and fluorescence due to bound fluorescent antibodies as single cells pass in a stream past photodetectors. The analysis of single cells in this way is called flow cytometry and the instruments that carry out the measurements and/or sort cells are called flow cytometers or cell sorters.

Fluorescent : Able to emit light of a certain wavelength when activated by light of a shorter wavelength.

Fluorescent antibody : Antiserum conjugated with a fluorescent dye, such as fluorescein or rhodamine. ***OR*** Immunoglobulin molecule (antibody) coupled with a fluorescent dye.

Fluorescent light : The light emitted by a substance when it is irradiated with light of a shorter wavelength.

Fluxes : Rate of emission, sorption, or deposition of a material from one pool to another. For example, the exchange of methane between the land and the atmosphere is a flux, while the production of methane within the soil is not.

FLX : Flexion.

FMD : Family medical doctor.

fMet : N-formylmethionine.

FMN : Flavin mononucleotide.

Foetid : With a strong offensive smell. (also spelled "fetid").

FOG : Fats, Oils and Greases. A measure of the non-petroleum based fats in waste treatment.

Folic acid : A B vitamin noted for its prevention of neural tube defects if taken in early pregnancy.

Follicle : Small cavity or sac in organ or tissue; e.g., hair follicle or ovarian follicle.

Follicle : A fluid-filled sac in the ovary where the egg grows before it is release during ovulation. ***OR*** Peripheral lymphoid tissues, such as lymph nodes and Peyer's patches, contain large areas of B cells called follicles, which are organized around follicular dendritic cells.

Follicular center cell lymphoma : A follicular center cell lymphoma is a type of B-cell lymphoma that tends to grow in the follicles of lymphoid tissues.

Follicular dendritic cells : The follicular dendritic cells of lymphoid follicles are cells of uncertain origin. They are characterized by long branching processes that make intimate contact with many different B cells. They have Fc receptors that are not internalized by receptor-mediated endocytosis and thus hold antigen:antibody complexes on the surface for long periods. These cells are crucial in selecting antigen-binding B cells during antibody responses.

Folliculits : Infection and inflammation of follicle(s).

Fomite : An object that is not in itself harmful but is able to harbor and transmit pathogenic organisms. Also called fomes. ***OR*** inanimate object capable of transmitting infectious organisms to a host, e.g., soiled bed linens, diapers, tissues and handkerchiefs, hospital respiratory equipment, dishrags, etc.

Food additive : A substance which is added to a food to add flavour, aroma or colour, alter the texture or nutritional content, or preserve the food. A food additive is not normally eaten or drunk by itself. All food additives are tested for safety before they can legally be used.

Food chain : The flow of energy and matter in living organisms through a producer-consumer sequence (See also food web). ***OR*** Movement of nutrients from one life form to another as a result of the different feeding habits and dietary requirements of organisms in an ecosystem. ***OR*** Very simple pathway of nutrient flow. Ex. Carnivore > herbivore > plant.

Food intoxication : Food poisoning caused by microbial toxins produced in a food prior to consumption. The presence of living bacteria is not required.

Food poisoning : A general term usually referring to a gastrointestinal disease caused by the ingestion of food contaminated by pathogens or their toxins. ***OR*** Usually caused by bacteria but some cases are caused by viruses, and symptoms include vomiting and diarrhoea. These symptoms are not only unpleasant but can sometimes be very serious and result in hospitalisation. In some instances the consequences of food poisoning can be life threatening

Food web : A network of many interlinked food chains, encompassing primary producers, consumers, decomposers, and detritivores. ***OR*** Diagram of the interconnections of nutrient flow through a food chain.

Food-borne infection : Gastrointestinal illness caused by ingestion of microorganisms, followed by their growth within the host. Symptoms arise from tissue invasion and/or toxin production.

Foot cell : A basal cell of a conidiophore as seen in *Aspergillus* and *Fusarium*.

Foot rot : Contagious disease of sheep caused by specific bacteria, which can spread rapidly through the flock. Indicated by inflammation between the claws of the hoof.

Fensic : Relating to the legal system. For example, forensic science is used to provide scientific evidence for court cases.

Formulating agent : A substance which alters the physical or chemical properties of a product; often used to improve the delivery of a drug to the target tissues.

Fossil : The preserved remains of a formerly-living (prehistoric) organism.

FPM : Full passive movements.

FPRL : Formyl peptide receptor like.

FR : Frequency of respiration; full range; fracture.

FR BB : Fracture of both bones.

Fragile-X syndrome : An inherited condition caused by a 'fragile' or broken site on the X chromosome, which can cause mental impairment.

Fragmentation : A type of asexual reproduction in which a thallus breaks into two or more parts, each of which forms a new thallus.

Frameshift mutations : Mutations arising from the loss or gain of a base or DNA segment, leading to a change in the codon reading frame and thus a change in the amino acids incorporated into protein. ***OR*** Insertions or deletions of genetic material that lead to a shift in the translation of the reading frame. The mutation usually leads to nonfunctional proteins.

Framework regions : The V domains of immunoglobulins and T-cell receptors contain relatively invariant framework regions that provide a protein scaffold for the hypervariable regions that make contact with antigen. ***OR*** gills do not attach to the stem.

Francesco Redi : Person who, in 1668, did an experiment with flies and meat that showed that larger organisms do not arise by spontaneous generation.

Francis Bacon : Person who,in about 1600, founded the empirical or experimental method of scientific inquiry and wrote *Novum Organum*.

Free energy : That part of the energy of a system that is available to do useful work. ***OR*** Intrinsic energy contained in a given substance that is available to do work, particularly with respect to chemical transformations; designated D G.

Free energy change : The total energy change in a system that is available to do useful work as the system goes from its initial state to its final state at constant temperature and pressure.

Free radical : An atom or group of atoms with an unpaired electron. Because they have a 'free' electron, free radicals are very unstable and highly reactive. The hydroxyl radical (OH.) is an example of a free radical found in cells.

Freeze-dry : Freeze drying (also known as Lyophili-zation) is a dehydration process typically used to preserve a perishable material, or to make the material more convenient for transport. Freeze drying works by freezing the material and then reducing the surrounding pressure to allow the frozen water in the material to sublimate directly from the solid phase to gas.

French polio : See Guillain-barrž syndrome.

FRET : Fluorescence resonance energy transfer.

FRJM : Full range of joint movement.

FROM : Full range of movement.

Fructose : A sugar found in fruits. ***OR*** A 6-carbon ketose with the formula $C_6H_{12}O_6$.

Fruiting body : A specialized structure that holds sexually or asexually produced spores; found in fungi and in some bacteria. ***OR*** Macroscopic reproductive structure produced by some fungi, such as mushrooms, and some bacteria, including myxobacteria. Fruiting bodies are distinctive in size, shape, and coloration for each species.

Frustule : Siliceous wall and protoplast of a diatom. ***OR*** A silicified cell wall in the diatoms.

Fry : Newly hatched baby fish.

FSC : Fracture, simple comminuted.

FSCC : Fracture, simple, complete, comminuted.

FSH : Follicle stimulating hormone.

FTIRS : Fourier transform infrared spectroscopy.

FTKA : Failed to keep appointment.

FTMS : Fourier transform mass spectrometry.

FTSG : Full thickness skin graft.

Fucoxanthin : Yellowish-brown pigment found in some members of the chromista, including kelps and diatoms.

Fugaceous : Soon disappearing e.g. Used of scales.

Fulvic acid : Yellow organic material that remains in solution after removal of humic acid by acidification.

Fumariaceae : A plant family(from genus *Fumaria*) which some botanists consider to be a subfamily within the Papaveraceae.

Functional food : A food which has a beneficial ingredient added to it so that it has health-giving properties over and above its usual nutritional value for example milk with added calcium or vitamins, breakfast cereals or probiotic yoghurts.

Functional group : An atom or group of atoms in an organic molecule that give it a particular property or characteristic.

Fungi : Fungi are single-celled and multicellular eukaryotic organisms, including the yeasts and molds, that can cause a variety of diseases. Immunity to fungi is complex and involves both humoral and cellmediated responses.

Fungicide : An agent that kills fungi. ***OR*** A chemical used to kill fungi, usually when it is growing on crops. For example, Shield is a fungicide used to kill black spot mould on roses.

Fungistasis : Suppression of germination of fungal spores or other resting structures in natural soils as a result of competition for available nutrients, presence of inhibitory compounds, or both.

Fungistatic : Inhibiting the growth and reproduction of fungi.

Fungivous : See mycophagous.

Fungus : A plant-like organism, examples include mushrooms, yeasts, and moulds. The plural is fungi. ***OR*** Achlorophyllous, heterotrophic, spore-bearing eucaryotes with absorptive nutrition; usually, they have a walled thallus. ***OR*** Nonphototrophic, eukaryotic microorganisms that contain rigid cell walls. ***OR*** A complex micro-organism that absorbs nutrients from either living organisms or the remains of dead organisms. Fungi include molds, mildew, yeasts, mushrooms and others that cause severe human illness (plural fungi).

FUO : Fever of unknown origin.

Furanose : A sugar that contains a five-membered ring as a result of intramolecular hemiacetal formation.

Furfuraceous : Covered in tiny particles like sand grains.

Furuncle : Boil; a painful nodule in the skin caused by inflammatory sequestration of *Staphylococci* and core formation.

Furunculosis : Perisistent sequential occurence of furuncles over a period of weeks or months.

Fusarium (Hyphomycetes) : Soil-borne fungi containing many plant pathogens that cause root rot, stem rot, fruit rot, and vascular wilt. Common on commodities, such as rice, bean, soybean, and other crops. Some species are important mycotoxin producers, and others notably *F. oxysporum, F. solani* and *F. moniliforme*, are recognized as opportunistic pathogens of man and animals.

Fusarium solani **:** Aw 0.90. Macroconidia dimensions 27-52 x 4.4-6.8; Microcondia dimensions 8-16 x 2-4 microns. Found in plants and soils. Can produce trichothecene toxins which may be associated with disease in humans and animals.

Fusarium sp. **:** Aw 0.90 A common soil fungus. It is found on a wide range of plants. It is often found in humidifiers. Several species in this genus can produce potent trichothecene toxins. The trichothecene (scirpene) toxin targets, the following systems: circulatory, alimentary, skin and nervous. Produces vomitoxin on grains during unusually damp growing conditions. Symptoms may occur either through ingestion of contaminated grains or possibly inhalation of spores. The genera can produce hemorrhagic syndrome in humans (alimentary toxic aleukia). This is characterized by nausea, vomiting, diarrhea, dermatitis, and extensive internal bleeding. Reported to be allergenic. Frequently involved in eye, skin and nail infections.

Fusiform : Spindleshaped; tapered at both ends.

Fusion protein : The protein(s) on the surface of a virus particle responsible for fusion of the virus envelope with cellular membranes.

Futile cycle : See pseudocycle.

Fv : See single-chain Fv.

FWB : Full weight bearing.

FX : Fracture.

FX BB : Fracture of both bones.

Fyn : See tyrosine kinase.

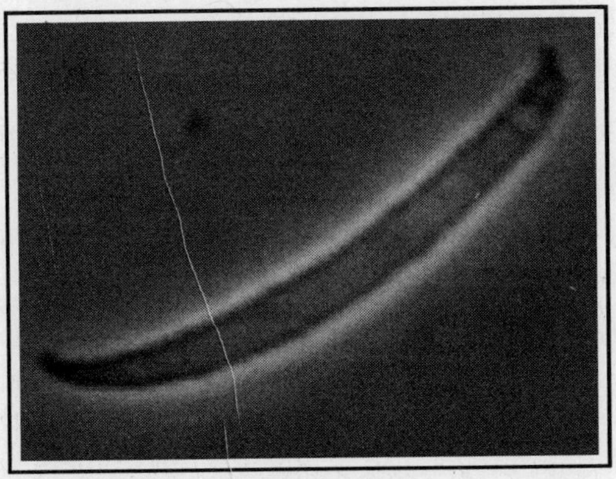

Fusarium

G

G : Gibbs free energy OR Guanine.

G proteins : G proteins are intracellular proteins that bind GTP and convert it to GDP in the process of cell signal transduction. There are two kinds of G protein, the heterotrimeric (a, b, g) receptor-associated G proteins, and the small G proteins, such as Ras and Raf, that act downstream of many transmembrane signaling events.

G1 phase : That period of the cell cycle in which preparations are being made for chromosome duplication, which takes place in the S phase.

G1P : Glucose-1-phosphate.

G2 phase : That period of the cell cycle between S phase and mitosis (M phase).

G3P : Glyceraldehyde-3-phosphate.

G6P : Glucose-6-phosphate.

GA : Gastric analysis; general anesthesia; general appearance.

GABA : Gamma-aminobutyric acid.

Gaia : Greek Mother Earth, "pre"-goddess.

Gal : Galactose.

Galactose : A 6-carbon, -aldose with the formula $C_6H_{12}O_6$.

Galen : Person who lived from 131 to 210 AD and is famous for his descriptions of human anatomy.

Galileo Galilei : Person who lived from 1564 to1642 and used a telescope to verify the idea that the planets move around the sun.

GalNAc : N-acetylgalactosamine.

GALT, gut-associated lymphoid tissues : See gut-associated lymphoid tissues.

Gametangium : Fungal structure that contains one or more gametes.

Gametangium : A structure that contains gametes or in which gametes are formed.

Gamete : Male or female reproductive cells - a sperm or egg. ***OR*** In eukaryotes, the haploid cell analogous to sperm and egg, which results from meiosis. ***OR*** A special sex cell, such as an egg or sperm, with one set of chromosomes.

Gametes : The ova and the sperm, haploid cells that unite during fertilization to generate a diploid zygote.

Gametocyte : The stage in the life cycle of some protozoa (Apicomplexa) which is destined to become a gamete (macro- or micro-).

Gametophyte : In plants and algae,their generation which produces gametes (eggs and sperm).

Gamma-proteobacteria : One of the five subgroups of proteobacteria, each with distinctive 16S rRNA sequences. This is the largest subgroup and is very diverse physiologically; many important genera are facultatively anaerobic chemoorganotrophs.

Gangrene : A condition where tissues dies due to lack of blood flow, which is often followed by infection.

Ganoderma (Basidiomycetes) : Large, very hard, woody bracket fungi that grow on living and dead trees. Some species are common on oaks, chestnuts, and conifers such as hemlock, spruce, and pine. Many species are being investigated for possible medicinal uses.

GAP : Glyceraldehyde-3-phosphate.

GAPDH : Glyceraldehyde-3-phosphate dehydro-genase.

Gas chromatograph : A laboratory instrument used to separate and analyse chemicals in a sample. Go to http://en.wikipedia.org/wiki/Gas_chromatography to find out more.

Gas chromatography : Chromatographic technique in which the stationary phase is a solid or an immobile liquid and the mobile phase is gaseous. The gaseous samples are separated based on their differential adsorption to the stationary phase.

Gas gangrene : A type of gangrene that arises from dirty, lacerated wounds infected by anaerobic bacteria, especially species of *Clostridium*. As the bacteria grow, they release toxins and ferment carbohydrates to produce carbon dioxide and hydrogen gas.

Gas vacuole : A subcellular organelle, found only in prokaryotes, which consists of clusters of hollow, cylindrical, gasfilled vesicles (gas vesicles). ***OR*** A gas-filled vacuole found in cyanobacteria and some other aquatic bacteria that provides flotation. It is composed of gas vesicles, which are made of protein.

Gasification : The conversion of soluble and suspended materials into gas during anaerobic decomposition. In clarifiers the resulting gas bubbles can become attached to the settled sludge and cause large clumps of sludge to rise and float on the water surface. In anaerobic sludge digesters, this gas is collected for fuel or disposed of using a waste gas burner.

Gastric juice : Collective name for the secretions of the stomach lining, containing HCl, pepsin, and other chemicals.

Gastrin : A hormone which triggers production of gastric juice.

Gastritis : Inflammation of the stomach.

Gastrocnemius : The calf muscle.

Gastroenteritis : An inflammation of the stomach and intestine resulting in diarrhea, with vomiting and cramps when irritation is excessive. When caused by an infectious agent, it is often associated with fever. ***OR*** An acute inflammation of the lining of the stomach and intestines, characterized by anorexia, nausea, diarrhea, abdominal pain, and weakness. It has various causes including food poisoning due to such organisms as *E. coli*, *S. aureus*, *Campylobacter* (campylobacteriosis), and *Salmonella* species; consumption of irritating food or drink; or psychological factors such as anger, stress, and fear. Also called enterogastritis.

Gastrointestinal tract : The long muscular tube from the mouth to the anus approximately 8.3 metres long. Food is ingested in the mouth and is digested mechanically and with t1he help of enzymes as it passes along the alimentary canal. Nutrients are absorbed by the body and waste is excreted through the anus as faeces approximately 24 hours after a meal is consumed.

GB : Gallbladder.

GBD : Gallbladder disease.

GBP : Glycan binding protein.

GC : Gas chromatography

GC : General condition; *Gonococcal*; *Gonococcus*.

GDNP : Glial derived neurotropic factor

GDP : Guanosine diphosphate

GE : Genetically engineered

GE : Gastroenteritis.

GEF, guanine-nucleotide exchange fact : See guanine-nucleotide exchange factor.

Gel : Inert polymer, usually made of agarose or polyacrylamide, that separates macromolecules such as nucleic acids or proteins during electrophoresis.

Gel electrophesis : A laboratory technique used to separate molecules, such as DNA, RNA, and proteins, according to size and (sometimes) charge. To find out more, get in-the-lab sheet: Gel electrophoresis.

Gel fitration chromatography : A technique that makes use of certain polymers that can form porous beads with varying pore sizes. In columns made from such beads, it is possible to separate molecules, which cannot penetrate beads of a given pore size, from small molecules that can. Also called gel-exclusion or molecular seive chromatography.

Gelatinous : Jelly like.

GEM : GE micro-organism.

Gene : A piece of DNA that determines a specific trait in an organism.OR Unit of heredity; a segment of DNA specifying a particular protein or polypeptide chain, a tRNA or an mRNA. ***OR*** A piece of DNA that carries the information needed to make a specific protein. Genes are sometimes called the basic unit of inheritance because the information they contain is transmitted from parents to their offspring. ***OR*** A DNA segment or sequence that codes for a polypeptide, rRNA, or tRNA.

Gene : A unit of genetic information on a chromosome.

Gene amplification : The duplication of a particular gene within a chromosome two or more times.

Gene cloning : Making an exact copy (or multiple copies) of a gene. ***OR*** Isolation of a desired gene from one organism and its incorporation into a suitable vector for the production of large amounts of the gene.

Gene conversion : In birds and rabbits, immunoglobulin receptor diversity is generated mainly by gene conversion, in which homologous inactive V gene segments exchange short sequences with an active, rearranged V-region gene.

Gene discovery : Identifying gene sequences for specific traits.

Gene expression : An important stage of viral replication at which virus genetic information is expressed one of the major control points in replication. ***OR*** The use of a gene to make a protein. Involves transcription of the DNA sequence into RNA, and translation of the RNA into an amino acid sequence (or protein). Only some genes are expressed in a particular cell at any one time.

Gene gun : A device that uses high-pressure gas or another propellant to shoot a spray of DNA-coated microprojectiles into cells and transform them. Sometimes it is called a biolistic device. ***OR*** A laboratory tool used to propel (or shoot) a gene (or other piece of DNA) into a cell at high energy. The aim is that the target cell will incorporate the new DNA and start using it.

Gene knockout : Gene knockout is jargon for gene disruption by homologous recombination.

Gene probe : A strand of nucleic acid which can be labeled and hybridized to a complementary molecule from a mixture of other nucleic acids.

Gene segments : The V domains of the polypeptide chains of antigen receptors are encoded in sets of gene segments that must first undergo somatic recombination to form a complete V-domain exon. There are three types of gene segment: V gene segments that encode the first 95 amino acids, D gene segments that encode about 5 amino acids, and J gene segments that form the last 10-15 amino acids of the V region. There are multiple copies of each type of gene segment in the germline DNA, but only one is expressed for each type of receptor chain in a receptor-bearing lymphocyte.

Gene sequencing : Determining the order (or sequence) of the DNA bases (A, G, C, or T) that make up a specific gene.

Gene splicing : The cutting and rejoining of DNA sequences.

Gene targeting : A gene can be specifically disrupted by a technique known as gene targeting or gene knockout. Usually this involves homologous recombination in embryonic stem cells followed by the preparation of chimeric mice by injection of these cells into the blastoeyst.

Gene therapy : Gene therapy is the correction of a genetic defect by the introduction of a normal gene into bone marrow or other cell types. It is also known as somatic gene therapy because it does not affect the germline genes of the individual.

Gene variant : A sequence in the genetic code that differs between people is called a gene variant.

Gene : A segment of the genome that codes for a functional product.

General recombination : Recombination involving a reciprocal exchange of a pair of homologous DNA sequences; it can occur any place on the chromosome. ***OR*** Recombination that occurs between homologous chromosomes at homologous sites.

Generalized transduction : The transfer of any part of a bacterial genome when the DNA fragment is packaged within a phage capsid by mistake.

Generation time : Time needed for a population to double in number or biomass. ***OR*** The time required for a microbial population to double in number. ***OR*** The time it takes for a cell to double its mass under specified conditions.

Genetic code : Information for the synthesis of proteins contained in the nucleotide sequence of a DNA molecule (or in certain viruses, of an RNA molecule).

Genetic engineering (GE) : *In vitro* techniques for the isolation, manipulation, recombination, and expression of DNA. ***OR*** The deliberate modification of an organism's genetic information by directly changing its nucleic acid genome. ***OR*** The artificial altering of a living organism's genetic material to produce new traits, species or products.

Genetic immunization : Genetic immunization is a novel technique for inducing adaptive immune responses. Plasmid DNA encoding a protein of interest is injected into muscle; for unknown reasons, it is expressed and elicits antibody and T-cell responses to the protein encoded by the DNA.

Genetic locus : A specific region on a chromosomes. Plural = loci.

Genetic makeup : See genetic profile.

Genetic map : The arrangement of genes or other identifiable sequences on a chromosome.

Genetic marker test : A test that can be done to see if an organisms carries a particular gene. As genes are too large to use to quickly identify themselves, shorter segments of DNA that are unique to, but always associated with the gene you are testing for are used instead. These are called gene markers.

Genetic modification (GM) : The process of changing the genetic material of an organism, either by removing part of the DNA, adding in new DNA, or changing the existing DNA in some other way is called genetic modification (GM).

Genetic mutation : A base change in the DNA strand.

Genetic profile : The unique sequence of bases in someone's DNA code is their genetic profile.

Genetically modified : All human-designed changes in a plant or animal, whether done through traditional breeding or genetic engineering.

Genetically modified organism : An organism with an altered genome, usually made by adding or removing genes.

Genetics : The unique sequence of bases in an organisms DNA.

Genetics : The study of genes and heredity.

Geniculate : Bent like a knee.

Genital herpes : A sexually transmitted disease caused by the herpes simplex virus type 2.

Genital ulcer disease : See chancroid.

Genome : Complete set of genes present in an organism. ***OR*** The genetic material of an organism (from gene + chromosome). ***OR*** The full set of genes present in a cell or virus; all the genetic material in an organism; a haploid set of genes in a cell.

Genome replication : The stage of viral replication at which the virus genome is copied to form new progeny genomes.

Genome : The total genetic content of a cell or a virus.

Genomics : The study of sets of genes and their interactions. ***OR*** The study of the molecular organization of genomes, their information content, and the gene products they encode.

Genotype : Precise genetic constitution of an organism. ***OR*** (Noun) The genetic information, or DNA sequence, of an individual organism. (Verb) To identify DNA sequence information. ***OR*** The genetic makeup of an organism that combined with environmental factors contributes to the phenotype of the organism.

Genus : A well-defined group of one or more species that is clearly separate from other genera. ***OR*** Taxonomic level below family and above species. ***OR*** A group of closely related species. For example humans are from the genus 'Homo' and species 'sapiens'.

GEO : GE organism.

Geographic information system (GIS) : A data management system that organizes and displays digital map data from remote sensing and aids in the analysis of relationships between mapped features.

Geology : The study of rocks and minerals.

Geotrichum (Ascomycetes yeast) : Teleomorph Dipodascus, Galactomyces (Ascomycetes) Commonly found in soil, water, air, decaying leaves, rotting paper, and textiles. Involved in spoilage of food like bakery products, dairy products, juices, fruits, and vegetables. Can be found in indoor environments with some species producing strong odors.

Geotrichum sp. **:** Conidia dimensions 6-12 x 3-6 microns. Aw 0.90 A common contaminant of grains, fruits, dairy products, paper, textiles, soil and water, and often present as part of the normal human flora. The species *Geotrichum candidum* can cause a secondary infection (geotrichosis) in association with tuberculosis. This rare disease can cause lesions of the skin, bronchi mouth, lung and intestine.

Germ : A microbe, usually one that causes disease.

Germ line : Cells giving rise to the gametes (eggs and sperm in animals, ovules and pollen in plants).

Germ tube : The initial hypha that develops from a conidium or spore.

German measles : See rubella.

Germ-free, gnotobiotic : Mice that are raised in the complete absence of intestinal and other flora are called germ-free or gnotobiotic mice. Such mice have very depleted immune systems, but they can respond virtually normally to any specific antigen, provided it is mixed with a strong adjuvant.

Germicide : An agent that kills pathogens and many nonpathogens but not necessarily bacterial endospores.

Germinal centers : Germinal centers in secondary lymphoid tissues are sites of intense B-cell proliferation, selection, maturation, and death during antibody responses. Germinal centers form around follicular dendritic cell networks when activated B cells migrate into lymphoid follicles. They can be divided by morphology into the dark zone, which is rich in proliferating B lymphocytes, and a light zone, which contains fdcs and centrocytes.

Germination : The stage following bacterial endospore activation in which the endospore breaks its dormant state. Germination is followed by outgrowth.

Germline configuration : Immunoglobulin and T-cell receptor genes are said to be in the germline configuration in the DNA of germ cells and in all somatic cells in which somatic recombination has, not occurred.

Germline diversity : The germline diversity of antigen receptors is due to the inheritance of multiple gene segments that encode V domains; such diversity is distinguished from the diversity that is generated during gene rearrangement or after receptor gene expression, which is somatically generated.

Germline theory : One theory of antibody diversity, the germline theory, proposed that each antibody was encoded in a separate germline gene. This is now known not to happen in people, mice, and most other organisms, but appears to happen in Elasmobranchs, which have rearranged genes in the germline.

Germplasm : The genetic material that carries the inherited characteristics of an organism.

GF : Growth factor.

GFP : Green fluorescent protein.

GFR : Glomerular filtration rate.

GG : Gamma globulin.

GGF2 : Glial growth factor 2.

GH : Growth hormone.

GH : General health.

Ghon complex (gon) : The initial focus of parenchymal infection in primary pulmonary tuberculosis.

GI : Gastrointestinal.

Giardiasis : A common intestinal disease caused by the parasitic protozoan Giardia lamblia.

Gibb's free energy : See free energy.

Gills : The plates of tissue bearing the hymenium in an Agaric

Gingivitis : Inflammation of the gingival tissue.

Gingivostomatitis : Inflammation of the gingiva and other oral mucous membranes.

Glabrous : Smooth.

Gland : A group of special cells that make substances so that other parts of the body can work. For example, the pancreas is a gland that releases insulin so that other body cells can use glucose (sugar) for energy.

Glaucous : A term used to describe plants which are green with a grayish or whitish "bloom"on them.

GLC : Gas-liquid chromatography.

Glc : Glucose.

Gleba : The tissues enclosed within the fruit bodies of the Gasteromycetes.

Gliding motility : A type of motility in which a microbial cell glides along when in contact with a solid surface.

Gliocladium sp. **:** A fungus which is structurally similar to *Penicillium sp*. It is reported to be allergenic.

Gln : Glutamine.

Global regulaty systems : Regulatory systems that simultaneously affect many genes and pathways.

Globular protein : A folded protein that adopts an approximately globular shape. May also be called soluble proteins.

Gloiospora : Conidia aggregated in slimy heads at the tip of an annellide or phialide.

Glomalin : An extracellular glycoprotein produced by arbuscular mycorrhizal fungi.

Glomerulonephritis : An inflammatory disease of the renal glomeruli. ***OR*** inflammation of the glomeruli in the kidney.

Glomerulus (singular) : Spherical cluster of capillaries in the kidney where plasma is filtered ; glomeruli (plural)

GLP : Glucagon like peptide or good laboratory practice.

Glucans : Polysaccharides composed of glucose units held together by glycosidic linkages. Some types of glucans have a (1_3) and a (1_6) linkages and bind bacterial cells together on teeth forming a plaque ecosystem.

Gluconeogenesis : The synthesis of glucose from noncarbohydrate precursors such as lactate and amino acids. ***OR*** The production of sugars from nonsugar precursors such as lactate or amino acids. Applies more specifically to the production of free glucose by vertebrate livers.

Glucose : A simple sugar. More complex sugars are first converted to glucose during digestion, and then it is the glucose that circulates through the blood. ***OR*** Simple sugar, and the primary product of photosynthesis. It is polymerized to ake cellulose and chitin. ***OR*** A 6-carbon aldose, a simple monosaccharide, with the formula $C_6H_{12}O_6$ and widely used by living organisms.

Gluteus maximus : Seat or rump muscle.

Glutinous : When cap surface is covered with a slimy layer of gelatinous hyphae

Gly : Glycine.

Glycaemic : The effect a food has on blood glucose over a period of time.

Glycaemic index : A measure of the speed with which glucose is released from a food.

Glycaemic load : A measure of the speed with which glucose is released from a food that takes is adjusted for the serving size of the food.

GlyCAM-1 : Glycam-1 is a mucinlike molecule found on the high endothelial venuies of lymphoid tissues. It is an important ligand for the L-selectin molecule expressed on naive lymphocytes, directing these cells to leave the blood and enter the lymphoid tissues.

Glyceraldehyde 3-phosphate : A 3-carbon carbohydrate.

Glycocalyx : A network of polysaccharides extending from the surface of bacteria and other cells.OR Alternative term for capsule or slime layer. ***OR*** Extracellular polymeric material produced by some bacteria. Term initially applied to the polysaccharide matrix excreted by epithelial cells forming a coating on the surface of epithelial tissue. General term for polysaccharide compounds outside the bacterial cell wall. Also called slime layer, EPS, or matrix polymer (See EPS).

Glycogen : A polymer of glucose residues in 1,4 linkage, with 1,6 linkages at branchpoints. ***OR*** A highly branched polysaccharide containing glucose, which is used to store carbon and energy. ***OR*** A storage polysaccharide found in the liver and muscles.

Glycogenic : Describing amino acids whose metabolism may lead to gluconeogenesis.

Glycolipid : A lipid containing a carbohydrate group.

Glycolysis : The catabolic conversion of glucose to pyruvate with the production of ATP. ***OR*** The anaerobic conversion of glucose to lactic acid by use of the Embden-Meyerhof pathway. ***OR*** Reactions of the embdenmeyerh of (glycolytic) pathway in which glucose is oxidized to pyruvate.

Glycolysis : The chemical process where by a 6-carbon glucose molecule is split in to two 3-carbon pyruvate molecules.

Glycolytic pathway : See Embden-Meyerhof pathway.

Glycoprotein : A membrane-bound protein which has attached branching carbohydrates. These may function in cell-cell recognition, such as in human blood groups and immune system response, as well as in resisting compression of cells.or A protein linked to an oligosaccharide or a polysaccharide. Glycosaminoglycans. Long, unbranched polysaccharide chains composed of repeating disaccharide subunits in which one of the two sugars is either N-acetylglucosamine or N-acetylgalactosa-mine.

Glycosidase : Enzyme that hydrolyzes a glucosidic linkage between two sugar molecules.

Glycosidic bond : The bond between a sugar and an alcohol. Also the bond that links two sugars in disaccharides, oligosaccharides, and polysaccharides.

Glyoxylate cycle : A pathway that uses some of the enzymes of the TCA cycle and some enzymes whereby acetate can be converted into succinate and carbohydrates. ***OR*** A modified tricarboxylic acid cycle in which the decarboxylation reactions are bypassed by the enzymes isocitrate lyase and malate synthase; it is used to convert acetyl-coa to succinate and other metabolites. ***OR*** A modification of the Krebs cycle, which occurs in some bacteria. Acetyl coenzyme A is generated directly from oxidation of fatty acids or other lipid compounds.

Glyoxysome : An organelle containing some enzymes of the glyoxylate cycle.

GM : Genetically modified. ***OR*** Gross motor.

gm : Gram.

GMA : Gross motor activity.

GMO : Genetically modified organism.

GMP : Guanosine monophosphate or good manufacturing practice.

Gnotobiotic : Animals that are germfree (microorganism free) or live in association with one or more known microorganisms.

Gnotobiotic, germ-free : See germ-free.

GOK : God only knows (diagnosis).

Goldman equation : An equation expressing the quantitative relationship between the concentrations of charged species on either side of a membrane and the resting transmembrane potential.

Golgi apparatus : A complex series of double-membrane structures that interact with the endoplasmic reticulum and that serve as a transfer point for proteins destined for other organelles, the plasma membrane, or extracellular transport. ***OR*** A membranous eucaryotic organelle composed of stacks of flattened sacs (cisternae), which is involved in packaging and modifying materials for secretion and many other processes.

Gonad : Sexual organ (testes and ovaries in animals).

Gonococci : Bacteria of the species *Neisseria gonorrhoeae*-the organism causing gonorrhea.

Gonorrhea : An acute infectious sexually transmitted disease of the mucous membranes of the genitourinary tract, eye, rectum, and throat. It is caused by *Neisseria gonorrhoeae*.

Good manufacturing practice (GMP) : Regulations that describe the methods, equipment, facilities, and controls for producing human pharmaceutical products, veterinary products, and biologically derived products. Also called current Good Manufacturing Practice or cgmp. Go to http://www.cgmp.com/overview.htm to find out more.

Goodpasture's syndrome : Goodpasture's syndrome is an autoimmune disease in which autoantibodies against basement membrane or type IV collagen are produced and cause extensive vasculitis. It is rapidly fatal.

GP : General practitioner.

GPCR : G-protein coupled receptor.

gr : Grain.

GR : Gastric resection; great, greater; gross, grossly; group.

Gracilis : The muscle along the thigh,between the legs and used to hold the legs together.

Gradualism : The theory that organisms evolve by slow, constant change.

Graft rejection : Tissue and organ grafts between genetically distinct individuals almost always elicit an adaptive immune response that causes graft rejection, the destruction of the grafted tissue by attacking lymphocytes.

Graft-versus-host disease (GVHD) : When mature T lymphocytes are injected into a nonidentical immunoincompetent recipient, they can attack the recipient, causing a graft-versus-host (GVH) reaction; in human patients, mature T cells in allogeneic bone marrow grafts can cause graft-versus-host disease (GVHD). ***OR*** A disease that results when mature post-thymic T cells in donor grafts (e.g., bone marrow) recognize the host as foreign and attack it.

Gram molecular weight : For a given compound, the weight in grams that is numerically equal to its molecular weight.

Gram negative : Bacteria cells which lose the crystal violet during the decolorizing step and are then colored by the counterstain. *Pseudomonas* and *Thiobacillus* are examples of Gram negative strains.

Gram positive : Bacterial cells which retain the crystal violet stain during a staining procedure. Most strains of bacilli are gram positive.

Gram stain : A differential staining procedure that divides bacteria into gram-positive and gram-negative groups based on their ability to retain crystal violet when decolorized with an organic solvent such as ethanol. ***OR*** Differential stain that divides bacteria into two groups, Grampositive and Gramnegative, based on the ability to retain crystal violet when decolorized with an organic solvent such as ethanol. The cell wall of Grampositive bacteria consists chiefly of peptidoglycan and lacks the outer membrane of Gramnegative cells.

Gramineae : Also known as Poaceae,the plant family (in Class Monocotyledonae) commonly Known as grasses.

Grana : A stack of thylakoids in the chloroplast stroma.

Granules : Reservoirs for hydrolytic and digestive enzymes in certain types of phagocytic cells; delivered to phagosomes to form phagolysosomes to digest phagocytosed (a.k.a., phagocytized) foreign material; appearance under microscopy makes cells looks "granulated" (see primary granules and secondary granules)

Granulocyte : (Polymorphonuclear leukocyte) leukocyte with intracellular granules and multilobed nucleus.

Granulocyte, polymphonuclear leukocyte : See polymorphonuclear leukocyte.

Granulocyte-macrophage colony-stimulating factor, GM-CSF : Granulocyte-macrophage colony-stimu-lating factor (GM-CSF) is a cytokine involved in the growth and differentiation of myeloid and monocytic lineage cells, including dendritic cells, monocytes and tissue macrophages, and cells of the granulocyte lineage.

Granuloma : Term applied to nodular inflammatory lesions containing phagocytic cells. ***OR*** A granuloma is a site of chronic inflammation usually triggered by persistent infectious agents such as mycobacteria or by a nondegradable foreign body. Granulomas have a central area of macrophages, often fused into multinucleate giant cells, surrounded by T lymphocytes. ***OR*** small growths of nodular granulation tissue formed at sites of chronic inflammation that are usually composed of epithelioid cells, lymphocytes, multinucleated giant cells, eosinophils, plasma cells that are "walled off" by an area of fibrosis.

Granulomatous : Tending to form granulomas; usually referring to diseases or pathogenic microorganisms.

Granulose : Covered with small granules.

Granum : A stack of thylakoids within a chloroplast.

Granzymes : Granzymes are serine proteases produced by cytotoxic T cells and are involved in inducing apoptosis in the target cell.

GRAS : Generally regarded as safe.

Graves' disease : Graves' disease is an autoimmune disease in which antibodies against the thyroid-stimulating hormone receptor cause overproduction of thyroid hormone and thus hyperthyroidism.

Gravitational water : Portion of total soil water potential due to differences in elevation.

Grazing : See predation.

Green (sulfur) bacteria : Anoxygenic phototrophs containing chlorosomes and bacteriochlorophyll c, cs, d or e and light harvesting chlorophyll.

Greenhouse gas : A gas, such as water vapour, carbon dioxide, methane, chlorofluorocarbons (cfcs) and hydrochlorofluorocarbons (hcfcs), that absorbs and re-emits infrared radiation, warming the earth's surface and contributing to climate change. ***OR*** Gases released from the earth's surface through chemical and biological processes that interact with the chemicals in the stratosphere to decrease the release of radiation from the earth. It is believed that this leads to global warming.

Greg Mendel : Person who,in 1865,published a paper which has served as the foundation for our modern understanding of genetics.

Griseofulvin : An antibiotic from *Penicillium griseofulvum* given orally to treat chronic dermatophytic infections of skin and nails.

Grit : The heavy material present in wastewater, such as sand coffee grounds, eggshells, gravel and cinders.

Gross domestic product : The total value of goods and services produced by a country in one year.

Ground state : The lowest electronic energy state of an atom or a molecule.

Ground water : Portion of the water below the surface of the ground at a pressure equal to or greater than atmospheric.

Group translocation : A transport process in which a molecule is moved across a membrane by carrier proteins while being chemically altered at the same time.

Growth : An increase in cellular constituents. ***OR*** In microbiology, an increase in both cell number and cellular constituents. ***OR*** In microbiology, an increase in both cell number and cellular constituents.

Growth factor : A substance that must be present in the growth medium to permit eucaryotic cell proliferation. ***OR*** Organic compound necessary for growth because it is an essential cell component or precursor of such components and cannot be synthesized by the organism itself. Usually required in trace amounts.

Growth factors : Organic compounds that must be supplied in the diet for growth because they are essential cell components or precursors of such components and cannot be synthesized by the organism.

Growth fork : The region on a DNA duplex molecule where synthesis is taking place. It resembles a fork in shape, since it consists of a region of duplex DNA connected to a region of unwound single strands.

Growth limiting nutrient : That parameter which controls the replication rate of a bacterium or other cell in the environment where that cell resides (i.e., temperature, hydrogen ion concentration, ionic strength, etc.). ***OR*** A compound or ion which, when added to the environment in which a population of cells resides, results in an increase in cell biomass. Also known as limiting substrate.

Growth medium : The nutrient-rich environment in which bacteria, yeasts, algae, or specific cell or tissue types are grown in a laboratory environment.

Growth rate : The rate at which growth occurs, usually expressed as the generation time.

Growth rate constant : Slope of the log of the number of cells per unit volume plotted against time.

Growth yield coefficient : Quantity of biomass carbon formed per unit of substrate carbon consumed.

GS : Glutamine synthetase.

GSH : Glutathione (reduced glutathione).

GSSG : Glutathione disulfide (oxidized glutathione).

GST : Glutathione-S-transferase.

GSW : Gunshot wound.

GT : Gait training.

GTL : Gas-to-liquid.

GTP : Guanosine triphosphate.

GTT : Glucose tolerance test.

GU : Gastric ulcer; genitourinary.

Guanine : A purine base found in DNA or RNA. ***OR*** A purine derivative, 2-amino-6-oxypurine, found in nucleosides, nucleotides, and nucleic acids.

Guanine-nucleotide exchange factors, (GEFs) : The guanine-nucleotide exchange factors (gefs) are proteins that can remove the bound GDP from small G proteins; this allows GTP to bind and activate the G protein.

Guanosine : A purine nucleoside found in DNA and RNA.

Guillain-Barr syndrome : A relatively rare disease affecting the peripheral nervous system, especially the spinal nerves, but also the cranial nerves. The cause is unknown, but it most often occurs after an influenza infection or flu vaccination. Also called French Polio.

Gumma : A soft, gummy tumor occurring in tertiary syphilis.

GURT : Genetic use restriction technology.

Gut-associated lymphoid tissue (GALT) : The defensive lymphoid tissue present in the intestines. See Peyer's patches. ***OR*** The gut-associated lymphoid tissues (GALT) are lymphoid tissues closely associated with the gastrointestinal tract, including the palatine tonsils, Peyer's patches, and intraepithelial lymphocytes. The GALT has a distinctive biology related to its exposure to antigens from food and normal intestinal microbial flora.

Guttulate : Containing one or more oil droplets.

GXT : Graded exercise test.

Gymnothecium (pl. gymnothecia) : A non-ostiolate ascocarp composed of loosely interwoven hyphae and containing randomly dispersed asci.

GYN : Gynecology.

H

H : Hour.

***H* :** Planck's constant.

H antigens, histocompatibility antigens : H antigens or histocompatibility antigens, are known as major histocompatibility antigens when they encode molecules that present foreign peptides to T cells and as minor H antigens when they present polymorphic self peptides to T cells. See also histocompatibility.

H&P : History and physical.

H-2 complex : Term for the MHC in the mouse.

H-2, histocompatibility-2 : The major histocompatibility complex of the mouse is called H-2 (for histocompatibility-2). Haplotypes are designated by a lower-case superscript, as in H-2b.

HA : Hemagglutinin.

HA : Headache; high anxiety; hospital admission.

HAA : Hepatitis associated antibodies.

Habitat : Place where an organism lives.

HAC : Human artificial chromosome.

Haemagglutination-inhibition : An assay used for certain types of viruses which are able to agglutinate red blood cells. Haemagglutination-inhibition records blocking of this process by antibodies, and thus, the presence of antibodies against the virus.

Haematopoietic : A type of stem cell found in the bone marrow, which can differentiate into blood cells, such as red blood cells, white blood cells, and platelets.

Haemoglobin : The part of the red blood cells that carries oxygen.

Haemophilia : An inherited disease where blood fails to clot.

Hairpin loop : A single-stranded complementary region of DNA or RNA that folds back on itself and base-pairs into a double helix.

Half-life : The time required for the disappearance of one half of a substance.

Halitosis : Bad breath caused by certain bacteria in the mouth breaking down proteins into smelly byproducts.

Halobacteria extreme halophiles : A group of archaea that have an absolute dependence on high NaCl concentrations for growth and will not survive at a concentration below about 1.5 M NaCl.

Halogen : Any of the five elements F, Cl, Br, I, and At that form part of group VII A of the periodic table.

Halophile : A microorganism that requires high levels of sodium chloride for growth. ***OR*** Organism requiring or tolerating a saline environment.

Halophilic halotolerant : Bacteria which thrive in a highly salt environment, up to 25% NaCl.

Halotolerant : An organism capable of growing in the presence of NaCl but not requiring it.

HAMA : Human antimouse antibody.

Hamstrings : Muscles that are the back of the thigh.

Hansen's disease : See leprosy.

Hantavirus pulmonary syndrome : The disease in humans caused by the pulmonary syndrome hantavirus. Deer mice shed the virus in their feces, humans inhale the virus and first develop ordinary flulike aches and pains. Within a few days the hantavirus causes lung damage and capillary leakage. After about a week the infected person enters a crisis phase and may die.

Haploid : In eukaryotes, an organism or cell containing one chromosome complement and the same number of chromosomes as the gametes.

Haploid : Having one set of chromosomes.

Haploid cell : A cell containing only one chromosome of each type.

Haplotype : A haplotype is a linked set of genes associated with one haploid genome. The term is used mainly in connection with the linked genes of the major histocompatibility complex, which are usually inherited as one haplotype from each parent. Some MHC haplotypes are overrepresented in the population, a phenomenon known as linkage disequilibrium.

Hapten : A substance not inducing antibody formation but able to combine with a specific antibody. ***OR*** A molecule not immunogenic by itself but that, when coupled to a macromolecular carrier, can elicit antibodies directed against itself.

Haptens : Haptens are molecules that can bind antibody but cannot by themselves elicit an adaptive immune response. Haptens must be chemically linked to protein carriers to elicit antibody and T-cell responses.

Harbage transmission : The mode of transmission in which an infectious organism does not undergo morphological or physiological changes within the vector.

Hashimoto's thyroiditis : Hashimoto's thyroiditis is an autoimmune disease characterized by persistent high levels of antibody against thyroid-specific antigens. These antibodies recruit NK cells to the tissue, leading to damage and inflammation.

Hay fever : Allergic rhinitis; a type of atopic allergy involving the upper respiratory tract.

Hazardous : Dangerous materials are hazardous materials.

Hb : Hemoglobin.

HBD : Has been drinking.

HBF : Hepatic blood flow.

HBP : High blood pressure.

HCA : High content analysis.

HCG : Human chorionic gonadotrophin.

HCR : Hydridization chain reaction.

HCS : High content screening.

HCTZ : Hydrochlorothiazide.

HCV : Hepatitis C virus.

HCVD : Hypertensive cardiovascular disease.

HD : High density or Huntington disease

HD : Heart disease; Hemodialysis; Hospital day.

HDL : High-density lipoprotein.

HDS : Herniated disc syndrome.

Headworks : The facilities where wastewater enters a wastewater treatment plant. The headworks may consist of bar screens, comminutors, a wet well and pumps.

Health : A state of optimal physical, mental, and social well-being, and not merely the absence of disease and infirmity.

Healthy carrier : An individual who harbors a pathogen, but is not ill.

Heart disease : A condition that affects the heart muscle or the blood vessels of the heart.

Heat-shock proteins : Proteins produced when cells are exposed to high temperatures or other stressful conditions. They protect the cells from damage and often aid in the proper folding of proteins.

Heavy chain, H chain : All immunoglobulin molecules have two types of chain, a heavy chain (H chain) of 50-70 kda and a light chain of 25 kda. The basic immunoglobulin unit consists of two identical heavy chains and two identical light chains. Heavy chains come in a variety of heavy-chain classes or isotypes, each of which confers a distinctive functional activity on the antibody molecule.

Heavy isotopes : Forms of atoms that contain greater numbers of neutrons than the most common form (e.g., 15N, 13C).

Heavy metals : Those metals which have densities > 5.0 Mg m^3. These include the metallic elements Cu, Fe, Mn, Mo, Co, Zn, Cd, Hg, Ni, and Pb. Al and Se have densities < 5 but are also considered heavy metals.

HEENT : Head, ears, eyes, nose, throat.

Helical : In virology this refers to a virus with a helical capsid surrounding its nucleic acid.

Helicases : Enzymes that use ATP energy to unwind DNA ahead of the replication fork.

Helix : A spiral structure with a repeating pattern.

Helminthospium sp. **:** Reported to be allergenic.

Helminths : A general term referring to the nematodes (roundworms), trematodes (flat worms, flukes), cestodes (tapeworms) and acanthocephalans (thorny-headed worms).

Helper CD4 T cells : Helper CD4 T cells are CD4 T cells that can help B cells make antibody in response to antigenic challenge. The most efficient helper T cells are also known as TH2 cells, which make the cytokines IL-4 and IL-5. Some experts refer to all CD4 T cells, regardless of function, as helper T cells; we do not accept this usage because function can be determined only in cellular assays, and some CD4 T cells kill the cells they interact with.

Helvella (Ascomycetes) : A saddle-shaped fruiting body of a fungus found on soil, or sometimes on rotting wood. Occasionally found growing on soil under houses.

Hemadsorption : The adherence of red blood cells to the surface of something, such as another cell or a virus.

Hemagglutination : The agglutination of red blood cells by antibodies.

Hemagglutinin : The antibody responsible for a hemaggluti-nation reaction.

Hemagglutinin, Hemagglutination : A hemagglutinin is any substance that causes red blood cells to agglutinate, a process known as hemagglutination. The hemagglutinins in human blood are antibodies that recognize the ABO blood group antigens. Influenza and some other viruses have hemagglutinin molecules that bind to glycoproteins on host cells to initiate the infectious process.

Hematoma : Local swelling,tumor,or bruise filled with blood.

Hematopoiesis, hematopoietic stem cells, hematopoietic growth factors: Formation and development of blood cells. ***OR*** Hematopoiesis is the generation of the cellular elements of blood, including the red blood cells, leukocytes, and platelets. These cells all originate from pluripotent hematopoietic stem cells whose differentiated progeny divide under the influence of hematopoietic growth factors.

Hematopoietic lineage : A hematopoietic lineage is any developmental series of cells that derives from hematopoietic stem cells and results in the production of mature blood cells.

Heme group : A porphyrin ring with iron in the center.

Heme : An iron-porphyrin complex found in hemoglobin and cytochromes.

Hemiacetal : The product formed by the condensation of an aldehyde with an alcohol; it contains one oxygen linked to a central carbon in a hydroxyl fashion and one oxygen linked to the same central carbon by an ether linkage.

Hemocyanin : Respiratory pigment containing a porphyrin ring with copper in the center (**heme**=blood; **cyano**=blue, dark blue)

Hemoglobin : The constituent of red blood cells that gives them their colour and carries oxygen.

Hemolysin : The substance produced by microorganisms that lyses red blood cells, liberating hemoglobin. Also can refer to type of antibody that acts in conjuction with complement to cause the lysis of red blood cells.

Hemolysis : The disruption of red blood cells and release of their hemoglobin. There are several types of hemolysis when bacteria such as *Streptococci* and *Staphylococci* grow on blood agar. In a-hemolysis, a narrow greenish zone of incomplete hemolysis forms around the colony. A clear zone of complete hemolysis without any obvious color change is formed during b-hemolysis. ***OR*** the lysis of red blood cells (erythrocytes); either *in vivo* in the host or *in vitro* on blood agar artificial medium

- Betahemolysis complete hemolysis; clear zone around colonies on blood agar medium
- Alphahemolysis partial hemolysis, usually with greening of the medium due to the reduction of hemoglobin in the erythrocytes to methemoglobin.
- Gamma hemolysis no hemolysis.

Hemolytic disease of the newborn : See erythroblastosis fetalis.

Hemolytic uremic syndrome : A kidney disease characterized by blood in the urine and often by kidney failure. It is caused by enterohemorrhagic strains of *Escherichia coli* O157:H7 that produce a Shiga-like toxin, which attacks the kidneys.

Hemorrhage : Abnormal bleeding,usually internal,and often profuse.

Hemorrhagic fever : A fever usually caused by a specific virus that may lead to hemorrhage, shock, and sometimes death.

Hemorrhoids : Dilated or varicose veins in the anal area,often due to difficulty passing feces due to a low fiber diet.

Henderson-Hasselbalch equation : An equation that relates the pka, to the pH and the ratio of the proton acceptor (A-) and the proton donor (HA) species of a conjugate acid base pair.

HEP : Home exercise program.

Hepatitis : Any infection that results in inflammation of the liver. Also refers to liver inflammation as such. ***OR*** Inflammation of the liver which may be caused by a virus e.g. Hepatitis A, Hepatits B.

Hepatitis A : A type of hepatitis that is transmitted by fecal-oral contamination; it primarily affects children and young adults, especially in environments where there is poor sanitation and overcrowding. It is caused by the hepatitis A virus, a single-stranded RNA virus.

Hepatitis B (formerly serum hepatitis; hep²ah-ti_tis) : This form of hepatitis is caused by a double-stranded DNA virus (HBV) formerly called the "Dane particle." The virus is transmitted by body fluids.

Hepatitis C : About 90% of all cases of viral hepatitis can be traced to either HAV or HBV. The remaining 10% is believed to be caused by one and possibly several other types of viruses. At least one of these is hepatitis C (formerly non-A, non-B).

Hepatitis D (formerly delta hepatitis) : The liver diseases caused by the hepatitis D virus in those individuals already infected with the hepatitis B virus.

Hepatitis E (formerly enteric-transmitted NANB hepatitis) : The liver disease caused by the hepatitis E virus. Usually, a subclinical, acute infection results; however, there is a high mortality in women in their last trimester of pregnancy.

Hepatocyte : Liver cell.

Heptamer : Recombination signal sequences (RSS) flanking gene segments consist of a seven-nucleotide heptamer and a nine-nucleotide nonamer of conserved sequence, separated by 12 or 23 nucleotides. Rsss form the target for the site-specific RAG-1:RAG-2 recombinase that joins the gene segments.

Heraclides : Person who lived from 388 to 315 BC and said that the apparent daily rotation of the heavenly bodies was not due to their motion, but rather, due to the rotation of the Earth.

Herbalist : A person who uses plants or plant extracts to heal.

Herbivorous : An animal that eats only plants.

Herd immunity : The resistance of a population to infection and spread of an infectious organism due to the immunity of a high percentage of the population.

Hereditary angioneurotic edema : Hereditary angioneurotic edema is the clinical name for a genetic deficiency of the Cl inhibitor of the complement system. In the absence of Cl inhibitor, spontaneous activation of the complement system can cause diffuse fluid leakage from blood vessels, the most serious consequence of which is epiglottal swelling leading to suffocation.

Herma phroditism : Having organs of both sexes in one body (this term is used primarily with animals—"monoecious" is used with plants).

Hermaphroditic : Containing both male and female sex organs.

Herpes labialis : See cold sore.

Herpetic keratitis : An inflammation of the cornea and conjunctiva of the eye resulting from a herpes simplex virus infection.

HESC : Human embryonic stem cell.

Heterochromatin : Highly condensed regions of chromosomes that are not usually transcriptionally active.

Heterocyst : Differentiated cyanobacterial cell that carries out dinitrogen fixation. ***OR*** Specialized cells produced by cyanobacteria that are the sites of nitrogen fixation.

Heteroduplex DNA : A double-stranded stretch of DNA formed by two slightly different strands that are not completely complementary. ***OR*** An annealed duplex structure between two DNA strands that do not show perfect complementarity. Can arise by mutation, recombination, or the annealing of complementary single-stranded DNAs.

Heterofermentation : Any fermentation in which there is more than one major endproduct. Synonym of heterolactic fermentation.

Heterogeneous nuclear RNA (hnRNA) : The RNA transcript of DNA made by RNA polymerase II; it is then processed to form mRNA.

Heterokaryon : Hypha that contains at least two genetically dissimilar nuclei.

Heterolactic fermentation : A type of lactic acid fermentation in which sugars (e.g. lactose, glucose) are fermented to a range of products.

Heterolactic fermenters : Microorganisms that ferment sugars to form lactate, and also other products such as ethanol and CO_2.

Heteropolymer : A polymer containing more than one type of monomeric unit.

Heteropolysaccharide : The class name for polysaccharides composed of two or more different kinds of monomeric units.

Heterothallic : A fungus that requires mating between two compatible strains for sexual reproduction to occur. ***OR*** Hyphae that are incompatible with each other each requiring contact with another hypha of compatible mating type which, upon fusion, forms a dikaryon or a diploid.

Heterotroph : An organism that uses reduced, preformed organic molecules as its principal carbon source. ***OR*** A microorganism which uses organic matter for energy and growth. ***OR*** Organism capable of deriving carbon and energy for growth and cell synthesis from organic compounds; generally also obtain energy and reducing power equivalents from organic compounds. ***OR*** Organism which ingests other food.

Heterotrophic bacteria : Those bacteria that require an organic carbon source for growth; i.e. Derive energy and carbon from organic compounds.

Heterotrophic nitrification : Nitrification carried out by chemohetero-trophic microorganisms. ***OR*** Biochemical oxidation of ammonium to nitrite and nitrate by heterotrophic microorganisms.

Heterotrophic plate count bacteria (HPC) : Those bacteria that will grow on nonselective heterotrophic medium plates.

Heterotrophs : Organisms that must have carbonenergy compounds.

Heterozygote : An individual with a pair of two different alleles.

Heterozygous : Individuals heterozygous for a particular gene have two different alleles of that gene. ***OR*** The two alleles for a particular gene within a diploid cell are different. ***OR*** Describing an organism (a heterozygote) that carries two different alleles for a given gene.

Hexon hexamer : A capsomer composed of six protomers.

Hexose monophosphate pathway : See pentose phosphate pathway. ***OR*** A metabolic pathway present in a wide range of prokaryotic and eukaryotic microorganisms as well as in plants and animals; it involves the oxidative decarboxylation of glucose 6phosphate, via 6phosphogluconate, to ribulose 5phosphate, followed by a series of reversible, nonoxidative interconversions whereby hexose and triose phosphates are formed from pentose phosphates. Also called: HMP pathway, HMP shunt; oxidative pentose phosphate pathway, pentose phosphate pathway/cycle, phosphogluconate pathway; warburgdickens pathway.

Hexose : A sugar with a six-carbon backbone.

Heymann's nephritis : An excellent model for membranous glomerulonephritis is Heymann's nephritis, a disease induced by injecting animals with tubular epithelial tissue.

HF : Hay fever; heart failure.

Hfr strain : A bacterial strain that donates its genes with high frequency to a recipient cell during conjugation because the F factor is integrated into the bacterial chromosome.

HGB : Hemoglobin.

hGH : Human growth hormone.

HGP : Human genome project.

HGPRT : Hypoxanthine-guanine phosphoribosyl transferase.

HH : Hard of hearing; hiatus hernia.

Hibernation : Along-term(seasonal) decrease in body functions.

High endothelial venuies, HEVs, high endothelial cells : High endothelial venuies (hevs) are specialized venules found in lymphoid tissues. Lymphocytes migrate from blood into lymphoid tissues by attaching to and migrating across the high endothelial cells of these vessels.

High oxygen diffusion environment : A microbial environment in close contact with air and through which oxygen can move at a rapid rate (in comparison with the slow diffusion rate of oxygen through water).

High-energy compound : A compound that undergoes hydrolysis with a high negative standard free energy change.

High-energy molecule : A molecule whose hydrolysis under standard conditions makes available a large amount of free energy (the standard free energy change is more negative than about 27 kcal/mole); a high-energy molecule readily decomposes and transfers groups such as phosphate to acceptors.

High-zone tolerance : Tolerance to injected protein antigens occurs at low or high doses of antigen. Tolerance induced by the injection of high doses of antigen is called high-zone tolerance, whereas tolerance produced with low doses of antigen is called low-zone tolerance.

Hilum : A scar at the base of a conidium.

Hinge region : The hinge region of antibody molecules is a flexible domain that joins the Fab arms to the Fc piece. The flexibility of the hinge region in IgG and IgA molecules allows the Fab arms to adopt a wide range of angles, permitting binding to epitopes spaced variable distances apart.

Hippocrates : Person who lived from about 400 to 300 BC and said that the human body was composed of four elements (earth, air, fire, and water) plus the four humors (sanguis, choler, melancholia, and phlegma).

Hirsute : Hairy.

Hispid : Covered with stiff bristle-like hairs usually visible to the naked eye.

Histamine : Histamine is a vasoactive amine stored in mast cell granules. Histamine released by antigen binding to IgE molecules on mast cells causes dilation of local blood vessels and smooth muscle contraction, producing some of the symptoms of immediate hypersensitivity reactions. Antihistamines are drugs that counter histamine action.

Histocompatibility : Histocompatibility is literally the ability of tissues (Greek: histos) to get along with each other. It is used in immunology to describe the genetic systems that determine the rejection of tissue and organ grafts resulting from immunological recognition of histocompatibility (H) antigens.

Histocompatibility-2 : See H-2.

Histology : The study of tissues.

Histone : A small basic protein with large amounts of lysine and arginine that is associated with eucaryotic DNA in chromatin.

Histones : Proteins attached to the DNA of eukaryotes which allows it to be packaged into chromosomes.or The family of basic proteins that is normally associated with DNA in most cells of eukaryotic organisms.

Histoplasma sp. **:** A fungus which has filamentous growth at 25 degrees C and yeast growth at 37 degrees C. It is reported to be a human pathogen. It may be associated with birds.

Histoplasmosis : A systemic fungal infection caused by *Histoplasma capsulatum* var *capsulatum*.

HIV : Human immunodeficiency virus

Hives : An eruption of the skin.

HL : Hearing loss; heart and lungs.

HLA : HLA, the acronym for Human Leukocyte Antigen, is the genetic designation for the human MHC. Individual loci are designated by upper-case letters, as in HLA-A, and alleles are designated by numbers, as in HLA-A*0201.

HLA : Heart, lungs, abdomen.

HLA-DM : The invariant HLA-DM molecule in humans is involved in loading peptides onto MHC class II molecules. It is encoded in the MHC within a set of genes resembling MHC class II genes. A homologous protein in mice is called H-2M.

HLK : Heart, liver, kidney.

HLV : Herpes like virus.

HM : Heart murmur.

HMG-CoA : ?-Hydroxy-?-methylglutaryl-coa

HMO : Health maintenance organization.

HMP : Hot moist packs.

HNP : Herniated nucleus pulposus.

hnRNA : Heterogeneous nuclear RNA

hnRNA : Heterogeneous RNA OR heterogeneous nuclear RNA

HNV : Has not voided.

HO : History of.

HOB : Head of bed.

Hodgkin's disease : Hodgkin's disease is an immune system tumor characterized by large cells called Reed-Sternberg cells, which derive from mutated B-lineage cells. It exists in at least two polar forms, Hodgkin's lymphoma and nodular sclerosis.

Holarctic realm : A term used to collectively refer to the Palearctic and Nearctic Realms—the Biogeographical realm corresponding to most of the continents in the northern hemisphere.

Holdfast : A structure produced by some bacteria and algae that attaches the cell to a solid object. ***OR*** Polymers excreted by living organisms that mediate a stable interaction between the organism and a solid substratum. Examples of holdfast molecules excreted by microorganisms include pili, fimbriae, stalks, and exopolysaccharides.

Holistic : Looking at the entire system. In healthcare, this means looking at all aspects of a patient's well-being and not just treating a particular pain or disease.

Holoblastic : A mode of blastic conidium ontogeny in which all the cell wall layers of the conidiogenous cell are involved in conidium development.

Holoenzyme : A complete enzyme consisting of the apoenzyme plus a cofactor. ***OR*** An intact enzyme containing all of its subunits and any necessary cofactors with full enzymatic activity.

Holomictic : In these lakes there is a periodic (usually once a year in temperate zone lakes) mixing of the zones

Holomorph : Whole fungus consisting of all sexual and asexual stages in its life cycle.

Holothallic : A mode of thallic conidium ontogeny in which all the cell wall layers of the conidiogenous cell are involved in conidium development.

Holozoic nutrition : In this type of nutrition, nutrients (such as bacteria) are acquired by phagocytosis and the subsequent formation of a food vacuole or phagosome.

Homeostasis : Homeostasis is a generic term describing the status of physiological normality. In the case of lymphocytes, homeostasis refers to an uninfected individual who has normal numbers of lymphocytes.

Homeostasis : The ability of an animal to maintain a constant internal environment.

Homeothermic : The ability of an organism to maintain a constant body temperature(usually because that organism is also endothermic).

Homofermentation : Any fermentation in which there is only one major endproduct. Synonym of homolactic fermentation.

Homokaryon : Fungal hypha in which all nuclei are genetically identical.

Homolactic fermentation : A type of lactic acid fermentation in which sugars (e.g. Glucose, lactose, etc) are converted entirely, or almost entirely, to lactic acid.

Homolactic fermenters : Organisms that ferment sugars almost completely to lactic acid.

Homologous chromosomes : The two matching chromosomes of a pair (one from the father and one from the mother).

Homologous chromosomes : Chromosomes that carry the same pattern of genes, but not necessarily the same alleles.

Homologous recombination : Cellular genes can be disrupted by homologous recombination with copies of the gene into which erroneous sequences have been inserted. When these exogenous DNA fragments are introduced into cells, they recombine selectively with the cellular gene through remaining regions of sequence homology, replacing the functional gene with a nonfunctional copy.

Homopolymer : A polymer composed of only one type of monomeric building block.

Homo sapiens **:** The scientific name for humans.

Homothallic : Hyphae that are selfcompatible in that sexual reproduction occurs in the same organism by meiosis and genetic recombination; fusion of hypha results in a dikaryon or diploid. ***OR*** A fungus capable of sexual reproduction on a single thallus. ***OR*** Hyphae that are selfcompatible in that sexual reproduction occurs in the same organism by meiosis and genetic recombination; fusion of hypha results in a dikaryon or diploid.

Homozygote : An individual with a pair of two of the same allele.

Homozygous : Describing an organism (a homozygote) that carries two identical alleles for a given gene. ***OR*** The two alleles for a particular gene within a diploid cell are identical.

Homunculus : A tiny, preformed human which people thought merely increased in size to Become a baby.

Hizon : See soil horizon.

Horizontal gene transfer : The process in which genes are transferred from one mature, independent organism to another. ***OR*** The process by which bacteria and Archaea transfer genes to other bacteria or Archaea, especially obvious when a sudden environmental change forces resident bacteria or Archaea to seek a faster method of evolution than will occur slowly due to mutations of native DNA. Methods used include.

Conjugation - One cell binds to another cell, often unrelated, opening a connection through which it can pass a strand of DNA, copied from its primary DNA, into a separate contained packet of DNA, called a plasmid, to the recipient cell, without risking spilling cell contents from either donor or recipient cell into the environment. Conjugation is often a two-way transfer that enhances survival of both participants. If bacteria develop antibiotic resistance due to improper application of antibiotics (incorrect selection of treatment antibiotic, insufficient dosage to kill all bacteria exposed to the antibiotic), escaping bacteria may choose to pass antibiotic resistance along in plasmids to other bacteria they encounter in their environment (whether they normally live in hospitals, sewage systems, lakes, ponds, rivers or turf that is grazed by livestock.

DNA scavenging - One cell samples DNA strands strewn about during cell wall breakdown following death of bacterial cells that were intolerant of primary environmental conditions (temperature, ph, salinity, available food sources, antibiotics produced by native bacteria to keep them dominant, etc.), sometimes delivering beneficial enzyme pathways to the resident species. When horizontal gene transfer happens naturally in the environment, huge leaps in adaptation can occur, creating new species very distinct from parental strains.

Hormogonia : Small motile fragments produced by fragmentation of filamentous cyanobacteria; used for asexual reproduction and dispersal.

Hormone : A chemical produced naturally by cells, which acts as a signal or message to cause widespread changes in the rest of the body. Examples or hormones are sex hormones, like oestrogen and progesterone; and insulin, which is needed to control the level of glucose sugar in the blood. Hormones are usually produced in one part of the body and travel in the blood to their points of action. ***OR*** A chemical substance made in one cell and secreted so as to influence the metabolic activity of a select group of cells located at other sites in the organism. ***OR*** One of the body's messenger molecules which affects the functioning of some other area of the body.

Hormone receptor : A protein that is located on the cell membrane or inside the responsive cell and that interacts specifically with the hormone.

Host : Organism capable of supporting the growth of a virus or other parasite. ***OR*** an organism (animal or plant) that harbors or provides nutrition for another organism. ***OR*** The body of an organism that harbors another organism. It can be viewed as a microenvironment that shelters and supports the growth and multiplication of a parasitic organism. (parasite; pathogen).

Host cell : A cell used for growth and reproduction of a virus.

Host immune system : The cells and tissues involved in recognizing and attacking foreign substances in the body.

Host range : The plants that are attacked by a particular pathogen. Plant viruses are typically specific to a few host plants (narrow host range) but tospoviruses (INSV and TSWV) has a very wide host range with many different plants susceptible to attack.

Host restriction : The degradation of foreign genetic material by nucleases after the genetic material enters a host cell.

Host-versus-graft disease, HVGD : Host-versus-graft disease (HVGD) is another name for the allograft rejection reaction. The term is used mainly in relation to bone marrow transplantation.

House sewer : The pipeline connecting the house and drain and the septic tank.

HP : Hot pack.

HPC : High performance computing.

HPE : History and physical examination.

HPI : History of present illness.

HPLC : High performance liquid chromatography. ***OR*** high pressure liquid chromatography.

HPV : Human papilloma virus.

HR : Heart rate; hospital record.

HRT : Hours of Retention Time.

HS : Bedtime; heart sounds.

HSA : Human serum albumin.

HSP : Heat shock protein (dict.) Or hematopoietic stem and progenitor cell.

HSPs : High scoring segment pairs.

HSV : Herpes simplex virus.

HT : Heart; hospital treatment; hypertension; hyperthyroidism.

hTEP : Human TEP.

HTGS : High throughput genomic sequencing.

HTN : Hypertension.

HTPS : High throughput screening.

HUGO : Human genome organization.

Hulle cells : Thick-walled cells with characteristic thin-walled pores, usually associated with cleistothecia of *Aspergillus*.

Human herpesvirus 6 (HHV-6, type A and B) : HHV-6 was discovered in 1986 and was initially called the human B-lymphotropic virus. The virus was later shown to have a marked tropism for CD41 T cells and was renamed HHV-6. HHV-6 is genetically similar to cytomegalovirus. HHV-6 causes exanthem subitum (roseola infantum or sixth disease) in infants and has been suspected of involvement in many conditions, including opportunistic infections in immunocompromised patients, hepatitis, lymphoproliferative diseases, synergistic interactions with HIV, lymphadenitis, and chronic fatigue syndrome.

Human immunodeficiency virus (HIV) : The virus that causes acquired immune deficiency syndrome (AIDS). ***OR*** The human immunodeficiency virus (HIV) is the causative agent of the acquired immune deficiency syndrome (AIDS). HIV is a retrovirus of the lentivirus family that selectively infects macrophages and CD4 T cells, leading to their slow depletion, which eventually results in immunodeficiency.

Human leukocyte antigen complex (HLA) : A class of proteins found on the surface of cells in the immune system, which help the immune system to distinguish 'self' from 'non-self'. Matching the type of human leukocyte antigen is important for transplantation to reduce the risk of organ rejection. ***OR*** An antigen on the surface of cells of human tissues and organs that is recognized by the immune system cells and therefore is important in graft rejection and regulation of the immune response. This is the same as MHC class II.

Humanization : Humanization is a term used to describe the genetic engineering of mouse hypervariable loops of a desired specificity into otherwise human antibodies. The DNA encoding hypervariable loops of mouse monoclonal antibodies or V regions selected in phage display libraries is inserted into the framework regions of human immuno-globulin genes. This allows the production of antibodies of a desired specificity that do not cause an immune response in humans treated with them.

Humerus : The bone in the upper arm.

Humic acid : Darkcolored organic material extracted from soil by various reagents (e.g., dilute alkali) and that is precipitated by acid (pH 1 to 2).

Humic substances : Series of relatively highmolecularweight, browntoblack substances formed by secondary synthesis reactions. The term is generic in a sense that it describes the colored material or its fractions obtained on the basis of solubility characteristics, such as humic acid or fulvic acid.

Humicuola sp. **:** Grow on products with a high cellulose content. These fungi are also found in soil and on plant debris.

Humidity : Water vapor in the air.

Humification : Process whereby the carbon of organic residues is transformed and converted to humic substances through biochemical and chemical processes.

Humor : The four fluids of which Hippocrates thought the body was composed (Sanguis, Choler, Melancholia, and Phlegma).

Humoral system : The portion of the immune system consisting of B- cells which develop In bone marrow.

Humoral immunity, humal immune response : The type of immunity that results from the presence of soluble antibodies in blood and lymph; also known as antibody-mediated immunity. ***OR*** Humoral immunity is the antibody-mediated specific immunity made in a humoral immune response. Humoral immunity can be transferred to unimmunized recipients by using immune serum containing specific antibody.

Humus : Total of the organic compounds in soil exclusive of undecayed plant and animal tissues, their "partial decomposition" products, and the soil biomass. The term is often used synonymously with soil organic matter. ***OR*** The dark organic material in soils, produced by the decomposition of soils. The matter that remains after the bulk of detritus has beenconsumed (leaves, roots). Humus mixes with top layers of soil (rock particles), supplies some of the nutrients needed by plants -increases acidity of soil; inorganic nutrients more soluble under acidic conditions, become more available, e.g. Wheat grows best at pH 5.5-7.0. Humus modifies soil texture, creates loose, crumbly texture, that allows water to soak in and nutrients retained; permits air to be incorporated into soil.

Huntington's disease : A genetic disorder of the nervous system, characterised by involuntary movements and progressive mental deterioration.

HUPO : Human proteome organization.

Husbandry : The breeding and raising of animals, or the cultivation of crops.

HV : Hepatic vein; herpes virus; home visit; hospital visit; hyperventilation.

HW : Healing well.

HWP : Hot wet pack.

HX : Hypoxanthine.

HX : History.

Hyaline : Clear and colourless. ***OR*** Colourless.

Hyaline mycelia : Sterile mycelia which is white or transparent. No fruiting structures are produced by the mycelia. Visual identification of these organism is not possible. Often associated with allergic symptoms.

Hyalo : A prefix meaning hyaline to lightly coloured.

Hyaluronic acid : A mucopolysaccharide that lubricates tissues and inhibits dissemination of invading microorganisms.

Hyaluronidase : Enzyme that breaks down hyaluronic acid; normally found in the spleen and the testicles and produced by species of *Streptococcus* and *Staphylococcus*.

Hybrid : An organism whose parents are from different species.

Hybrid (chimeric) plasmid : A plasmid that contains DNA from two different organisms.

Hybridization : Natural formation or artificial construction of a duplex nucleic acid molecule by complementary base pairing between two nucleic acid strands derived from different sources.

Hybridoma : A fast-growing cell line produced by fusing a cancer cell (myeloma) to another cell, such as an antibody-producing cell.

Hybridomas : Monoclonal antibodies are most commonly produced from hybridomas. These are hybrid cell lines formed by fusing a specific antibodyproducing B lymphocyte with a myeloma cell that is selected for its ability to grow in tissue culture and for an absence of immunoglobulin chain synthesis.

Hydatid cyst : Larva of *Echinococcus* spp. Which is a large fluid filled cyst containing many invaginated scolecies and daughter cysts that also contain many invaginated scolecies. There are two types of hydatid cysts *E. granulosus* has uniloculular (one-chambered) cyst and *E. multilocularis* has a multi-locular (many chambered) cyst.

Hydraulic loading : Hydraulic loading refers to the flows (MGD or m^3/day) to a treatment plant or treatment process.

Hydrocarbon : Any chemical compound containing only carbon and hydrogen elements. Some simple examples of hydrobarbons are: methane (CH_4), ethylene (C_2H_4), ethane (C_2H_6), etc.

Hydrocarbon : A molecule made of only carbon and hydrogen atoms.

Hydrochloric Acid : (HCl), an acid made from a hydrogen ion and a chloride ion and found in the stomach.

Hydrogen : The first element on the periodic table, containing one proton and one electron.

Hydrogen bond : Chemical bond between a hydrogen atom of one molecule and two unshared electrons of another molecule. ***OR*** A weak attraction which holds two molecules containing hydrogen together. For example, in water two hydrogen atoms are bonded to one oxygen atom.

Hydrogen peroxide : A substance made up of hydrogen and oxygen. It is produced by cells, but can be toxic in large amounts and is broken down rapidly by enzymes.

Hydrogen sulfide gas : Hydrogen sulfide is a gas with a rotten egg odor. This gas is produced under anaerobic conditions. Hydrogen sulfide is particularly dangerous because it dulls your sense of smell so that you don't notice it after you have been around it for a while and because the odor is not noticeable in high concentrations. The gas is very poisonous to your respiratory system, explosive, flammable, and colorless.

Hydrogenosome : A microbodylike organelle that contains a unique electron transfer pathway in which hydrogenase transfers electrons to protons (which act as the terminal electron acceptors) and molecular hydrogen is formed.

Hydrogenoxidizing bacterium : Facultative lithotrophs that, in the absence of an oxidizable organic source, oxidize H2 for energy and synthesize carbohydrates with carbon dioxide as their source of carbon.

Hydrolysis : The process in which carbohydrates and starches are simplified into organic soluble organics, usually by facultative anaerobes. ***OR*** The cleavage of a molecule by the addition of water. Hydrophilic. Preferring to be in contact with water. ***OR*** Splitting of one large molecule in to two smaller ones through the addition of a water molecule.

Hydrophilic : Compatible with water; "waterloving" OR A polar substance that has a strong affinity for water (or is readily soluble in water). ***OR*** "Water loving". Hydrophilic compounds dissolve easily in water, and are usually polar. ***OR*** Term used to describe substances which stick to or mix with water.

Hydrophobic : A nonpolar substance lacking affinity for water (or which is not readily soluble in water). ***OR*** Incompatible with water; "waterhating". ***OR*** Term used to describe substances which are water-repellant and can not mix With water.

Hydrophobic effect : The noncovalent association of nonpolar groups with each other in aqueous solution. ***OR*** "Water fearing". Hydrophobic compounds do not dissolve easily in water, and are usually non-polar. Oils and other long hydrocarbons are hydrophobic. ***OR*** Preferring not to be in contact with water, as is the case with the hydrocarbon portion of a fatty acid or phospholipid chain.

Hydroxyapatite : A calcium phosphate gel used, in the case of nucleic acids, to selectively absorb duplex DNA-RNA from a mixture of single-stranded and duplex nucleic acids.

Hygrophanous : Of water-soaked appearance and translucent, changing colour on drying (usually paler) and more opaque.

Hygroscopic : Absorbing or attracting moisture from the air.

Hygroscopic water : Water adsorbed by a dry soil from an atmosphere of high relative humidity.

Hyl : 5-hydroxylysine.

Hymen : A membrane which partially covers the opening of the vagina in a virgin.

Hymenium : The fertile sexual spore-bearing tissues (In an Agaric this covers the gills.) ***OR*** Layer of hyphae which are fertile in producing asci (fungi in the phylum Ascomycota) or basidia (fungi in the phylum Basidiomycota) from the process of meiosis.

Hyp : 4-hydroxyproline.

Hypanthium : In a flower, an enlarged receptacle which had grown to surround the ovary (for example, an apple).

Hyperacute graft rejection : Hyperacute graft rejection of an allogenic tissue graft is an immediate reaction caused by natural preformed antibodies that react against antigens on the graft. The antibodies bind to endothelium and trigger the blood clotting cascade, leading to an engorged, ischemic graft and rapid loss of the organ.

Hypercalcemia : Having too much calcium in one's system.

Hyperendemic disease : A disease that has a gradual increase in occurrence beyond the endemic level, but not at the epidemic level, in a given population; also may refer to a disease that is equally endemic in all age groups.

Hypereosinophilia : Hypereosinophilia is an abnormal state in which there are extremely large numbers of eosinophils in the blood.

Hyperglycemia : High blood sugar,another name for diabetes,due either to not enough insulin being produced by the pancreas or faulty insulin receptors in the liver.

Hyperimmunization : Repetitive immunization to achieve a heightened state of immunity is called hyperimmunization.

Hypermutation : A rapid production of multiple mutations in a gene or genes through the activation of special mutator genes. The process may be deliberately used to maximize the possibility of creating desirable mutants.

Hyperparasite : Parasite that feeds on another parasite.

Hypersensitivity : A condition of increased immune sensitivity in which the body reacts to an antigen with an exaggerated immune response that usually harms the individual. Also termed an allergy.

Hypersensitivity reactions, hypersensitivity diseases, hypersensitivity : Immune responses to innocuous antigens that lead to symptomatic reactions upon reexposure are called hypersensitivity reactions. These can cause hypersensitivity diseases if they occur repetitively. This state of heightened reactivity to antigen is called hypersensitivity. Hypersensitivity reactions are classified by mechanism: type I hypersensitivity reactions involve IgE antibody triggering of mast cells; type II hypersensitivity reactions involve IgG antibodies against cellsurface or matrix antigens; type III hypersensitivity reactions involve antigen:antibody complexes; and type IV hypersensitivity reactions are T cell-mediated.

Hypertension : High blood pressure.

Hyperthermophile : An organism that has its growth optimum above 80°C. ***OR*** A prokaryote having a growth temperature optimum of 80°C or higher. ***OR*** A bacterium that has its growth optimum between 80ºC and about 113ºC. Hyperthermophiles usually do not grow well below 55ºC.

Hypertonic : Having a solute concentration greater than that of the solution to which it is being compared.

Hypervariable regions, HV : The hypervariable regions (HV) of immunoglobulin and T-cell receptor V domains are small regions that make contact with the antigen and differ extensively from one receptor to the next. Cf. Framework regions.

Hypha : A single filament of a fungus. ***OR*** The unit of structure of most fungi and some bacteria; a tubular filament. ***OR*** Long and often branched tubular filament that constitutes the vegetative body of many fungi and funguslike organisms. Bacteria of the order Actinomycetes also produce branched hyphae. ***OR*** The individual filament or thread that make up a fungus.

Hyphal fragment : A portion of the fungal mycelium that does not have any spores or other diagnostic fungal structures, and therefore, could not be identified.

Hyphomycetes : A class of mycelial moulds which reproduce asexually by conidia on hyphae or aggregations of hyphae. ***OR*** A group of fungi in which asexual spores called conidia are produced from special conidiogenous cells.

Hypobiosis : In parasitology A temporary halt in nematode development within the host at an early stage in the parasitic phase of the life cycle in nematodes with direct life cycles only. Nematodes arrest as immature forms in the definitive hosts at a time when conditions in the external environment pose a hazard to survival of free living pre-parasitic stages (usually in winter or dry seasons). Once conditions are more favorable, the arrested development resume their development to adults the life cycle continues.

Hypochlhydria (hyperchlhydric) : Abnormally low production of gastric acids

Hypoferremia : Deficiency of iron in the blood.

Hypogeous : A fruit-body produced below ground – like truffles

Hypoglycemia : Abnormally low glucose levels. ***OR*** Low blood sugar level due to over secretion of insulin.

Hypogynous : A term referring to plants with flower parts (petals, sepals, etc.) Attached below the ovary (so the ovary is superior).

Hypolimnion : Abnormally low blood pressure. ***OR*** deep part of lake low temperature and low oxygen concentrations because of low light penetration and consequent low photosynthetic activity. ***OR*** The deep, bottom layer of water in a lake or an ocean.

Hypotheca : The smaller half of a diatom frustule.

Hypothesis : A tentative assumption or educated guess developed to explain a set of observations. ***OR*** A tentative explanation for something that happens that can be tested by scientific investigation.

Hypothesis : A testable,tentative answer to a question.

Hypotonic : Having a solute concentration less than that of the solution to which it is being compared.

Hypoxic : Having a low oxygen level. ***OR*** Insufficient availability of oxygen in an environment to support aerobic respiration.

Hythergraph : A graph of temperature vs. rainfall.

I

I&D : Incision and drainage.

I&E : Internal and external.

I&O : Intake and output.

I.P. (IP) : Intraperitoneal.

I.V. (IV) : Intravenous.

IA : Incurred accidentally; intra-arterial.

IAC : Internal acoustic canal.

IBW : Ideal body weight.

IC : Inspiratory capacity; intensive care; intracerebral; between meals.

iC3b : The inactive complement fragment iC3b is produced by cleavage of C3b and is the first step in C3b inactivation.

ICAMs, ICAM-1, ICAM-2, ICAM-3 : The ICAMs (intercellular adhesion molecules) are cell-surface ligands for the leukocyte integrins and are crucial in the binding of lymphocytes and other leukocytes to certain cells, including antigenpresenting cells and endothelial cells. They are members of the immuno-globulin superfamily. ICAM-1 is the most prominent ligand for the integrin CD11a: CD18 or LFA-1. It is rapidly inducible on endothelial cells by infection, and plays a major role in local inflammatory responses. ICAM-2 is constitutively expressed at relatively low levels by endothelium. ICAM-3 is expressed only on leukocytes and is thought to play an important part in adhesion between T cells and antigen-presenting cells, particularly dendritic cells.

ICAT : Isotope-coated affinity tags.

ICC : Intensive coronary care.

Iccosomes : Iccosomes are small fragments of membrane coated with immune complexes that fragment off the processes of follicular dendritic cells in lymphoid follicles early in a secondary or subsequent antibody response.

ICCU : Intensive coronary care unit.

ICDA : International Classification of Diseases, Adapted.

ICF : Intensive care facility.

ICFX : Intracapsular fracture.

ICM : Inner cell mass.

ICM : Intercostal margin.

ICN : Intensive care nursery.

ICOS : ICOS is a CD28-related protein that is induced on activated T cells and can enhance T-cell responses. It binds a ligand known as LICOS, which is distinct from the B7 molecules.

Icosahedral : In virology this term refers to a virus with an icosahedral capsid, which has the shape of a regular polyhedron having 20 equilateral triangular faces and 12 corners.

Icosahedral symmetry : The symmetry displayed by a regular polyhedron that is composed of 20 equilateral triangular faces with 12 corners.

ICP : Intracranial pressure.

ICS : Intercostal space.

ICT : Information & communication technologie

ICT : Intermittent cervical traction.

ICU : Intensive care unit.

ID : Identification; ill defined; ineffective dose; infectious disease; during the day.

IDA : Iron deficiency anemia.

IDD : Insulin dependent diabetes.

Identification : The process of determining that a particular isolate or organism belongs to a recognized taxon.

Idiotype : A set of one or more unique epitopes in the variable region of an immunoglobulin that distinguishes it from immunoglobulins produced by different plasma cells.

Idiotype, idiotopes : Each immunoglobulin molecule has the potential of binding a variety of antibodies directed at its unique features or idiotype. An idiotype is made up of a series of idiotopes from idiotype epitopes.

Idiotypic network : Lymphocyte antigen receptors can recognize one another through idiotope - anti-idiotope interactions, forming an idiotypic network of receptors that may be important for the generation and maintenance of the receptor repertoire. The proposed components of idiotype networks exist, but their functional significance is uncertain.

IDK : Internal derangement of the knee.

IDL : Intermediate-density lipoprotein.

IE : Inner ear.

IF : Initiation factor.

IFC : Integrated fluidic circuits.

IFN : See interferons.

IFR : Inspiratory flow rate.

Ig, immunoglobulin : Standard abbreviation for immunoglobulin. Different immunoglobulin isotypes are called igM, IgD, IgG, IgA, and IgE.

IgA : A type of immunoglobulin present in blood and body secretions which may aid in fighting infections. ***OR*** Immunoglobulin A; the class of immunoglobulins that is present in dimeric form in many body secretions (e.g., saliva, tears, and bronchial and intestinal secretions) and protects body surfaces. IgA also is present in serum. ***OR*** igA is the class of immunoglobulin characterized by á heavy chains. IgA antibodies are secreted mainly by mucosal lymphoid tissues.

IgD : IgD is the class of immunoglobulin characterized by ä heavy chains. It appears as surface immunoglobulin on mature naive B cells but its function is unknown. ***OR*** Immunoglobulin D; the class of immuno-globulins found on the surface of many B lymphocytes; thought to serve as an antigen receptor in the stimulation of antibody synthesis.

IgE : IgE is the class of immunoglobulin characterized by d-heavy chains. It is involved in allergic reactions. ***OR*** Immunoglobulin E; the immunoglobulin class that binds to mast cells and basophils, and is responsible for type I or anaphylactic hypersensitivity reactions such as hay fever and asthma. IgE is also involved in resistance to helminth parasites.

IgG : Immunoglobulin G. ***OR*** IgG is the class of immunoglobulin characterized ã heavy chains. It is the most abundant class of immunoglobulin found in the plasma.OR Immunoglobulin G; the predominant immunoglobulin class in serum. Has functions such as neutralizing toxins, opsonizing bacteria, activating complement, and crossing the placenta to protect the fetus and neonate.

IgM : IgM is the class of immunoglobulin characterized by ì heavy chains. It is the first immunoglobulin to appear on the surface of B cells and the first to be secreted. ***OR*** Immunoglobulin M; the class of serum antibody first produced during an infection. It is a large, pentameric molecule that is active in agglutinating pathogens and activating complement. The monomeric form is present on the surface of some B lymphocytes.

IGS : Intergenic spacer or internal guide sequence.

IH : Infectious hepatitis; interval history.

IHC : Immunohistochistochemistry.

IHD : Ischemic heart disease.

IHP : Inositol hexaphosphate.

IHR : Intrinsic heart rate.

IID : Insulin dependent diabetes.

IL : See interleukins. ***OR*** Interleukin. ***OR*** Ilio-lumbar.

Ile : Isoleucine.

Ileocecal junction : The site of transition from the terminal ileum of the small intestine to the cecum of the large intestine.

Ileum : The lower (distal) portion of the small intestine extending from the jejunum to the ileocecal junction and encompassing approximately twelve feet in length (~3/5 of the length of the small intestine). ***OR*** The third (last) section of the small intestine.

Ilium : The large, flared bone forming the top of the "hipbones".

Illicit : Unlawful acts are illicit acts.

Illuviation : Deposition of soil material removed from one horizon to another in the soil.

IM : Infectious mononucleosis; internal medicine; intramuscular.

IMBC : Indirect maximum breathing capacity.

Imbricate : Overlapping like roof tiles.

IME : Independent medical examiner.

IMI : Intramuscular injection.

Imine : A molecule containing a nitrogen atom attached to a carbon atom by a double bond. The nitrogen is also covalently linked to a hydrogen.

IMM : Immediate; immobilize; immunize.

Immature B cells : Immature B cells are B cells that have rearranged a heavy- and a light-chain V-region gene and express surface IgM, but have not yet matured sufficiently to express surface IgD as well.

Immature dendritic cells : Tissues throughout the body contain immature dendritic cells, which only leave the tissues in response to an inflammatory mediator or an infection. See also dendritic cells.

Immediate reaction, immediate hypersensitivity reactions : During allergic reactions, there are normally two phases: the first happens almost immediately and is called the immediate reaction. Hypersensitivity reactions that occur within minutes of exposure to antigen are called immediate hypersensitivity reactions; such reactions are antibody mediated. Cf. Delayed-type hypersensitivity.

Immigration : Moving in to a new location.

Immobilization : Conversion of an element from the inorganic to the organic form in microbial or plant biomass. ***OR*** The incorporation of a simple, soluble substance into the body of an organism, making it unavailable for use by other organisms.

Immtal : Live forever.

Immune : Having the ability to resist a specific disease.

Immune clearance : When late amounts of antigen are injected into the blood, they are initially removed slowly by normal catabolic processes that also degrade plasma proteins. However, if the antigen elicits an antibody response, then antigen is removed at an accelerated rate as antigen:antibody complexes, a process known as immune clearance.

Immune complex : The product of an antigen-antibody reaction, which may also contain components of the complement system. ***OR*** The binding of antibody to a soluble antigen forms an immune complex. Large immune complexes form when sufficient antibody is available to cross-link the antigen; these are readily cleared by the reticuloendothelial system of cells bearing Fc and complement receptors. Small, soluble immune complexes that form when antigen is in excess can be deposited in and damage small blood vessels.

Immune deviation : Immune deviation is a term used to describe the polarization of an immune response to one dominated by TH1 or TH2 by the injection of antigen.

Immune modulation : Immune modulation is a general term encompassing various alterations in an immune response.

Immune response : The immune response is the response made by the host to defend itself against a pathogen.

Immune response genes, Ir genes, immune response gene defect : Immune response genes (Ir genes) are genetic polymorphisms that control the intensity of the immune response to a particular antigen. Virtually all Ir phenotypes are due to the differential binding of peptide fragments of antigen to MHC molecules, especially MHC class II molecules. The term is little used now. An immune response gene defect is usually, but not always, due to failure to bind an immunogenic peptide, so that no T-cell response is observed.

Immune surveillance : It has been proposed that most tumors that arise are detected and eliminated by immune surveillance mediated by lymphocytes specific for tumor antigens. There is little evidence for the efficacy of this proposed process, but it remains an important concept in tumor immunology. ***OR*** The still somewhat hypothetical process in which lymphocytes such as natural killer (NK) cells recognize and destroy tumor cells; other cells with abnormal surface antigens (e.g., virus-infected cells) also may be destroyed.

Immune system : The defensive system in a host consisting of the nonspecific and specific immune responses. It is composed of widely distributed cells, tissues, and organs that recognize foreign substances and microorganisms and acts to neutralize or destroy them. ***OR*** The body's defensive system which recognises foreign substances and microorganisms and acts to neutralis or destroy them. It is composed of widely distributed cells, tissues, and organs. ***OR*** The immune system is the name used to describe the tissues, cells, and molecules involved in adaptive immunity, or sometimes the totality of host defense mechanisms. ***OR*** The system which protects the body from foreign invaders (viruses, bacteria, cancer, etc.).

Immunity : Immunity is the ability to resist infection.OR The ability of a human or animal body to resist infection by microorganisms or their harmful products such as toxins. ***OR*** The ability of a human or animal body to resist infection by microorganisms or their harmful products such as toxins.

Immunization : Immunization is the deliberate provocation of an adaptive immune response by introducing antigen into the body. See also active immunization; passive immunization.

Immunobiology : Immunobiology is the study of the biological basis for host defense against infection.

Immunoblot (western blot) : Detection of proteins immobilized on a filter by complementary reaction with specific antibody.

Immunoblotting : The electrophoretic transfer of proteins from polyacrylamide gels to nitrocellulose sheets to demonstrate the presence of specific proteins through reaction with labeled antibodies. ***OR*** Immunoblotting is a common technique in which proteins separated by gel electrophoresis are blotted onto a nitrocellulose membrane and revealed by the binding of specific labeled antibodies.

Immunocompromised : Incapable of developing a normal immune response, usually as a result of disease (lupus, HIV), malnutrition, or immunosuppressive therapy (chemo-therapy, corticosteroids).

Immunocontraceptive : A molecule that partially or completely suppresses the reproductive system of an animal by stimulating an immune response, for example against sperm, is an immunocontraceptive.

Immunodeficiencies : The inability to produce a normal complement of antibodies or immunologically sensitized T cells in response to specific antigens.

Immunodeficiency diseases : Immunodeficiency diseases are a group of inherited or acquired disorders in which some aspect or aspects of host defense are absent or functionally defective.

Immunodiffusion : A technique involving the diffusion of antigen and/or antibody within a semisolid gel to produce a precipitin reaction where they meet in proper proportions. Often both the antibody and antigen diffuse through the gel; sometimes an antigen diffuses through a gel containing antibody.

Immunodiffusion : Immunodiffusion is the detection of antigen or antibody by the formation of an antigen:antibody precipitate in a clear agar gel.

Immunoelectrophoresis : Immunoelectrophoresis is a technique in which antigens are first separated by their electrophoretic mobility and are then detected and identified by immunodiffusion. ***OR*** The electrophoretic separation of protein antigens followed by diffusion and precipitation in gels using antibodies against the separated proteins.

Immunofluorescence : A technique used to identify particular antigens microscopically in cells or tissues by the binding of a fluorescent antibody conjugate. ***OR*** Technique to visualize specific antibodies and any attached homologous antigens by means of conjugating the antibodies to a fluorescent dye. ***OR*** A cytological technique in which a specific fluorescent antibody is used to label an antigen. Frequently used to determine the location of an antigen in a tissue or a cell.

Immunofluorescent microscopy, immunohistochemistry, immuno-electronmicroscopy : There are three ways of detecting molecules in tissues: immunofluorescent microscopy that reveals the presence of any molecule against which you have a specific antibody; immunohistochemistry, in which one links an enzyme that produces a change in a molecule that is visible under the microscope; and immunoelectron-microscopy, in which different sized gold particles are linked to antibodies and detected as bound gold particles.

Immunogenetics : Immunogenetics was originally the analysis of genetic traits by means of antibodies against genetically polymorphic molecules such as blood group antigens or MHC proteins. Immunogenetics now refers to the genetic analysis, by any technique, of molecules important in immunology.

Immunogenic immunogen : Substance which is capable of eliciting immune response. An immunogen usually has a fairly high molecular weight (usually greater than 10,000), thus, a variety of macromolecules such as proteins, lipoprotein, polysaccharides, and some nucleic acids can act as immunogens. ***OR*** Any molecule that can elicit an adaptive immune response on injection into a person or animal is called an immunogen. In practice, only proteins are fully immunogenic because only proteins can be recognized by T lymphocytes.

Immunoglobulin : Antibody.

Immunoglobulin A : See IgA.

Immunoglobulin D : See IgD.

Immunoglobulin domains, Ig domains, immunoglobulin fold, immunoglobulin-like domains : Many proteins are partly or entirely composed of protein domains known as immunoglobulin domains or Ig domains because the were first described in antibody molecules. Immunoglobulin domains are characteristic of proteins of the immunoglobulin superfamily, which includes antibodies, T-cell receptors, MHC molecules, and many other proteins described in this book. The immunoglobulin domain consists of a sandwich of two â sheets held together by a disulfide bond and called the immunoglobulin fold. There are two main types of immunoglobulin domain: C domains and V domains. Domains less closely related to the canonical Ig domains are sometimes also called immunoglobulin-like domains.

Immunoglobulin E : See IgE.

Immunoglobulin G : See IgG.

Immunoglobulin M : See IgM.

Immunoglobulin repertoire : The immunoglobulin repertoire, also known as the antibody repertoire, is the total variety of immunoglobulin molecules in the body of an individual.

Immunoglobulin superfamily, Ig superfamily : Many proteins involved in antigen recognition and cell-cell interaction in the immune system and other biological systems are members of a protein family called the immunoglobulin superfamily, or Ig superfamily, because their shared structural features were first defined in immuno-globulin molecules. All members of the immunoglobulin superfamily have at least one immunoglobulin or immunoglobulin-like domain.

Immunoglobulins, Ig : A protein made in a B plasma cell and usually secreted; it interacts specifically with a foreign agent. Synonymous with antibody. It is composed of two heavy and two light chains linked by disulfide bonds. Immunoglobulins can be divided into five classes (IgG, IgM, IgA, IgD, and IgE) based on their heavy-chain component. ***OR*** All antibody molecules belong to a family of plasma proteins called immunoglobulins (Ig). Membrane-bound immunoglobulin serves as the specific antigen receptor on B lymphocytes.

Immunohistochemisty : The detection of antigens in tissues by means of visible products produced by the degradation of a colorless substrate by antibodylinked enzymes is called immunohistochemistry. This technique has the advantage that it can be combined with other stains to be viewed in the light microscope, whereas immunofluorescence microscopy requires a special dark-field or UV microscope.

Immunological ignorance : Immunological ignorance describes a form of self tolerance in which reactive lymphocytes and their target antigen are both detectable within an individual, yet no autoimmune attack occurs. Most autoimmune diseases probably reflect the loss of other lymphocytes known as regulatory or suppressor T cells.

Immunological memory : When an antigen is encountered more than once, the adaptive immune response to each subsequent encounter is speedier and more effective, a crucial feature of protective immunity known as immunological memory. Immunological memory is specific for a particular antigen and is long-lived.

Immunologically privileged sites : Allogeneic tissue placed in certain sites in the body, such as the brain, does not elicit graft rejection. Such sites are called immunologically privileged sites. Immunological privilege results from the effects of both physical barriers to cell and antigen migration, and soluble immunosuppressive mediators such as certain cytokines.

Immunology : Immunology is the study of all aspects of host defense against infection and of adverse consequences of immune responses. ***OR*** The branch of science that deals with the immune system and attempts to understand the many phenomena that are responsible for both acquired and innate immunity. It also includes the use of antibody-antigen reactions in other laboratory work (serology and immunochemistry).

Immunopathology : The study of diseases or conditions resulting from immune reactions.

Immunophilins : Immunophilins are proteins with peptidyl-prolyl cis-trans isomerase activity that bind the immunosuppressive drugs cyclosporin A, tacrolimus, and rapamycin.

Immunoprecipitation : A reaction involving soluble antigens reacting with antibodies to form a large aggregate that precipitates out of solution.

Immunoprecipitation analysis : Soluble proteins, or membrane proteins solubilized in detergents, can be labeled and then detected by immunoprecipitation analysis using specific antibodies. The immunoprecipitated labeled protein is usually detected by SDS-PAGE followed by autoradiography. When proteins that do not react directly with the antibody used are nevertheless precipitated, they are said to co-immunoprecipitate.

Immunoreceptor tyrosine-based activation motifs, (ITAMs), immunoreceptor tyrosine-based inhibity motifs, (ITIMs) : The T and B cell antigen receptors are associated with transmembrane molecules with immunoreceptor tyrosine-based activation motifs (itams) in their cytoplasmic domains. These tyrosine-containing motifs are sites of tyrosine phosphorylation and of association with tyrosine kinases and other phosphotyrosine-binding moieties involved in receptor signaling. Related motifs with opposing effects are immunoreceptor tyrosine-based inhibitory motifs (itims), which recruit phosphatases to the receptor site that remove the phosphate groups added by tyrosine kinases.

Immunoregulation : The ability of the immune system to sense and regulate its own responses is called immunoregulation.

Immunosuppress : Suppress, or reduce, a response from the immune system.

Immunosuppressive drugs : Compounds that inhibit adaptive immune responses are called immunosuppressive drugs. They are used mainly in the treatment of graft rejection and severe autoimmune disease.

Immunotherapy : A treatment that uses the body's natural defence system, the immune system, to fight disease. For example, using a vaccine is used to kill a cancer cell - see the Hub's information sheet on melanoma research.

Immunotoxin : A monoclonal antibody that has been attached to a specific toxin or toxic agent (antibody 1 toxin 5 immunotoxin) and can kill specific target cells.

Immunotoxins : Immunotoxins are antibodies that are chemically coupled to toxic proteins usually derived from plants or microbes. The antibody targets the toxin moiety to the required cells. Immunotoxins are being tested as anticancer agents and as immunosuppressive drugs.

IMP : Inosine monophosphate.

Impatiens necrotic spot virus (INSV) : Impatiens necrotic spot virus is a member of the tospovirus group and was first called TSWV impatiens. In 1991, researchers determined that this thripstransmitted virus commonly found destroying impatiens, cyclamen and other greenhouse crops, was distinct from TSWV (not just a special strain) and named it INSV.

Imperfect flower : A term referring to a flower which has either male or female flower parts but not both—note that this would,therefore, also be an incomplete flower.

Impetigo : Pyoderma with yellowish weeping vesicular lesions that crust over when the pustule erupts historically seen on the face around the nose and mouth or on the limbs in children. ***OR*** This superficial cutaneous disease, most commonly seen in children, is characterized by crusty lesions, usually located on the face; the lesions typically have vesicles surrounded by a red border. It is the most frequently diagnosed skin infection caused by *S. pyogenes* (impetigo can also be caused by *S. aureus*).

In vitro* :** Literally meaning "in glass" and referring to assays performed in a test tube or another artificial laboratory environment. ***OR processes or reactions that occur outside of the body and in an artificial environment e.g. Test-tube or laboratory environment, are referred to as *in vitro* (literally meaning: in glass).

***In vitro* fertilisation :** A technique where an egg is fertilised by sperm in the laboratory, before being transferred into the womb.

***In vitro* :** Literally, "in glass," describing whatever happens in a test tube or other receptacle, as opposed to what happens in whole cells of the whole organism (*in vivo*).

In vivo* :** Literally meaning "in living" and referring to assays performed in a living animal. ***OR in the body, in a living organism, as opposed to *in vitro*; when a study or an experiment is done in the living organism, it is done *in* vivo.

INC : Incomplete; inconclusive; increase.

Incidence : Rate of occurrence of an event; number of new cases of disease occuring over a specified period of time; may be expressed per a known population size (see also prevalence)

Incineration : The conversion of dewatered wastewater solids by combustion (burning) to ash, carbon dioxide, and water vapor.

Inclusion bodies : Granules of organic or inorganic material lying in the cytoplasmic matrix of bacteria.

Inclusion conjunctivitis : An acute infectious disease that occurs throughout the world. It is caused by *Chlamydia trachomatis* that infects the eye and causes inflammation and the occurrence of large inclusion bodies.

Incomplete flower : A term referring to a flower which is missing one or more of the four layers of flower parts.

Incubation period (latent period) : The period after pathogen entry into a host and before signs and symptoms appear. ***OR*** between the time that infecting organisms enter the host and onset of signs and symptoms of disease.

Incubaty carrier : An individual who is incubating a pathogen in large numbers but is not yet ill.

Incus : One of the bones in the middle ear.

IND : Investigational new drug. ***OR*** Independent.

Independence : An ecological relationship in which an organism is living free from the influence, guidance, or control by another organism.

Index case : The first disease case in an epidemic within a given population.

Indicat organism : An organism whose presence indicates the condition of a substance or environment, for example, the potential presence of pathogens. Coliforms are used as indicators of fecal pollution.

Indicator plants : Plants that react with characteristic symptoms are one of the most reliable ways to detect a virus problem. Indicator plants, such young petunia plants, or fava bean seedlings, may be placed throughout a greenhouse range to monitor for the spread of tospovirus by thrips. If thrips feed on the leaves and spread virus, necrotic spots result. Indicator plants are also used in plant virology to indicate if a virus is present in a plant by grinding sample leaves then rubbing the extract onto test plants such as cucumber seedlings, Nicotiana, tomato and beans to see if lesions, mosaic or other symptoms result.

Indigestion : (Or dyspepsia), is a general term covering a group of symptoms in the digestive tract including a feeling of fullness, bloating, nausea, heartburn, or gassy discomfort in the chest or abdomen. The symptoms develop during meals or shortly afterward. In most cases, indigestion is a minor problem that often clears up without professional treatment.

Indirect coombs test : The indirect coombs test is a variation of the direct coombs test in which an unknown serum is tested for antibodies against normal red blood cells by first mixing the two and then washing out the serum from the red blood cells and reacting them with anti-immunoglobulin antibody. If antibody in the unknown serum binds to the red blood cells, agglutination by anti-immunoglobulin occurs.

Indirect immunofluorescence : See immunofluorescence.

Indirect life cycle : A life cycle of a parasite in which the pre-parasitic larvae develop inside an appropriate intermediate host.

Induced fit : A change in the shape of an enzyme that results from the binding of substrate.

Induced systemic resistance (ISR) : The process of active resistance dependent on the host's physical or chemical barriers, activated by biotic or abiotic inducing agents.

Inducer : A small molecule that stimulates the synthesis of an inducible enzyme. ***OR*** Molecules that cause an increase in a protein activity when added to cells.

Inducible : Enzyme not synthesized or activated until needed.

Inducible enzyme : An enzyme whose level rises in the presence of a small molecule that stimulates its synthesis. ***OR*** Enzyme synthesized (induced) in response to the presence of an external substance (the inducer).

Inducible NO synthase, (INOS) : Macrophages and many other cells have an inducible NO synthase, that is induced by many different stimuli to activate NO synthesis. This is a major mechanism of host resistance to intracellular infection in mice, and probably in humans as well.

Inducible proteins : Those which are synthesized in different amounts depending on cellular signals.

Industrial ecology : The study of the ecology of industrial societies with a major focus on material cycling, energy flow, and the ecological impacts of such societies.

Inert : A substance that tends to be stable and does not react chemically with other substances.

INF : Infant; infarction; infected; inferior.

Infantile paralysis : See poliomyelitis.

Infection : Invasion of the body by microorganisms, like a throat or ear infection. ***OR*** Growth of an organism within another living organism. ***OR*** An invasion of the body by a micro-organism with subsequent establishment and multiplication of the micro-organism.

Infection thread : Cellulosic tube in a root hair through which rhizobia can travel to reach and infect root cells. ***OR*** A tubular structure formed during the infection of a root by nitrogen-fixing bacteria. The bacteria enter the root by way of the infection thread and stimulate the formation of the root nodule.

Infectious : Quickly spreading to others, for example, an infectious disease.

Infectious disease : A disease caused by an infectious agent; contagious disease. ***OR*** Any change from a state of health in which part or all of the host's body cannot carry on its normal functions because of the presence of an infectious agent or its products.

Infectious dose : The number of infecting organisms generally required to cause disease, in this case, via consumption of food.

Infectious dose 50 (ID_{50}) : Refers to the dose or number of organisms that will infect 50% of an experimental group of hosts within a specified time period.

Infectious mononucleosis : Infectious mononucleosis, or glandular fever, is the common form of infection with the Epstein-Barr virus. It consists of fever, malaise, and swollen lymph nodes. ***OR*** An acute, self-limited infectious disease of the lymphatic system caused by the Epstein-Barr virus and characterized by fever, sore throat, lymph node and spleen swelling, and the proliferation of monocytes and abnormal lymphocytes.

Infectivity : Infectiousness; the state or quality of being infectious or communicable.

Infiltration : The seepage of groundwater into a sewer system, including service connections. Seepage frequently occurs through defective or cracked pipes, pipe joints, connections or manhole walls.

Inflammation (Inflammatory response) : The first response of tissues to irritation or infection, which results in swelling, redness, heat, and pain in the affected tissue. ***OR*** A localized protective response to tissue injury or destruction. Acute inflammation is characterized by pain, heat, swelling, and redness in the injured area.

Inflammation, inflammatory response, inflammatory cells, inflammatory infiltrate : Inflammation is a general term for the local accumulation of fluid, plasma proteins, and white blood cells that is initiated by physical injury, infection, or a local immune response. This is also known as an inflammatory response. Acute inflammation is the term used to describe early and often transient episodes, whereas chronic inflammation occurs when the infection persists or during autoimmune diseases. Many different forms of inflammation are seen in different diseases. The cells that invade tissues undergoing inflammatory responses are often called inflammatory cells or an inflammatory infiltrate.

Inflorescence : A cluster of individual flowers in aspecific arrangement on a plant.

Influent : The liquid-raw (untreated) or partially treated - flowing into a reservoir, basin, treatment process or treatment plant.

Influenza flu : An acute viral infection of the respiratory tract, occurring in isolated cases, epidemics, and pandemics. Influenza is caused by three strains of influenza virus, labeled types A, B, and C, based on the antigens of their protein coats.

Infrared (IR) : The portion of the electromagnetic spectrum with wavelengths from about 0.75 μm to 1 mm.

Infrastructure : The basic facilities and services needed to support a community, such as transport, water, and power, and public institutions including schools, post offices, and prisons.

Infundibulifm : Funnel-shaped.

Ingoldian fungi : Aquatic hyphomycetes that often have a characteristic tetraradiate hyphal development form and which sporulate under water. Discovered by the British mycologist, C. T. Ingold, in the 1940s.

Inhibit : Reduce or prevent.

Inhibition : Prevention of growth or function.

Initial body : See reticulate body (RB).

Initiation factors : Those protein factors that are specifically required during the initiation phase of protein synthesis.

INJ : Inject; injection; injury.

Injury : Reversible phenotypic response of bacteria exposed to sublethal stressors.

INK : Injury not known.

Innate : Of scales not readily detached as in Lepiota

Innate immunity : The early phases of the host response to infection depend on innate immunity in which a variety of innate resistance mechanisms recognize and respond to the presence of a pathogen. Innate immunity is present in all individuals at all times, does not increase with repeated exposure to a given pathogen, and discriminates between a group of related pathogens.

Innate natural immunity : See nonspecific resistance.

Innovation : A new process or product.

Inoculate : To treat with microorganisms for the purpose of creating a favorable response. For example, treatment of legume seeds with rhizobia to stimulate N_2 fixation. ***OR*** Infect one living organism with a small sample of another living organism or part of a living organism. ***OR*** To deliver organisms such as bacteria or protests in to a growth medium.

Inoculation : The introduction of disease-causing microbes into the body to produce or boost immunity to a specific disease.

Inoculum : Material used to introduce a microorganism into a suitable situation for growth.

Inoculum size : See dosage.

Inoculums : Material used to introduce a microorganism into a suitable situation for growth.

Inganic waste : Waste material such as sand, salt, iron, calcium, and other mineral materials which are only slightly affected by the action of organisms. Inorganic wastes are chemical substances of mineral origin; whereas organic wastes are chemical substances usually of animal or plant origin.

Inositol trisphosphate, IP : When inositol phospholipid is cleaved by phospholipase C-γ, it yields inositol trisphosphate (IP3) and diacylglycerol. Inositol trisphosphate releases calcium ions from intracellular stores in the endoplasmic reticulum.

INR : Intermittent; interval.

INS : Insulin; insurance.

Insecticide : A chemical (synthetic or organic) used to kill or repel insects.

Insertion : Genetic mutation in which one or more nucleotides are added to DNA.

Insertion sequence : A simple transposon that contains genes only for those enzymes, such as the transposase, that are required for transposition.

Insertion sequence (IS element) : Simplest type of transposable element. Has only genes involved in transposition.

Insoluble : A substance that does not dissolve in liquid.

InsP$_3$: Inositol 1,4,5-trisphosphate.

Instar : An insect or a stage in an insect's life in between two molts.

Insulin : A hormone produced by the Beta-cells in the pancreas. Insulin plays a central role in the control of blood glucose levels in the body.

Insulin-dependent diabetes mellitus, IDDM : In insulin-dependent diabetes mellitus (IDDM), the b cells of the pancreatic islets of Langerhans are destroyed so that no insulin is produced. The disease is believed to result from an autoimmune attack on the P cells.

Integrated pest management (IPM) : IPM is a system that uses monitoring pest and environmental conditions and using the best combinations of pest control options such as sanitation, biocontrol, and pesticides to prevent unacceptable lines of damage.

Integration : Process by which a DNA molecule becomes incorporated into another genome. ***OR*** The incorporation of one DNA segment into a second DNA molecule to form a new hybrid DNA. Integration occurs during such processes as genetic recombination, episome incorporation into host DNA, and prophage insertion into the bacterial chromosome.

Integrin : Adhesive protein of the extracellular matrix in animals.

Integrins : A large and broadly distributed family of a/b heterodimers. Integrins are cellular adhesion receptors that mediate cell-cell and cell-substratum interactions. Integrins usually recognize linear amino acid sequences on protein ligands.

Integrins, β : Integrins are heterodimeric cell-surface proteins involved in cell-cell and cell-matrix interactions. They are important in adhesive interactions between lymphocytes and antigen-presenting cells and in lymphocyte and leukocyte migration into tissues. The b1-integrins, or very late antigens (VLA), are a family of integrins with shared β1 chains and different á chains that mediate adhesion to other cells and to extracellular matrix proteins.

Inteins : Internal intervening sequences of precursor self-splicing proteins that separate exteins and are removed during formation of the final protein.

Intellectual property : An idea or invention that is protected by law, such as a patent, copyright, or trademark.

Intercalary : Within a hyphal element.

Intercalating agent : A chemical, usually containing aromatic rings, that can sandwich in-between adjacent base pairs in a DNA duplex. The intercalation leads to an adjustment in the DNA secondary structure, as adjacent base pairs are usually close-packed.

Intercalating agents : Molecules that can be inserted between the stacked bases of a DNA double helix, thereby distorting the DNA and inducing insertion and deletion mutations.

Intercellular adhesion molecules : See ICAMs.

Interdigitating dendritic cell : Special dendritic cells in the lymph nodes that function as potent antigen-presenting cells and develop from Langerhans cells.

Interdigitating reticular cells : See dendritic cells.

Interface : The common boundary layer between two substances such as between water and a solid (metal) or between water and a gas (air) or between a liquid (water) and another liquid (oil).

Interfering RNA : A single-stranded RNA molecule introduced into a cell to bind by complementary base pairing to a specific RNA molecule in order to prevent its translation. This stops the protein that is coded from being formed. Sometimes called inhibitory RNA. See also RNA interference (RNAi).

Interferon (IFN) : One of a family of proteins that are liberated by special host cells in the mammal in response to viral infection. The interferons attach to an infected cell, where they stimulate antiviral protein synthesis. ***OR*** A glycoprotein that has nonspecific antiviral activity by stimulating cells to produce antiviral proteins, which inhibit the synthesis of viral RNA and proteins. Interferons also regulate the growth, differentiation, and/or function of a variety of immune system cells. Their production may be stimulated by virus infections, intracellular pathogens (chlamydiae and rickettsias), protozoan parasites, endotoxins, and other agents.

Interferons, Interferon-a, IFN-a, interteron-b, IFN-b, interferon-g, IFN-g : Interferons are cytokines that can induce cells to resist viral replication. Interferon-a (IFN-a) and interteron-b (IFN-b) are produced by leukocytes and fibroblasts, respectively, as well as by other cells, whereas interferon-g (IFN-g is a product of CD4 TH1 cells, CD8 T cells, and NK cells. IFN-g has as its primary action the activation of macrophages.

Interleukin : A glycoprotein produced by macrophages and T cells that regulates growth and differentiation, particularly of lymphocytes. Interleukins promote cellular and humoral immune responses. ***OR*** Interleukin, abbreviated IL, is a generic term for cytokines produced by leukocytes. We use the more general term cytokine, but the term interleukin is used in the naming of specific cytokines such as IL-2.

Interleukin-2, IL-2 : Interleukin-2 (IL-2) is the cytokine that is most central to the development of an adaptive immune response.

Intermediate : Used of gills that do not reach the stem.

Intermediate filaments : Small protein filaments, about 8 to 10 nm in diameter, in the cytoplasmic matrix of eucaryotic cells that are important in cell structure.

Intermediate host : The host that serves as a temporary but essential environment for development of a parasite and completion of its life cycle. ***OR*** A stage in the life cycle that is essential to parasite development (with some exceptions), and the mode of transmission of the parasite to the definitive host is usually by predation. Intermediate hosts are not required for all parasite life cycles.

Interphase : Non-dividing stage in a cell's lifecycle.

Interspecies : Between two different species, such as tomato and weeds.

Interspecies hydrogen transfer : The process in which organic matter is degraded anaerobically by the interaction of several groups of microorganisms in which hydrogen production and hydrogen consumption are closely coupled among species. ***OR*** The linkage of hydrogen production from organic matter by anaerobic heterotrophic microorganisms to the use of hydrogen by other anaerobes in the reduction of carbon dioxide to methane. This avoids possible hydrogen toxicity.

Interspecific interactions : Interactions among organisms of different species.

Intertriginous candidiasis : A skin infection caused by *Candida* species. Involves those areas of the body, usually opposed skin surfaces, that are warm and moist (axillae, groin, skin folds).

Intervening sequence : See intron.

Intestine : Portion of the digestive tract between the pyloric sphincter of the stomach and the anus, composed of the small intestine and the large intestine.

Intoxicant : An alcoholic substance that can make a person intoxicated (drunk) is an intoxicant.

Intoxication : A disease that results from the entrance of a specific toxin into the body of a host. The toxin can induce the disease in the absence of the toxin-producing organism.

Intracellular : Inside the cell.

Intracellular cytokine staining : Staining for cytokines in cells that produce them can be achieved by permiablizing the cell and reacting it with a labelled fluorescent anticytokine antibody. This procedure is called intracellular cytokine staining.

Intracutaneous, subcutaneous, intramuscular, intranasal, intravenous : Injections can be administered by a number of routes: intracutaneous (intradermal) - entering the skin or dermis; subcutaneous - entering below the skin or dermis; intramuscular - entering the muscle; intranasal - by way of the nose; and intravenous - entering a vein.

Intraepidermal lymphocytes : T cells found in the epidermis of the skin that express the gd T-cell receptor.

Intranuclear inclusion body : A structure found within cells infected with the cytomegalovirus.

Intraspecies : Within same species; Elk vs. Elk.

Intraspecific interactions : Interactions among organisms of the same species.

Intrathymic dendritic cells : See dendritic cells.

Intrinsic factors : Food-related factors such as moisture, pH, and available nutrients that influence microbial growth.

Intron : A noncoding intervening sequence in a split or interrupted gene, which codes for RNA that is missing from the final RNA product. ***OR*** A DNA sequence which lies within a coding sequence but is not itself used to code for a protein. Although it may initially be transcribed into RNA, it is removed during RNA processing. The resulting mRNA does not contain any of the intron information.

Invariant chain : The major histocompatibility complex (MHC) class II proteins are assembled in the endoplasmic reticulum with the invariant chain (Ic), which is involved in shielding the MHC class II molecules from binding peptides and in delivering them to cellular vesicles. There Ii is degraded, leaving the MHC class II molecules able to bind peptide fragments of antigen.

Invasins (invasive factors) : Structure or macromolecule that enables a pathogenic microorganism to enter and spread throughout the cells and/or tissues of the host body; specific recognition of receptor sites on target cells enhances pathogenic advantage.

Invasiveness : The ability of a microorganism to enter a host, grow and reproduce within the host, and spread throughout its body.

Invertebrate : An animal without a backbone, for example an insect, snail, or worm.

Inverted repeat : A chromosome segment that is identical to another segment on the same chromosome except that it is oriented in the opposite direction.

Involute : Of the cap rolled inwards at the margin.

IO : Intestinal obstruction.

IOFB : Intraocular foreign body.

Ion : An atom or group of atoms that has gained or lost electrons and has an electrical charge.

Ion exchange : A reversible process where ions are swapped between different substances on the basis of their electrical charge.

Ion : An atom or small molecule which carries a positive or negative charge.

Ion-exchange resin : A polymeric resinous substance, usually in bead form, that contains fixed groups with positive or negative charge. An anion exchange resin has positively-charged groups and is therefore useful in exchanging the anionic groups in a test sample; a cation exchange resin is itself negatively charged, and has the opposite application. The resin is usually used in a column chromatographic procedure.

Ionised : Convert (an atom, molecule or substance) into an ion or ions, typically by removing one or more electrons.

Ionization : The process of adding electrons to, or removing electrons from, atoms or molecules, thereby creating ions. High temperatures, electrical discharges, and nuclear radiation can cause ionization.

Ionizing radiation : Radiation of very short wavelength or high energy that causes atoms to lose electrons or ionize.

Ions : Atoms, groups of atoms, or compounds, that are electrically charged as a result of the loss of electrons (cations) or the gain of electrons (anions).

IOP : Intraocular pressure.

IOU : Intensive therapy observation unit.

IP : Interphalangeal; intra-peritoneal.

IP 1 : Inositol-1-phosphate.

IP 3 : Inositol-1,4,5 –phosphate.

IP_1 : Inositol-1-phosphate.

IP_3 : Inositol 1,4,5-triphosphate.

IPPA : Inspection, palpation, percussion, auscultation.

IPPB : Intermittent positive-pressure breathing.

IPPO : Intermittent positive-pressure inflation with oxygen.

IPPR : Intermittent positive pressure respiration.

IPPV : Intermittent positive pressure ventilation.

IPR : Intellectual property rights.

IPTG : Isopropylthiogalactoside.

IR : Infrared.

IR : Internal rotation.

IRE : Internal rotation in extension.

IRF : Internal rotation in flexion.

IRGSP : International rice genome-sequencing programme.

Iris : The colored ring of eye muscles,behind the cornea,that control the size of the pupil.

IRP : Iron-responsive protein.

Irradiated : When objects are exposed to radiation, like X-rays, they are irradiated.

Irregular flower : A term referring to a flower which has bilateral symmetry, also known as a Zygomorphic flower.

Irrigation : Intentional watering of the soil.

IS : Insertion sequence. ***OR*** In situ; intercostal space.

Isaac Newton : Person who lived from 1643 to1727 and based his laws of gravity not only on the legendary falling apple,but also on the motion of the planets.

Ischemia (ischemic) : Deficiency of oxygenation of a body part caused by obstruction in or constriction of blood vessel(s).

Ischium : The "bottom" hip bones on which you sit.

ISCOMs, immune stimulaty complexes : ISCOMS are immune stimulatory complexes of antigen held within a lipid matrix that acts as an adjuvant and enables the antigen to be taken up into the cytoplasm after fusion of the lipid with the plasma membrane.

ISH : In situ hybridization.

ISL : Interscapular line.

Islet cells : Cells in the pancreas that secrete compounds to help control blood glucose, like insulin and glucagon.

Isoelectric focusing : Isoelectric focusing is an electrophoretic technique in which proteins migrate in a pH gradient until they reach the place in the gradient at which their net charge is neutral-their isoelectric point. Uncharged proteins no longer migrate; thus each protein is focused at its isoelectric point.

Isoelectric point pH : The pH at which a protein has no net charge.

Isoenzyme : When two different enzymes catalyze the same reaction(s), they are isoenzymes of each other. Isoenzymes could differ from each other in their primary structure or electrophoretic mobility.

Isolation : Any procedure in which an organism present in a particular sample or environment, is obtained in pure culture.

Isomerase : An enzyme that catalyzes an intra-molecular rearrangement.

Isomerization : Rearrangement of atomic groups within the same molecule without any loss or gain of atoms.

Isomorphous substitution : Substitution in a crystalline clay sheet of one atom by a similarly sized atom of lower valence.

Isotonic : Having a solute concentration equal to that of the solution to which it is being compared.

Isotope : One or more forms of a chemical element that differ in atomic mass. ***OR*** Different form of the same element containing the same number of protons and electrons, but differing in the number of neutrons.

Isotopes : Atoms of the same element with different numbers of neutrons.

Isotype switching : The first antibodies produced in a humoral immune response are IgM, but activated B cells subsequently undergo isotype switching or class switching to secrete antibodies of different isotypes: IgG, IgA, and IgE. Isotype switching does not affect antibody specificity significantly, but alters the effector functions that an antibody can engage. Isotype switching occurs by recombination involving the deletion of DNA between the rearranged V region and the selected C-region exon at so-called S regions.

Isotypes : Immunoglobulins are made in several distinct isotypes or classes-IgM, IgG, IgD, IgA and IgE-each of which has a distinct heavy-chain C region encoded by a distinct C-region gene. The isotype of an antibody determines the effector mechanisms that it can engage on binding antigen. The different heavy-chain C regions are encoded in exons 3' to the V(D)J rearrangement site. This allows the same antibody heavy-chain V region to link up with different heavy-chain C-region isotypes as a result of somatic recombination.

Isotypic exclusion : Isotypic exclusion describes the use of one or other of the lightchain isotypes, ê or ë, by a given B cell or antibody.

Isozymes : Multiple forms of an enzyme that differ from one another in one or more of the properties.

ISSR : Inter-simple sequence repeat.

IST : Insulin shock therapy.

IT : Individual therapy; inhalation therapy; intensive therapy.

ITF : Innovation task force.

ITP : Inosine triphosphate.

ITS : Internal transcribed spacer.

ITT : Internal tibial torsion.

IU : Immunizing unit; international unit; intrauterine.

IUP : Intrauterine pregnancy.

IUT : Intrauterine transfusion.

IV : Intervertebral; intravascular; intravenous; *in vitro*; *in vivo*.

IVD : Intervertebral disc.

IVEP : *In vitro* embryo production.

IVF : *In vitro* fertilization.

IVJC : Intervertebral joint complex.

IVM : *In vitro* maturation.

IVT : *In vitro* transcription.

Ingoldian fungi

J

J : Joule.

J : Joining segment.

J chain : A polypeptide present in polymeric IgM and IgA that links the subunits together.

J gene segments, joining gene segments : The J gene segments, or joining gene segments, are found some distance 5' to the C genes in immunoglobulin and T-cell receptor loci. A V and in the case of IgH chains, tcrâ chains and tcrä chains, a D gene segment must also rearrange to a J gene segment to form a complete V-region exon.

Jaccard coefficient (SJ) : An association coefficient used in numerical taxonomy; it is the proportion of characters that match, excluding those that both organisms lack.

JALMA : Japanese Leprosy Mission for Asia.

JAMA : Journal of the American Medical Association.

James Hutton : Person who lived from 1726 to 1797 and said that land forms can be accounted for by current mechanisms.

James Watson and Francis Crick : People who, in 1953, published a paper describing a theoretical structure and means of replication for DNA.

Janus kinases, JAKs : Cytokine receptors signal via Janus kinases (jaks)-tyrosine kinases that are activated by the aggregation of cytokine receptors. These kinases phosphorylate proteins known as STATS, for Signal Transducers and Activators of Transcription. STATS are normally found in the cytosol, but move to the nucleus on phosphorylation and activate a variety of genes.

Jaundice : Jaundice is when there is a yellowing of the skin and whites of the eyes because of release of pigments from the liver into the blood.

JCT : Junction.

Jean Baptiste Lamarck : Person who published a theory of evolution in the early 1800s saying that organisms could inherit acquired traits.

Jejunum : The middle portion of the small intestine, approximately eight feet in length (~2/5 of the length of the small intestine), extending from the duodenum to the ileum. ***OR*** The second (middle) section of the small intestine.

JIVET : Juvenile *in vitro* embryo technology.

JJ : Jaw jerk.

JND : Just noticeable difference.

Jock itch : See tinea cruris.

Johannes Kepler : Person who lived from 1571 to 1630 and further refined Copernicus' concept of a sun-centeredsystem.

John Needham : Person who,in the1740s,claimed that there was a "life force" in the air that caused bacteria to appear by spontaneous generation.

JRA : Juvenile rheumatoid arthritis.

JT : Joint.

Junctional diversity : Junctional diversity is the diversity present in antigen-specific receptors that is created during the process of joining V, D, and J gone segments.

JUV : Juvenile.

JV : Jugular vein.

JVD : Jugular-venous distension.

JVP : Jugular venous pulse.

K

K : Kilo (10^3g).

K : Potassium.

KA : Ketoacidosis.

Karlvon Baer : Person who, in 1828, published on the developmental stages in mammalian eggs and showed that an undifferentiated, single-celled egg grows into a many-celled embryo.

Karlvon Linné : Person who lived from 1707 to1778 and developed the concepts of scientific names and the taxonomic hierarchy.

Karyogamy : Fusion in a cell of haploid (n) nuclei to form a diploid (2n).

Karyotype : Obtaining blood cells, growing, staining, and photographing them, cutting out chromosomes from the picture, and lining them up in pairs to check for Abnormalities.

kb : Kilobase.

KB : Knee bearing amputation.

k_{cat} **:** Turnover number.

KCL : Potassium chloride.

kDa : Kilodalton.

kelp (kelp) : A common name for any of the larger members of the order Laminariales of the brown algae.

Keratinocyte : Cell found in skin-associated lymphoid tissue; secretes cytokines that may induce an inflammatory response.

Keratitis : Inflammation of the cornea of the eye.

Ketogenic : Describing amino acids that are metabolized to acetoacetate and acetate.

Ketone : A functional group of an organic compound in which a carbon atom is double-bonded to an oxygen. Neither of the other substituents attached to the carbon is a hydrogen. Otherwise the group would be called an aldehyde. ***OR*** An organic molecule with a carbony group (–C=O) attached in the middle of the molecule.

Ketone bodies : Refers to acetoacetate, acetone, and ?-hydroxybutyrate made from acetyl-coa in the liver and used for energy in nonhepatic tissue.

Ketosis : A condition in which the concentration of ketone bodies in the blood or urine is unusually high.

KFT : Kidney function test.

KG : Kilogram.

KI : Potassium iodide.

Kick net : 500 micron white mesh net is designed to meet the requirements of groups performing USEPA Rapid Bioassessment Protocols for Benthic Invertebrates. (Benthic = bottom dwelling).

Killer activaty receptors, KARs : Killer activatory receptors (kars) are cell-surface receptors on NK cells or cytotoxic T cells; they are receptors that can activate killing by these cells.

Killer T cell : Killer T cell is a commonly used term for cytotoxic T cell.

Kilobase : One thousand bases in a DNA molecule.

Kinase : An enzyme catalyzing phosphorylation of an acceptor molecule, usually with ATP serving as the phosphate (phosphoryl) donor.

Kinetoche : A structure that attaches laterally to the centromere of a chromosome; it is the site of chromosome tubule attachment.

Kinetoplast : A special structure in the mitochondrion of kinetoplastid protozoa. It contains the mitochondrial DNA.

Kingdom animalia : Kingdom of organisms, known as "animals," which ingest their food.

Kingdom monera : The Kindgom to which prokaryotes such as bacteria and blue-green algae belong.

Kingdom plantae : Group of multi cellular eukaryotes, most of which have chlorophyll and do photosynthesis to make their own food.

Kingdom protista : Simple eukaryotes including protozoa and algae.

Kingdom fungi : Kingdom of organisms which obtain their nutrition by secreting chemicals in to their environment to digest the food, then absorbing the nutrients.

Kinin system : The kinin system is an enzymatic cascade of plasma proteins that is triggered by tissue damage to produce several inflammatory mediators, including the vasoactive peptide bradykinin.

Kirby-Bauer method : A disk diffusion test to determine the susceptibility of a microorganism to chemotherapeutic agents.

KJ : Knee jerk.

KK : Knee kick.

KLS : Kidney, liver, spleen.

Km : See Michaelis constant.

Koch's Postulates : Set of laws formulated by Robert Koch to prove that an organism is the causal agent of disease. ***OR*** A set of rules for proving that a microorganism causes a particular disease. ***OR*** Of laws formulated by Robert Koch to prove that an organism is the causal agent of disease.

Koplik's spots : Lesions of the oral cavity caused by the measles (rubeola) virus that are characterized by a bluish white speck in the center of each.

Kean hemorrhagic fever : An acute infection caused by a virus that produces varying degrees of hemorrhage, shock, and sometimes death. Krebs cycle See tricarboxylic acid (TCA) cycle.

Krebs cycle : The oxidative process in respiration by which pyruvate (via acetyl coenzyme A) is completely decarboxylated to CO_2. The pathway yields 15 moles of ATP (150,000 calories).

Krebs cycle : The second stage incellular respiration in which pyruvic acid is converted to CO_2, also known as the citric acid cycle.

K strategy : Ecological strategy where organisms depend on physiological adaptations to environmental resources. K strategists are usually stable and permanent members of the community.

KUB : Kidney, ureter, and bladder x-ray.

Kumarahou : New Zealand native shrub, also known as gumdigger's soap or poverty weed.

Kupffer cells : Kupffer cells are phagocytes lining the hepatic sinusoids; they remove debris and dying cells from the blood, but are not known to elicit immune responses.

KVO : Keep vein open.

Kyoto protocol : An international agreement, reached in 1997 in Kyoto, Japan, which extends the commitments of the United Nations Framework Convention on Climate Change. In particular, it sets targets for future emissions by each developed country.

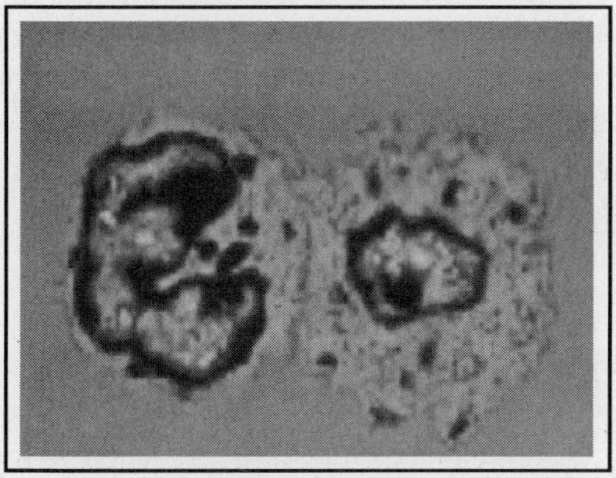

Killer T-cell

L

L : Liter.

L chain : See light chain.

L&H : Lungs and heart.

L&S : Liver and spleen.

Labia mina : The inner set of folds of tissue surrounding the vestibule area, part of the external female genitalia.

Labiatae : Plant family characterized by lipped flowers and square stems, commonly known as the mint family (which includes many frequently-used herbs such as peppermint, basil, lavender, and oregano).

LAC : Laceration; long arm cast.

Lactase : An enzyme used to digest the b-1-4 glycosidicl inkage in milk sugar.

Lactation : The production of milk.

Lactic acid bacteria : A large group of beneficial bacteria that have similar properties and all produce lactic acid as an end product of the fermentation process. They are widespread in nature and are also found in our digestive systems. Lactic acid bacteria have been used to ferment foods for at least 4000 years e.g. Yoghurt, cheese. They may also be responsible for food spoilage.

Lactic acid fermentation : A fermentation that produces lactic acid as the sole or primary product.

Lactic acid : A 3-carbon molecule, $CH_3CHOHCOOH$, which results from lactic acid fermentation, gives yogurt its tangy and sour taste

Lactobacillus acidophilus **:** One of the main "good" bacteria in yogurt.

Lactose : The naturally occurring type of sugar found in milk and milk products is called lactose. It is less sweet than other types of sugars. Enzymes in a mammal's digestive system break the lactose down into glucose and galactose, simple sugars that can be more easily absorbed than lactose. ***OR*** "Milk sugar"—a disaccharide made from glucose +galactose and found in milk.

Lactose intolerance : The inability to digest lactose, resulting in discomfort and bloating, is referred to as lactose intolerance. ***OR*** Caused when your intestines lack a certain enzyme, called lactase, needed to digest lactose, the sugar in milk and other dairy products. Within 30 minutes to two hours of eating these foods, you may suffer cramps, gas, bloating, and diarrhoea. Lactase tablets and drops can be used to treat the symptoms.

Lacunae : Spaces in which osteocytes or chondrocytes are found.

Lacunose : A surface with wide sunken pits like an irregular honey comb.

LAE : Long above elbow (cast).

Lag phase : Period after inoculation of fresh growth medium during which population numbers do not increase. ***OR*** A period following the introduction of microorganisms into fresh culture medium when there is no increase in cell numbers or mass during batch culture.

Lageniform : Flask-shaped.

Lager : Pertaining to the process of aging beers to allow flavor development.

LAL : Limulus amebocyte lysate.

LAM : Laminectomy.

Lamella : (i) A thin layer, platelike arrangement or membrane. (ii) Layers of protoplasmic membranes within the chloroplast that contain the photosynthetic pigments.

Lamellae : Other name for gills. ***OR*** Having layers (such as the concentric rings of the matrix in bone tissue).

Lamellate : With gills.

Lamellulae : Gills that don't reach the stem interspersed between lamellae.

Lamina propria : Thin layer of fibrous connective tissue immediately beneath the surface epithelium of mucous membranes.

Laminaria **:** A type of brown algae that is a good source of iodine when eaten and is used gynecologically.

Laminarin : A beta-glucan polysaccharide produced by many chromists through photosynthesis. ***OR*** One of the principal storage products of the golden-brown algae; a polymer of glucose.

Lampbrush chromosome : Giant diplotene chromosome found in the oocyte nucleus. The loops that are observed are the sites of extensive gene expression.

LANC : Long arm navicular cast.

Lancefield system : One of the serologically distinguishable groups (as group A, group B) into which *Streptococci* can be divided.

Lanceolate : Lance-shaped.

Land farming : The addition of waste material, such as a hydrocarbon waste, to the soil surface so that it will be degraded. The soil may be moistened or mixed to stimulate the desired degradation process.

Langerhans cell : Cell found in the skin that internalizes antigen and moves in the lymph to lymph nodes where it differentiates into a dendritic cell.

Langerhans' cells : Langerhans' cells are phagocytic dendritic cells found in the epidermis. They can migrate from the epidermis to regional lymph nodes via the afferent lymphatics. In the lymph node they differentiate into dendritic cells.

Lanose : Woolly.

Laparoscopy : An operation that involves a tiny camera (laporoscope) being inserted into the body of an organism for the duration of the operation.

Large intestine : The portion of the digestive tract that extends from the terminal ileum (ileocecal junction) of the small intestine to the anus (approximately 5 feet in length), composed of the cecum, colon and rectum and containing >1011 bacteria per gram of feces with anaerobes 1000fold more common than other microbes.

Large pre-B cells, pre-B-cell receptor : The large pre-B cells have a cell-surface pre-B-cell receptor, which is lost on the transition to small pre-B cells, in which light-chain gene rearrangement occurs.

Larvae : The newly hatched, wingless, often worm-like form of many insects before metamorphosis into the winged form.

Larynx : The widened area at the top of the trachea,which contains the vocal cords.

LAT : See linker of activation in T cells.

Late mRNA : Messenger RNA produced later in a virus infection, which codes for proteins needed in capsid construction and virus release.

Late pro-B cell : The late pro-B cell is the stage in B-cell development in which VH to DJH joining occurs.

Latency : Some viruses can enter a call but not replicate, a state known as latency. Latency can be established in various ways; when the virus is reactivated and replicates, it can produce disease.

Latent Infection : Viruses which are able to down-regulate their gene expression can establish a truly latent state, i.e. With strictly limited gene expression and without ongoing virus replication. Latent virus infections typically persist for the entire life of the host.

Latent period : The initial phase in the one-step growth experiment in which no phages are released.

Latent virus infections : Virus infections in which the virus quits reproducing and remains dormant for a period before becoming active again.

Late-phase reaction : In type I immediate hypersensitivity reactions, the late-phase reaction persists and is resistant to treatment with antihistamine.

Lateral : On the side.

Latex : Fluid often milky that oozes from cut surfaces in some species. Characteristic of lactarius.

Latin binomial : See binomial nomenclature.

Latissimus Dsi : The muscles on the back,over the kidneys.

Law of mass action : The finding that the rate of a chemical reaction is a function of the product of the concentrations of the reacting species.

Laxative : A substance which stimulates evacuation of the bowels.

Lazzaro Spallanzani : Person who,in the late1760s,demonstrated that broth which was boiled in flasks,which were then sealed,stayed sterile and did not spoil.

LBB : Low back bending.

LBP : Low back pain; low blood pressure.

LBT : Low back tenderness.

LC : Liguid chromatography

LCAT : Lecithin:cholesterol acyl transferase.

Lck : The tyrosine kinase Lck associates most strongly with the cytoplasmic tails of CD4 and CD8. It plays a central role in signal transduction and activation of the T-cell receptor.

LCM : Laser capture microdissection.

LCMV : See lymphocyte choriomeningitis virus.

LCR : Ligase chain reaction or low complexity region.

LDH : Lactate dehydrogenase.

LDL : Low-density lipoprotein.

LDP : Lumbo-dorsal pain.

Leaching : (i) Removal of valuable metals from ores by microbial action. (ii) The removal of materials in solution from the soil.

Leader region : The region of an mRNA between the 5' end and the initiation codon for translation of the first polypeptide chain.

Leader sequence : A nontranslated sequence at the 5_ end of mRNA that lies between the operator and the initiation codon; it aids in the initiation and regulation of transcription.

Leader sequence : An N-terminal signal sequence that directs secretion and processing of proteins.

Lecithin : A phospholipids found in egg yolk (used as an emulsifying agent).

Lectin : Adhesin specific for polysaccharide target receptor (sugar residues)

Lectin complement pathway : The lectin pathway for complement activation is triggered by the binding of a serum lectin (mannan-binding lectin; MBL) to mannose-containing proteins or to carbohydrates on viruses or bacteria.

Lectins : Plant proteins with a high affinity for specific sugar residues. ***OR*** Agglutinating proteins usually extracted from plants.

Leghemoglobin : Ironcontaining, red pigment(s) produced in root nodules during the symbiotic association between rhizobia and leguminous plants. The pigment is similar but not identical to mammalian hemoglobin.

Legionellosis : See Legionnaires' disease.

Legionnaires' disease : A pulmonary form of legionellosis, resulting from infection with *Legionella pneumophila*.

Leguminosae : Plant family characterized by a pod-shaped ovary surrounding its seeds (peas,beans).

Leishmanias : Zooflagellates, members of the genus *Leishmania*, that cause the disease leishmaniasis.

Leishmaniasis : The disease caused by the protozoa called leishmanias.

Lens : The clear structure,behind the iris and pupil,that focuses the light on the retina.

Lenticular : Shaped like a double convex lens.

Lentiviruses : Lentiviruses are a group of retroviruses that include the human immunodeficiency virus, HIV-1. They cause disease after a long incubation period and can take years to become apparent.

Lepromatous (progressive) leprosy : A relentless, progressive form of leprosy in which large numbers of *Mycobacterium leprae* develop in skin cells, killing the skin cells and resulting in the loss of features. Disfiguring nodules form all over the body.

Leprosy Hansen's disease : A severe disfiguring skin disease caused by *Mycobacterium leprae*.

Leprosy, lepromatous leprosy, tuberculoid leprosy : Leprosy is caused by *Mycobacterium leprae* and occurs in a variety of forms. There are two polar forms: lepromatous leprosy, which is characterized by abundant replication of leprosy bacilli and abundant antibody production without cell-mediated immunity; and tuberculoid leprosy, in which few organisms are seen in the tissues, there is little or no antibody, but cell-mediated immunity is very active. The other forms of leprosy are intermediate between the polar forms.

Lesion : Pathological or traumatic discontinuity of tissue or loss of function.

Lethal dose 50 (LD_{50}) : Refers to the dose or number of organisms that will kill 50% of an experimental group of hosts within a specified time period.

Leu : Leucine.

Leucocytosis : An increase in the number of WBCs.

Leucopenia : A decrease in the number of WBCs.

Leukaemia : A cancer of the blood or bone marrow, which normally causes excess production of white blood cells. ***OR*** Leukemia is the unrestrained proliferation of a malignant white blood cell characterized by very high numbers of the malignant cells in the blood. Leukemias can be lymphocytic, myelocytic, or monocytic. ***OR*** A progressive, malignant disease of blood-forming organs, marked by distorted proliferation and development of leukocytes and their precursors in the blood and bone marrow. Certain leukemias are caused by viruses (HTLV-1, HTLV-2).

Leukocidin : A microbial toxin that can damage or kill leukocytes.

Leukocyte : White blood cell. ***OR*** Leukocyte is a general term for a white blood cell. Leukocytes include lymphocytes, polymorphonuclear leukocytes, and monocytes. ***OR*** Any colorless white blood cell. Can be classified into granular and agranular lymphocytes.

Leukocyte adhesion deficiency : Leukocyte adhesion deficiency is an immunodeficiency disease in which the common P chain of the leukocyte integrins is not produced. This mainly affects the ability of leukocytes to enter sites of infection with extracellular pathogens, so that infections cannot be effectively eradicated.

Leukocyte common antigen : See CD45.

Leukocyte functional antigens, LFAs, LFA-1, LFA-2, LFA-3 : The leukocyte functional antigens or lfas, are cell adhesion molecules initially defined with monoclonal antibodies: LFA-1 is a â2 integrin; LFA-2 is a member of the immunoglobulin superfamily, as is LFA-3, now called CD58. LFA-1 is particularly important in T-call adhesion to endothelial cells and antigen-presenting cells.

Leukocyte integrins : See leukocyte functional antigens.

Leukocytosis : An abnormally large number (or an abnormal increase in the number (cytosis)) of leukocytes in the blood, as during hemorrhage, infection, inflammation, or fever, specifically a count of 12,000 or more per cubic millimeter. ***OR*** Leukocytosis is the presence of increased numbers of leukocytes in the blood. It is commonly seen in acute infection.

Leukopenia : An abnormally low number (or an abnormal reduction in the number (penia)) of leukocytes in the blood, specifically a count of 5000 or less per cubic millimeter.

Leukotrienes : Leukotrienes are lipid mediators of inflammation that are derived from arachidonic acid. They are produced by macrophages and other cells.

Levulose : Another name for fructose because it rotates polarized light to the left.

LF : Left foot; linear fracture; low frequency.

LFA : Left forearm.

LFA-1, LFA-2, LFA-3 : See leukocyte functional antigens.

LHG : Left hand grip.

LHRH : Luteinizing hormone releasing hormone.

LHS : Left hand side.

LIC : Left iliac crest.

License : A person or organisation deemed competent is given the right to manufacture a product designed by someone else by paying them a licensing fee.

Lichen : An organism composed of a fungus and either green algae or cyanobacteria in a symbiotic association. ***OR*** Fungus and an alga or a cyanobacterium living in symbiotic association.

LICOS : LICOS is the ligand for ICOS, a CD28-related protein that is induced on activated T cells and can enhance T-cell responses. LICOS is produced on activated dendritic cells, monocytes, and B cells.

Liebig's law of the minimum : Living organisms and populations will grow until lack of a resource begins to limit further growth.

LIF : Leukaema inhibitory factors.

Life cycle : The stages an organism goes through over time containing a possible pre-parasitic and a parasitic phase. Usually represented as a circle beginning with the formation of a individual and ending with the production of the next generation. In parasitology the life cycle of the parasite gives the investigator predictive value for understanding pathogenesis and clinical signs, and for understanding the epidemiology and control of the parasite. Life cycles can either be direct or indirect.

Ligand : Molecule, ion, or group bound to the central atom in a chelate or a coordination compound. ***OR*** A (usually small) molecule that binds to another, such as oxygen when it binds to myoglobin.

Ligase : An enzyme that joins DNA fragments together. ***OR*** An enzyme that catalyzes the joining of two molecules together. In DNA it joins 5'-OH to 3' phosphates.

Ligation : The joining together of DNA fragments using ligases.

Light chain, L chain : The immunoglobulin light chain (L chain) is the smaller of the two types of polypeptide chain that make up all immunoglobulins. It consists of one V and one C domain, and is disulfide-bonded to the heavy chain. There are two classes of light chain, known as ê and ë.

Light compensation point : Where the rate of photosynthesis is lower than the rate of respiration usually about 1% of the light intensity of sunlight

Light microscope : A microscope that uses light to make a specimen visible.

Light zone : See germinal centers.

Lignicolous : Growing on wood.

Lignin : A natural substance found in the cell walls of wood, which binds the wood fibres together making them rigid and strong. ***OR*** A complex polymer that occurs in woody material of higher plants. It is highly resistant to chemical and enzymatic degradation. Lignin is composed of various intermolecular linkages between phenylpropanes and guaiacyl, syringyl, p-hydroxyphenyl, and biphenyl nuclei. Analysis of the metabolic pathway has shown that the dimeric lignin compounds are degraded to protocatechuate or 3OMGA (3-O-methylgallic acid) and that these compounds are cleaved by protocatechuate 4,5-dioxygenase encoded by ligab. Among the dimeric lignin compounds, the degradation of ß-aryl ether and the biphenyl structure is the most important, because ß-aryl ether is most abundant in lignin (50%) and the biphenyl structure is so stable that its decomposition is rate-limiting in lignin degradation . *Sphingomonas paucimobilis* is able to degrade a wide variety of dimeric lignin compounds, including ß-aryl ether, biphenyl, and diarylpropane.

Ligulate flowers : Irregular flowers with five,united petals.

Liliaceae : Family of monocots in which the sepals and petals are very similar in appearance.

Lime, agricultural : Soil amendment containing calcium carbonate, magnesium carbonate or other materials to neutralize soil acidity and furnish calcium or magnesium or both for plant growth.

Limiting substrate : See growth limiting nutrient for basic definition. If all nutrients are in perfect proportion there is balanced growth. If all nutrients are balanced but not present in sufficient quantity, growth is substrate limited. If there is an excess of nutrient in the bulk fluid, but it is not reaching biofilm bacteria because of transport limitations, growth is transport limited.

LIMS : Laboratory information management system.

LINE : Long interspersed nuclear element.

Linear epitope : See continuous epitopes.

Linkage disequilibrium : Alleles at linked loci within the major histocompatibility complex are said to be in linkage disequilibrium if they are inherited together more frequently than predicted from their individual frequencies.

Linkage : The tendency of markers to be inherited together. Linkage of two markers is an indication that they are close to one another in the genome.

Linked recognition : Epitopes recognized by B cells and helper T cells must be physically linked for the helper T cell to activate the B cell. This is called linked recognition.

Linker of activation in T cells, LAT : The adaptor protein known as linker of activation in T cells (LAT) is a cytoplasmic protein with several tyrosines that become phosphorylated by the tyrosine kinase ZAP-70. It becomes associated with membrane lipid rafts and coordinates downstream signaling events in T-cell activation.

Linkers : Short oligonucleotides that can be ligated (connected) to larger DNA fragments, then cleaved (cut) to yield overlapping cohesive (sticky) ends, suitable for ligation to other DNAs that contain comparable cohesive ends.

Linking number : The net number of times one polynucleotide chain crosses over another polynucleotide chain. By convention, right-handed crossovers are given a plus designation.

Lipid : A fat or fat-like substance found in body tissues, examples include cholesterol, triglycerides, and fatty acids. ***OR*** A biological molecule that is soluble in organic solvents. Lipids include steroids, fatty acids, prostaglandins, terpenes, and waxes. ***OR*** A class of biochemical compounds which includes fats, oils, and waxes.

Lipid bilayer : Model for the structure of the cell membrane based on the interaction between the hydrophobic regions of phospholipids.

Lipophilic : Having an affinity for fat.

Lipopolysaccharide (LPS) : Complex lipid structure containing unusual sugars and fatty acids found in many Gramnegative bacteria. ***OR*** A molecule containing both lipid and polysaccharide, which is important in the outer membrane of the Gram-negative cell wall. ***OR*** Amphiphilic molecule integrated in (hydrophobic end) and extending outward from (hydrophilic end) the outer membrane of Gramnegative bacterial cells; structurally composed of hydrophobic lipid A (responsible for endotoxin activity) + core polysaccharide + hydrophilic osomatic antigen polysaccharide side chain; provides protection from environmental insults and acts as virulence factor. ***OR*** Complex lipid structure containing unusual sugars and fatty acids found in many Gramnegative bacteria. ***OR*** Usually refers to a unique glycolipid found in Gram negative bacteria.

Liposome : A spherical particle formed by a lipid bilayer enclosing an aqueous solution. It may be used to administer chemotherapeutic agents or in diagnostic testing.

Lipoteichoic acid : Amphiphilic molecule integrated in and extending outward from the cell wall (peptidoglycan layer) of Grampositive bacterial cells; structurally composed of hydrophobic glycolipid and hydrophilic polyglycosylpolyglycerophosphate; often serves as an antiphagocytic adhesin.

LIQ : Left inner quadrant.

Liquefaction : The conversion of large solid particles of sludge into very fine particles which either dissolve or remain suspended in wastewater.

LIR : Left iliac region.

Listeriosis : A sporadic disease of animals and humans, particularly those who are immuno-compromised or pregnant, caused by the bacterium *Listeria monocytogenes*.

Lithosphere : The rocky layers that make up the Earth's crust.

Lithotroph : An organism that uses reduced inorganic compounds as its electron source. ***OR*** Organism that uses an inorganic substrate such as ammonia or hydrogen as an electron donor in energy metabolism. There are two types of lithotrophs: chemolithotroph and photolithotroph.

Litter : Surface layer of the forest floor consisting of freshly fallen leaves, needles, twigs, stems, bark, and fruits.

Littal Zone : Zone of emergent vegetation in a lake or ocean (**litt**=the seashore)

LK : Left kidney.

LKS : Liver, kidney, spleen.

LL : Left lateral; left leg; left lower; left lung.

LLB : Left lateral bending; long leg brace.

LLC : Long leg cast.

LLCC : Long leg cylinder cast.

LLD : Leg length discrepancy.

LLE : Left lower extremity

LLF : Left lateral flexion; left little finger.

LLL : Left lower leg; left lower lid; left lower lung.

LLN : Lower limit of normal.

LLQ : Left lower quadrant.

LLR : Left lumbar region.

LLS : Long leg splint.

LLWC : Long leg walking cast.

LLX : Left lower extremity.

LMO : Living modified organism.

LN : Lymph node.

LNAPL : See NAPL

Lobeliaceae : A plant family within class Dicotyledones in which flowers are characterized by two petals up and three petals down.

LOC : Lab-on-a-chip.

LOC : Level of consciousness; loss of consciousness.

Log growth : A growth phase in which cell production is maximum.

Log phase : See exponential phase.

LOM : Left otitis media; limitation of movement; loss of movement.

Lophotrichous : Having a tuft of polar flagella. ***OR*** A cell with a cluster of flagella at one or both ends.

LOQ : Left lower quadrant; left outer quadrant.

LOS : Length of stay; loss of sight.

Louis Pasteur : Person who, in 1864, disproved spontaneous generation in microscopic organisms such as bacteria.

Low oxygen diffusion environment : An aquatic environment in which microorganisms are surrounded by deep water layers that limit oxygen diffusion to the cell surface. In contrast, microorganisms in thin water films have good oxygen transfer from air to the cell surface.

Low-zone tolerance : See high-zone tolerance.

LPS : Lipopolysaccharide.

LPS-binding protein : A special plasma protein that binds bacterial lipopolysaccharides and then attaches to receptors on monocytes, macrophages, and other cells. This triggers the release of IL-1 and other cytokines that stimulate the development of fever and additional endotoxin effects.

LPS-binding protein, LBP : A molecule of bacterial lipopolysaccharide (LPS) has first to be bound by the LPS-binding protein (LBP) before it can interact with CD14, an LPS:LBP-binding protein on cells such as macrophages.

LQ : Lower quadrant.

LRQ : Lower right quadrant.

LRT : Lower respiratory tract.

LS : Left side; liver and spleen; lumbar spine; lumbosacral.

L-selectin : L-selectin is an adhesion molecule of the selectin family found on lymphocytes. L-selectin binds to CD34 and glycam-1 on high endothelial venules to initiate the migration of naive lymphocytes into lymphoid tissue. Also called CD62L.

LSK : Liver, spleen, kidneys.

LSP : Lumbar spine.

LTR : Long terminal repeat.

LUA : Left upper arm.

Luciferase : Enzyme which catalyzes production of light from luciferin. ***OR*** Enzyme which activates luciferin to produce bioluminescence.

Luciferin : Compound whose activated form emits light.

LUE : Left upper extremity.

Lumbar vertebrae : The vertebrae in the lower back which are not attached to ribs.

Lumen : The space inside a tubular organ, e.g., a blood vessel or intestine

Luminescence : Production of light.

Luminescent : Emission of light, glows in the dark. Characteristic of some fungi e.g. Omphalotus.

LUOQ : Left upper outer quadrant.

LUQ : Left upper quadrant.

Luxury uptake : The absorption by plants of nutrients in excess of their need for growth. Luxury contents accumulated during early growth may be used for later growth.

LX : Lower extremity.

Lyase : An enzyme that catalyzes the removal of a group to form a double bond, or the reverse reaction.

Lyme disease (LD) : A tick-borne disease caused by the spirochete *Borrella burgdorferi.* ***OR*** Lyme disease is a chronic infection with *Borrelia burgdorferi*, a spirochete that can evade the immune response.

Lymph : Clear extracellular fluid that is derived from tissue fluids, circulates in the lymphatic vessels and is filtered by the lymph nodes. ***OR*** Lymph is the extracellular fluid that accumulates in tissues and is carried by lymphatic vessels back through the lymphatic system to the thoracic duct and into the blood.

Lymph node : A small secondary lymphoid organ that contains lymphocytes, macrophages, and dendritic cells. It serves as a site for (1) filtration and removal of foreign antigens and (2) the activation and proliferation of lymphocytes. ***OR*** mass of lymphatic tissue located in lymphatic vessels and responsible for filtration of lymph.

Lymphangitis : Inflammation of lymphatic vessel(s).

Lymphatic system : The lymphatic system is the system of lymphoid channels and tissues that drains extracellular fluid from the periphery via the thoracic duct to the blood. It includes the lymph nodes, Peyer's patches, and other organized lymphoid elements apart from the spleen, which communicates directly with the blood. ***OR*** System of vessels, nodes, and masses of tissue, e.g., spleen, responsible for filtering lymphatic fluid from the tissues and recirculating it back to the venous system; also transports lipids and proteins to the circulatory system and returns fluid that leaks out of capillaries.

Lymphatic tissue (lymphoid tissue) : Tissue consisting of lymphocytes enclosed in network of reticular and collagenous fibers.

Lymphatic vessels, lymphatics : Lymphatic vessels, or lymphatics, are thin-walled vessels that carry lymph through the lymphatic system.

Lymphoblast : Lymphoid tissue cell that is the immature precursor (progenitor) of a mature lymphocyte. ***OR*** A lymphoblast is a lymphocyte that has enlarged and increased its rate of RNA and protein synthesis.

Lymphocyte : A nonphagocytic, mononuclear leukocyte (white blood cell) that is an immunologically competent cell, or its precursor. Lymphocytes are present in the blood, lymph, and lymphoid tissues. See B cell and T cell.

Lymphocyte function-associated antigens : See leukocyte functional antigens.

Lymphocyte receptor repertoire : The lymphocyte receptor repertoire is the totality of the highly variable antigen receptors carried by B and T lymphocytes.

Lymphocyte(s) : Mononuclear, nonphagocytic, immuno-logically competent leukocytes with agranular cytoplasm (agranulocyte) and large nucleus concentrated in blood, lymph and lymphoid tissue that are divided into two classes B lymphocytes (B cells) and T lymphocytes (T cells); active in immunological responses in the body, including the production of lymphokines (cytokines) and antibodies.

Lymphocytes : All adaptive immune responses are mediated by lymphocytes. Lymphocytes are a class of white blood cells that bear variable cellsurface receptors for antigen. These receptors are encoded in rearranging gene segments. There are two main classes of lymphocyte-B lymphocytes (B cells) and T lymphocytes (T cells)-which mediate humoral and cell-mediated immunity, respectively. Small lymphocytes have little cytoplasm and condensed nuclear chromatin; on antigen recognition, the cell enlarges to form a lymphoblast and then proliferates and differentiates into an antigenspecific effector cell.

Lymphocytic chiomeningitis virus, LCMV : Lymphocytic choriomeningitis virus (LCMV) is a virus that causes a nonbacterial meningitis in mice and occasionally in humans. It is used extensively in experimental studies.

Lymphogranuloma venereum (LGV) : A sexually transmitted disease caused by *Chlamydia trachomatis* serotypes L1-L3, which affect the lymph organs in the genital area.

Lymphoid dendritic cells : Dendritic cells can arise from myeloid cells, in which case they are called myeloid dendritic cells, or from lymphoid tissues, in which case they are called lymphoid dendritic cells. Functional differences exist between these two lineages.

Lymphoid organs : Lymphoid organs are organized tissues characterized by very large numbers of lymphocytes interacting with a nonlymphoid stroma. The central or primary lymphoid organs, where lymphocytes are generated, are the thymus and bone marrow. The main peripheral or secondary lymphoid organs, in which adaptive immune responses are initiated, are the lymph nodes, spleen, and mucosal-associated lymphoid tissues such as tonsils and Peyer's patches.

Lymphoid tissue : See lymphatic tissue.

Lymphokine : Extracellular soluble protein produced by Tlymphocytes that activate cellmediated immunity and inflammation; type of cytokine.

Lymphokine : A biologically active glycoprotein (e.g., IL-1) secreted by activated lymphocytes, especially sensitized T cells. It acts as an intercellular mediator of the immune response and transmits growth, differentiation, and behavioral signals.

Lymphokines : Lymphokines are cytokines produced by lymphocytes.

Lymphomas : Lymphomas are tumors of lymphocytes that grow in lymphoid and other tissues but do not enter the blood in large numbers. There are many types of lymphoma, which represent the transformation of various developmental stages of B or T lymphocytes.

Lymphopoiesis : Lymphopoiesis is the differentiation of lymphoid cells from a common lymphoid progenitor.

Lymphotoxin, LT : Lymphotoxin (LT) is also known as tumor necrosis factor-â (TNF-â), a cytokine secreted by inflammatory CD4T cells that is directly cytotoxic for some cells.

Lyn : See tyrosine kinase.

Lyophilization, Lyophilize : Freeze-drying, (sometimes spelled incorrectly as liophilization).

Lys : Lysine.

Lysing : A disintegration or breakdown of cells which releases organic matter.

Lysis : The rupture or physical disintegration of a cell. ***OR*** Rupture of a cell, resulting in loss of cell contents.

Lysogenic : See lysogens.

Lysogenic virus : A virus that can adopt an inactive (Iysogenic) state, in which it maintains its genome within a cell instead of entering the Iytic cycle. The circumstances that determine whether a lysogenic (temperate) virus will adopt an inactive state or an active lytic state are often subtle and depend upon the physiologic state of the infected cell.

Lysogens : Bacteria that are carrying a viral prophage and have the potential of producing bacteriophages under the proper conditions.

Lysogeny : An association where a prokaryote contains a prophage and the virus genome is replicated in synchrony with the host chromosome. ***OR***

The state in which a phage genome remains within the bacterial host cell after infection and reproduces along with it rather than taking control of the host and destroying it.

Lysosome : An organelle that contains hydrolytic enzymes designed to break down proteins that are targeted to that organelle. ***OR*** Cell organelle containing digestive enzymes.

Lysosomes : Lysosomes are acidified organelles that contain many degradative hydrolytic enzymes. Material taken up into endosomes is eventually delivered to lysosomes.

Lysozyme : An enzyme that degrades peptidoglycan by hydrolyzing the b(1 _ 4) bond that joins N-acetylmuramic acid & N-acetylglucosamine.

Lytic cycle : A virus life cycle that results in the lysis of the host cell.

Lytic granules : Lytic granules containing perforin and granzymes are a defining characteristic of armed effector cytotoxic cells.

Lytic infection : A virus infection that leads to the Iysis of the host cell, yielding progeny virus particles.

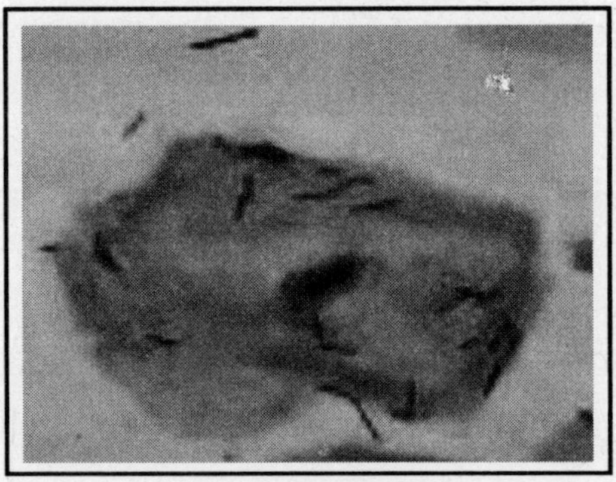

Lactobacillus

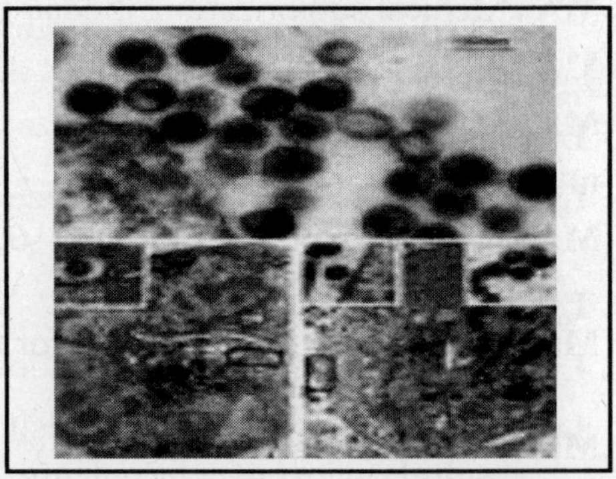

Lentivirus

M

M : Milli (10^{-3}l).

M : Molar (mol/L).

M : Microorganisms - small organisms which require a microscope to be seen. M represents the SS in the mixed liquor and is part of the F/M ratio.

M cells (microfold cells) : Specialized cell of the intestinal mucosa and other sites, such as the urogenital tract, that delivers the antigen from the apical face of the cell to lymphocytes clustered within the pocket in its basolateral face. ***OR*** Specialized cells in the Peyer's patches that sample the microenvironment and uptake foreign antigens for processing by underlying macrophages.

M phase : That period of the cell cycle when mitosis takes place.

M&F : Mother and father.

MA : Medical authorization; moderately advanced; muscle activity.

MAA : Market authorization application.

MAAP : Multiple arbitrary amplicon profiling.

mAb : Monoclonal antibody.

MAC : Memmary alveolar cell. ***OR*** Maximum allowable concentration; macerated.

Mac-1 : Mac-1 is another name for the leukocyte integrin CD11b:CD18 (or complement receptor 2 (CR2).

MacConkey streak : Laboratory test for the presence of Gram negative bacteria. We use this test to detect contamination of Bacillus products such as CF 1000, 1002, 4002 and some of the Enz-odor® products.

Macroconidium (pl. macroconidia) : The larger of two different types of conidia produced by a fungus in the same manner.

Macrofauna : Soil animals that are > 1000 micrometers in length (e.g., vertebrates, earthworms, and large arthropods).

Macroglobulin : Macroglobulin describes plasma proteins that are globulins of high molecular weight, including immunoglobulin M (IgM) and a2-macroglobulin.

Macrolide (macrolide antibiotic) : An antibiotic containing a macrolide ring, a large lactone ring with multiple keto and hydroxyl groups, linked to one or more sugars. ***OR*** class of antibiotics produced by species of *Streptomyces* and containing a large oxygenated lactone ring (e.g., erythromycin, azithromycin).

Macromolecule : Large molecule formed from the connection of a number of small molecules. ***OR*** A large molecule that is a polymer of smaller units joined together. ***OR*** A giant molecule formed by joining smaller molecules together is a macromolecule. For example, DNA and RNA are macromolecules formed by joining nucleotides together. Proteins are formed by joining amino acids together.

Macromolecule : A large, complex molecule.

Macromolecule vaccine : A vaccine made of specific, purified macromolecules derived from pathogenic microorganisms.

Macronematous : Having a conidiophore that is morpho-logically different from the vegetative hyphae.

Macronucleus : The larger of the two nuclei in ciliate protozoa. It is normally polyploid and directs the routine activities of the cell.

Macronutrient : A substance required in large amounts for growth, usually attaining a concentration of > 500 mg kg^{-1} in mature plants. Usually refers to N, P, K, Ca, Mg, and S. ***OR*** An element required in large proportion by plants and other life forms for survival and growth. Macronutrients include nitrogen (N), potassium (K), and phosphorous (P).

Macrophage : Monocytederived large mononuclear phagocytic cell arising from hematopoietic stem cells in the bone marrow and mainly localized in the tissues; performs an array of immune response functions including nonspecific phagocytosis and pinocytosis, antigen presentation to T cells to initiate specific immune responses, secretion of lymphokines (cytokines).

Macrophage (a.k.a., mononuclear phagocyte) : The name for a large mononuclear phagocytic cell, present in blood, lymph, and other tissues. Macrophages are derived from monocytes. They phagocytose and destroy pathogens; some macrophages also activate B cells and T cells.OR phagocytic cell (mononuclear leukocytes) active in tissues; characteristics distinctive in different tissues; may become activated macrophages upon appropriate signals (e.g., cytokines like interferongamma) with increased efficiency and new functions.

Macrophage activation : Resting macrophages will not destroy certain intracellular bacteria unless the macrophage is activated by a T cell. Macrophage activation is important in controlling infection and also causes damage to neighbouring tissues.

Macrophage mannose receptor : The macrophage mannose receptor is highly specific for certain carbohydrates that occur on the surface of some pathogens but not on host cells.

Macropinocytosis : Dendritic cells are unique in being able to carry out macropinocytosis, a process in which large amounts of extracellular fluid are taken up in single vesicles. This is one means of antigen uptake.

Macropore : Larger soil pores, generally having a minimum diameter between 30 and 100 micrometers, from which water drains readily by gravity.

MadCAM-1 : Madcam-1 is the mucosal cell adhesion molecule-1 or mucosal addressin that is recognized by the lymphocyte surface proteins L-selectin and VLA-4, allowing specific homing of lymphocytes to mucosal tissues.

Maduromycosis : A subcutaneous fungal infection caused by *Madurella mycetoma*; also termed an eumycotic mycetoma.

Madurose : The sugar derivative 3-O-methyl-D-galactose, which is characteristic of several actinomycete genera that are collectively called maduromycetes.

Magnetic resonance imaging : A method used to image the inside of tissue, which can help to diagnose medical conditions.

Magnetosome : Magnetite particles in magnetotactic bacteria that are tiny magnets and allow the bacteria to orient themselves in magnetic fields. ***OR*** Small particle of Fe_3O_4 present in cells that exhibit magnetotaxis.

Magnetotactic bacteria : Bacteria that can orient themselves in the earth's magnetic field due to the presence of magnetosomes.

Maintenance energy : The energy a cell requires simply to maintain itself or remain alive and functioning properly. It does not include the energy needed for either growth or reproduction.

MAJ : Major; majority.

Major basic protein : Eosinophils can be triggered to release their major basic protein, which can then act on mast cells to cause their degranulation.

Major histocompatibility complex (MHC) : A large set of cell surface molecules in each individual, encoded by a family of genes, that serves as a unique biochemical marker of individual identity. It can trigger T-cell responses that may lead to rejection of transplanted tissues and organs. MHC molecules are also involved in the regulation of the immune response and the interactions between immune cells.

MAL : Malignant.

Malaria : A serious infectious illness caused by the parasitic protozoan Plasmodium. Malaria is characterized by bouts of high chills and fever that occur at regular intervals.

Malaria : Disease caused when the parasite, *Plasmodium vivax* invades and kills RBCs.

MALDI-TOF : Matrix assisted laser desorption/ionisation time-of-flight.

Malignant : Referring to a disease tending to become progressively worse (morbidity = illness) and potentially result in death (mortality = death).

Malleus : One of the bones in the middle ear.

MALT : See mucosal-associated lymphoid tissue.

Malt (mawlt) : Grain soaked in water to soften it, induce germination, and activate its enzymes. The malt is then used in brewing and distilling.

Maltose : Malt sugar found in sprouting grains.

Mammal : An animal,such as humans and cats,that has fur/hair,provides milk for its young, and has a four-chambered heart.

Man : Mannose.

Mandible : The jaw bone.

Mannan-binding lectin, MBL, mannan-binding lectin pathway, MB-lectin pathway : The mannan-binding lectin (MBL), also called mannose-binding protein, is an acute-phase protein that binds to mannose residues. It can opsonize pathogens bearing mannose on their surfaces and can activate the complement system via the mannan-binding lectin pathway (MB-lectin pathway) an important part of innate immunity.

Mantle zone : The follicular mantle zone is a rim of B lymphocytes that surrounds lymphoid follicles. The precise nature and role of mantle zone lymphocytes have not yet been determined.

Manuka : The manuka or tea tree (*Leptospermum scoparium*) is a shrub or small tree native to New Zealand and southeast Australia.

Manure : Excreta of animals, with or without an admixture of bedding or litter, fresh or at various stages of decomposition or composting. In some countries the term may denote any fertilizer material.

MAP kinases : See mitogen-activated protein kinases.

Marburg viral hemorrhagic fever : An acute infection caused by a virus that produces varying degrees of hemorrhage, shock, and sometimes death.

Margin : The edge or rim of a mushroom cap.

Marginal zone, marginal zone B cells : The marginal zone of the lymphoid tissue of the spleen lies at the border of the white pulp. It contains a unique population of B cells, the marginal zone B cells, which do not circulate and are distinguished by a distinct set of surface proteins.

Marginate : When there is a small circular ridge around the bulb at the base of the stem where the universal veil was attached.

Marker : A "landmark" that can be localized to a specific region of the genome.

Marsupial : A mammal,usually with a pouch,whose young are born extremely immature.

Maryland Department of Agriculture, Plant Protection & Weed Management Section : The Plant Protection and Weed Management Section of the Maryland Department of Agriculture is based in Annapolis, with field offices in Cheltenham, Riverdale and Salisbury. The Section serves Maryland Agriculture through programs including Plant Certification & Virusfree Plant Programs, Nursery Inspection and Licencing, Apiary Inspection, Pest Survey, IPM and Biocontrol, and Entomology and Plant Pathology Laboratories. For more information call 4108415920.

MAS : Marker-assisted selection. ***OR*** Manifest anxiety scale.

Mash : The soluble materials released from germinated grains and prepared as a microbial growth medium.

Mashing : The process in which cereals are mixed with water and incubated in order to degrade their complex carbohydrates (e.g., starch) to more readily usable forms such as simple sugars.

Masking agent : Substance used to cover up or disguise unpleasant odors. Liquid masking agents are dripped into wastewater, sprayed into the air, or evaporated (using heat) with the unpleasant fumes or odors and then discharged into the air by blowers to make an undesirable odor less noticeable.

MASP-1, MASP-2 : The components of the MB-lectin pathway of complement activation include two serine proteases, MASP-1 and MASP-2, that bind to mannan-binding lectin and play the same role in cleaving C4 as do Cl r and Cl s in the classical pathway.

MASS : Massage; massive.

Mass flow (nutrient) : Movement of solutes associated with net movement of water.

Mass spectrometer : A instrument that can measure the mass of atoms.

Mast cells, Mastocytosis : A bone marrow-derived cell present in a variety of tissues that resembles peripheral blood-borne basophils and contains an

Fc receptor for IgE. It undergoes IgE-mediated degranulation. ***OR*** "tissuebased basophils;" contain histamine; function in allergic reactions, inflammation, other immune responses. ***OR*** Mast cells are large cells found in connective tissues throughout the body, most abundantly in the submucosal tissues and the dermis. They contain large granules that store a variety of mediator molecules including the vasoactive amine histamine. Mast cells have high-affinity Fcå receptors (fcåri) that allow them to bind IgE monomers. Antigenbinding to this IgE triggers mast-cell degranulation and mast-cell activation, producing a local or systemic immediate hypersensitivity reaction. Mast cells have a crucial role in allergic reactions. Mastocytosis indicates an overproduction of mast cells.

MAT : Maternal.

Matric potential : Portion of the total soil water potential due to the attractive forces between water and soil solids as represented through adsorption and capillarity.

Matrix Protein : A structural protein of a virus particle which underlies the envelope and links it to the core.

Maturation : The stage of viral replication at which a virus particle becomes infectious.

Mature B cells : Mature B cells are B cells that have acquired surface IgM and IgD and have become able to respond to antigen.

MAX : Maxilla; maximal; maximum.

Mb : Myoglobin.

MB : Muscle balance.

MBA : Masters of Business Administration – an advanced university degree in business studies.

MBC : Maximal breathing capacity.

MBH : Maximum benefit from hospitalization.

MBL, MB-lectin pathway : See mannan-binding lectin.

MCB : Master cell bank.

MCD : Magnetic circular dichroism.

MCP : Macrophage chemotatic protein ***OR*** monocyte chemotactic protein.

MCRT : Mean Cell Retention Time - days. An expression of the average time that a microorganism will spend in an activated sludge process.

MCS : Multiple cloning site.

M-CSF : Macrophage-colony stimulating factor.

MD : Molecular dynamics.

MDA : Microdroplet array or multiple drop array.

MDC : Minimum detectable concentration.

MDV : Multiple dose vial.

ME : Medical examiner.

Mean growth rate constant (k) : The rate of microbial population growth expressed in terms of the number of generations per unit time.

Measles : A highly contagious skin disease that is endemic throughout the world. It is caused by a morbilli virus in the family Paramyxoviridae, which enters the body through the respiratory tract or through the conjunctiva.

Mechanical aeration : The use of machinery to mix air and water so that oxygen can be absorbed into the water. Some examples are paddle wheels, mixers, rotating brushes to agitate the surface of an aeration tank; pumps to create fountains; and pumps to discharge water down a series of steps forming falls or cascades.

Mechano receptors : Sensory neurons stimulated by physical means such as pressure, motion, touch,or stretching'.

MED : Minimum effective dose; medial.

Media : The material in the trickling filter on which slime accumulates and organisms grow. As settled wastewater trickles over the media, organisms in the slime remove certain types of wastes thereby partially treating the wastewater. Also the material in a rotating biological contactor (RBC) or in a gravity or pressure filter.

Medical mycology : The discipline that deals with the fungi that cause human disease.

Medium : Any liquid or solid material prepared for the growth, maintenance, or storage of microorganisms.

MEDLARS : Medical Literature Analysis and Retreival System.

MEDS : Medications, medicines.

Medulla : The medulla is generally the central or collecting point of an organ. The thymic medulla is the central area of each thymic lobe, rich in bone marrow-derived antigen-presenting cells and the cells of a distinctive medullary epithelium. The medulla of the lymph node is a site of macrophage and plasma cell concentration through which the lymph flows on its way to the efferent lymphatics.

Medulla : The farther, smaller bulge between the pons and the top of the spinal cord.

MEF : Maximal expiratory flow.

MEFR : Maximal expiratory flow rate.

MEH : Medical eye history.

Meiosis : A special cell division resulting in sex cells with half as many chromosomes. ***OR*** The sexual process in which a diploid cell divides and forms two haploid cells. ***OR*** In eukaryotes, reduction division, the process by which the change from diploid to haploid occurs. ***OR*** Process in which diploid cells undergo division to form haploid sex cells.

MEK : Methyl ethyl ketone.

MEL : Melanoma.

Melancholia : Black bile: one of the four humors of which people used to think the body was composed and which they thought was produced by the spleen.

Melanistic : Occurrence of an individual with a black phenotype in a normally light-colored species.

Melanoma : A cancer of a particular type of skin cell, called a melanocyte. Melanocytes are responsible for skin colour.

Melting temperature (Tm) : The temperature at which double-stranded DNA separates into individual strands; it is dependent on the G + C content of the DNA and is used to compare genetic material in microbial taxonomy.

Membrane : A sheet-like composite of protein and lipid that is the boundary of cells and organelles.

Membrane attack complex (MAC) : The complex complement components (C5b-C9) that create a pore in the plasma membrane of a target cell and leads to cell lysis. C9 probably forms most of the actual pore.

Membrane cofactor of proteolysis : Membrane cofactor of proteolysis (MOP or CD46) is a host-cell membrane protein that acts in conjunction with factor I to cleave C3b to its inactive derivative iC3b and thus prevent convertase formation.

Membrane filter technique : The use of a thin porous filter made from cellulose acetate or some other polymer to collect microorganisms from water, air, and food.

Membrane immunoglobulin : B cells carry on their surfaces many molecules of membrane immunoglobulin (mig) of a single specificity, which acts as the receptor for antigen.

Membrane protein : A protein that is associated with a membrane, rather than found free in the cell. A membrane protein may be integral (embedded or buried) in the membrane, or peripheral (attached more loosely, by interactions with either lipid or intergral membrane proteins).

Membrane transport : The facilitated transport of a molecule across a membrane.

Membrane-attack complex : The membrane-attack complex is made up of the terminal complement components, which assemble to generate a membrane-spanning hydrophilic pore, damaging the membrane.

Membrane-disrupting exotoxin : A type of exotoxin that lyses host cells by disrupting the integrity of the plasma membrane.

Membranous glomerulonephritis : Membranous glomerulonephritis is a disease of the kidneys characterized by proteinuria and heavy deposits of antibody and complement.

Memnoniella (Hyphomycetes) : Mainly isolated from soils and dead plant material in tropical countries but has also been isolated from indoor sources such as paper, wallpaper, and textiles. Exposure to this genus should be avoided as they can produce griseofulvins, a potentially toxic metabolite. Emerging research has proposed that *Memnoniella* species actually belong to *Stachybotrys*.

Memory B cell : A lymphocyte capable of initiating the antibody-mediated immune response upon detection of a specific antigen molecule for which it is genetically programmed. It circulates freely in the blood and lymph and may live for years.

Memory cells, memory B cells, memory T cells : Memory cells are the lymphocytes that mediate immunological memory. They are more sensitive to antigen than are naive lymphocytes and respond rapidly on reexposure to the antigen that originally induced them. Both memory B cells and memory T cells have been defined.

MEMS : Micro electro mechanical systems.

Meninges : Three membranes covering the brain and spinal cord; Dura mater (outermost), pia mater (middle), and arachnoid (innermost).

Meninges : The three membranes covering the brain and spinal cord.

Meningitis : A condition that refers to inflammation of the brain or spinal cord meninges (membranes). The disease can be divided into bacterial (septic) meningitis (caused by bacteria) and aseptic meningitis syndrome (caused by nonbacterial sources). ***OR*** inflammation of the meninges.

Meningitis : Inflammation (infection) of the meninges.

Meniscus : The curved surface of a liquid(such as water) in a glass container,or one of two crescent-shaped pieces of cartilage in our knee joint.

Menstrual cycle : Cyclical sexual receptiveness in a female of some species of mammals, followed by shedding of the unused endometrium.

Mercaptans : Compounds containing sulfur which have an extremely offensive skunk-like odor. Also sometimes described as smelling like garlic or onions.

Merino : A breed of sheep that produces very fine wool.

Merodiploid : An organism that is diploid for some but not all of its genes.

Meromictic : Lakes do not exhibit mixing of layers and the layers are usually renamed to Monolimnion (similar to the hypolimnion) and Mixolimnion (similar to the epilimnion) with the dividing zone being the chemolimnion containing the chemocine.

Meront : See schizont.

Merosporangium (pl. merosporangia) : A small cylindrical sporangium with the sporangiospores aligned in a row.

Merozoite : The invasive stage that is the end result of schizogony.

Mesofauna : Soil animals between 200 to 1000 micrometers in length, including nematodes, oligochaete worms, smaller insect larvae, and small arthropods.

Mesophile : Liking moderate temperatures; optimal growth at 20 to 45 ºC. ***OR*** Organism whose optimum temperature for growth falls in an intermediate range of approximately 15 to 40 ºC. ***OR*** A microorganism with a growth optimum around 20 to 45°C, a minimum of 15 to 20°C, and a maximum about 45°C or lower.

Mesophilic bacteria : A group of bacteria that grow and thrive in a moderate temperature range between 68 F (20 ºC) and 113 F (45 ºC).

Mesophyll : The middle layer of a leaf.

Mesopore : Soil pores ranging in diameter from 0.2 to 50 (Micron).

Mesopotamia : The area between the Tigris and Euphrates Rivers, also known as Babylon, Persia, and now Iran/Iraq.

Mesosome : An invagination of the bacterial cell membrane.

Mesozoic era : A geological period around 180 million years.

Messenger RNA : RNA molecule transcribed from DNA, which contains the information to direct the synthesis of a particular protein. ***OR*** Single-stranded RNA synthesized from a DNA template during transcription that binds to ribosomes and directs the synthesis of protein. ***OR*** molecule transcribed from DNA, which contains the information to direct the synthesis of a particular protein.

Met : Methionine.

Metabolic channeling : The localization of metabolites and enzymes in different parts of a cell.

Metabolic control engineering : Modification of the controls for biosynthetic pathways without altering the pathways themselves in order to improve process efficiency.

Metabolic pathway : A series of reactions in a cell, where the product from one reaction is used as the reactant for the next reaction, and so on.

Metabolic pathway engineering (MPE) : The use of molecular techniques to improve the efficiency of pathways that synthesize specific products.

Metabolic turnover : A measure of the rate at which already existing molecules of the given species are replaced by newly-synthesized molecules of the same type. Usually isotopic labeling is required to measure turnover.

Metabolism : All biochemical reactions in a cell, both anabolic and catabolic. ***OR*** All biochemical reactions in a cell, both anabolic and catabolic. ***OR*** Sum total of all the chemical and physical processes within a living organism. ***OR*** The total of all chemical reactions in the cell; almost all are enzyme catalyzed.

Metabolite : A cellular chemical produced as part of metabolism.

Metabolized : Broken down or otherwise transformed by the body.

Metabolome : All the natural metabolites, or small molecules, expressed by an individual cell or organism at one particular point in time. Does not include DNA, RNA, enzymes, or structural molecules.

Metabolomics : The study of the metabolome (all the small molecules produced in a cell as a result of metabolism).

Metacarpels : The bones in the palm of the hand.

Metacercaria : Stage of trematode life cycle that the cercaria after invading the second intermediate host or attaching itself to vegetation, develop into. When ingested by the definitive host, the metacercaria will develop to the adult stage.

Metacestode : Mature tapeworm larvae.

Metachromatic granules : Granules of polyphosphate in the cytoplasm of some bacteria that appear a different color when stained with a blue basic dye. They are storage reservoirs for phosphate. Sometimes called volutin granules.

Metalimnion : The middle layer of water in a lake or ocean where light does not penetrate and which has a thermocline (gets colder as it gets deeper).

Metamorphosis : A change of form, especially the conversion of a larval form to an adult form.

Metaphase : That stage in mitosis or meiosis when all of the chromosomes are lined up on the equator (i.e., an imaginary line that bisects the cell).

Metastasis : The transfer of a disease like cancer from one organ to another not directly connected with it.

Metatarsals : The bones in the arch of the foot,the bones that make up the sole of the foot.

Methane : CH_4. A gas that contributes to the greenhouse effect.

Methanogenesis : Biological production of methane.

Methanogenic bacterium (methanogen) : Methaneproducing prokaryote; member of the Archaea.

Methanogens : Strictly anaerobic archaeons that derive energy by converting CO_2, H_2, formate, acetate, and other compounds to either methane or methane and CO_2.

Methanotroph : Organism capable of oxidizing methane. ***OR*** A bacterium that uses reduced one-carbon compounds such as methane and methanol as its sole source of carbon and energy.

Metula (pl. Metulae) : A sterile cell below the phialides of some *Aspergillus* and *Penicillium* species.

MFD : Midforceps delivery; minimum fatal dose.

MFT : Muscle function test.

MFU : Medical follow up.

MFW : Multiple fragment wounds.

MG : Muscle group.

mg/l : Milligrams per liter = ppm (parts per million) - expresses a measure of the concentration by weight of a substance per unit volume.

MGD : Million gallons daily - refers to the flow through a waste treatment plant.

MH : Marital history; medical history; mental health.

MHB : Maximum hospital benefit.

MHC : Major histocompatability complex.

MHC class IB : The MHC class IB molecules encoded within the MHC are not highly polymorphic like the MHC class I and MHC class II molecules, and present a restricted set of antigens.

MHC class II compartment, MIIC : The MHC class II compartment is a site in the cell where MHC class II molecules accumulate, encounter HLA-DM, and bind antigenic peptides, before migrating to the surface of the cell.

MHC class II transactivator, CIITA : The protein that activates the transcription of MHC class II genes, the MHC class II transactivator (CIITA), is one of several defective genes in the disease bare lymphocyte syndrome, in which MHC class II molecules are lacking on all cells.

MHC congenic, MHC recombinant, MHC mutant : Various specialized strains of mice are used to explore the role of MHC polymorphism *in vivo*. These are called MHC congenic, meaning that the mice differ only at the MHC complex, MHC recombinant, meaning that the mice have a crossover within the MHC, or MHC mutant, meaning that they are mutant at one or more loci.

MHC haplotype : MHC genes are inherited in most cases as an MHC haplotype, the set of genes in a haploid genome inherited from one parent. Thus, if the parents are designated as ab and cd, then the offspring are most likely to be ac, ad, bc, or bd.

MHC molecules : MHC molecules is the general name given to the highly polymorphic glycoproteins encoded by MHC class I and MHC class II genes, which are involved in the presentation of peptide antigens to T cells. They are also known as histocompatibility antigens.

MHC, MHC class I, MHC class II : See major histocompatibility complex.

MHC:peptide tetramers : The development of MHC:peptide tetramers held together by fluorescent streptavidin, which has four binding sites for the biotin attached to the tail of the MHC molecule, has made it possible to stain specific T cells in any species.

MHC-restricted antigen recognition, MHC restriction : MHC-restricted antigen recognition, or MHC restriction, refers to the fact that a given T cell will recognize a peptide antigen only when it is bound to a particular MHC molecule. Normally, as T cells are stimulated only in the presence of self MHC molecules, antigen is recognized only as peptides bound to self MHC molecules.

MHR : Maximal heart rate.

MI : Mental illness; mitral insufficiency; myocardial infarction.

MIC : Minimal inhibitory concentration.

MIC molecules : MIC molecules are MHC class I-like molecules that are expressed in the gut under conditions of stress and are encoded within the class I region of the human MHC. They are not found in mice.

Micelle : An aggregate of lipids in which the polar head groups face outward and the hydrophobic tails face inward; no solvent is trapped in the center.

Michaelis constant (Km) : A kinetic constant for an enzyme reaction that equals the substrate concentration required for the enzyme to operate at half maximal velocity. ***OR*** The substrate concentration at which an enzyme-catalyzed reaction proceeds at one-half of the maximum velocity.

Michaelis-Menten equation Henri-Michaelis-Menten equation : An equation relating the reaction velocity to the substrate concentration of an enzyme.

Microaerophile (microaerophilic) : A microorganism that requires low levels of oxygen for growth, around 2 to 10%, but is damaged by normal atmospheric oxygen levels. ***OR*** Organism that require or prefer a low concentration of oxygen for growth. Sometimes indicates an organism that will carry out its metabolic activities under aerobic conditions but will grow much better under anaerobic conditions.

Microaggregate : Clustering of clay packets stabilized by organic matter and precipitated inorganic materials.

Microarray : A technique for determining which genes are actively being used in a cell by checking which mRNA molecules are present. ***OR*** Profiling of gene expression by measuring binding of RNA from growing cells to an array of function-specific oligonucleotides attached to an inert surface.

Microarthropod : Invertebrate with jointed legs and exoskeleton; includes insects and arachnids. They form part of the soil mesofauna.

Microbe, Microbes : Another word for a microorganism, which is an organism that is too small to be seen clearly with the naked eye. It includes bacteria, viruses and fungi. ***OR*** An organism so small you can see it only with a microscope; a minute living organism. ***OR*** Microscopic living organisms, including bacteria, viruses, and fungi.

Microbial : Anything of, or related to microorganisms.

Microbial biomass : Total mass of microorganism alive in a given volume or mass of soil.

Microbial count : The number of microorganisms which grow on an agar plate. The number is used to indicate whether a substance is contaminated.

Microbial dietary adjuvant : A substance added to the diet to stimulate specific microbial processes and populations.

Microbial ecology : The study of microorganisms in their natural environments, with a major emphasis on physical conditions, processes, and interactions that occur on the scale of individual microbial cells.

Microbial interactions : Complex relationships among species; Neutral, Antagonistic, or Synergistic.

Microbial loop : The mineralization of organic matter synthesized by photosynthetic phytoplankton through the activity of microorganisms such as bacteria and protozoa. This process "loops" minerals and carbon dioxide back for reuse by the primary producers and makes the organic matter unavailable to higher consumers.

Microbial mat : A firm structure of layered microorganisms with complementary physiological activities. ***OR*** A firm structure of layered microorganisms with complementary physiological activities that can develop on surfaces in aquatic environments.

Microbial population : Total number of living microorganisms in a given volume or mass of soil.

Microbial transformation : See bioconversion.

Microbiology : Study of microorganisms. ***OR*** The study of organisms that are usually too small to be seen with the naked eye. Special techniques are required to isolate and grow them. ***OR*** The science and study of microorganisms, including protozoans, algae, fungi, bacteria, and viruses.

Microbiology : The study of bacteria and other microscopic organisms.

Microbiota : Microscopic flora.

Microbivous : Feeding on microorganisms, including bacteria and fungi.

Microbivy : The use of microorganisms as a food source by organisms that can ingest or phagocytose them.

Microconidium (pl. microconidia) : The smaller of two different types of conidia produced by a fungus in the same manner.

Microcosm : A community or other unit that is representative of a larger unity.

Microenvironment : The immediate environment surrounding a microbial cell or other structure, such as a root. ***OR*** Immediate physical and chemical surroundings of a microorganism.

Microfauna : Protozoa, nematodes and arthropods generally < 200 micrometers long.

Microfilaments : Protein filaments, about 4 to 7 nm in diameter, that are present in the cytoplasmic matrix of eucaryotic cells and play a role in cell structure and motion.

Microflora : Bacteria (including actinomycetes), fungi, algae, and viruses.

Microhabitat : Clusters of microaggregates with associated water within which microbes function. May be composed of several microsites (e.g., aerobic and anaerobic).

Micrometer : Onemillionth of a meter, or 10^{-6} meter, the unit usually used for measuring microorganisms.

Micron : A unit of length. One millionth of a meter or one thousandth of a millimeter. One micron equals 0.00004 of an inch.

Micronematous : The smaller of the two nuclei in ciliate protozoa. Micronuclei are diploid and involved only in genetic recombination and the regeneration of macronuclei.OR Having a conidiophore that is not morphologically different from the vegetative hyphae.

Micronutrient : Nutrients such as zinc, manganese, and copper that are required in very small quantities for growth and reproduction. Also called trace elements. ***OR*** An element required by plants and bacteria, in proportionately smaller amounts, for survival and growth. Micronutrients include Iron (Fe), Managanese (MN), Zinc (Zn), Boron (B), and Molybdenum (Mo). ***OR*** Chemical element necessary for growth found in small amounts, usually < 100 mg kg^{-1} in a plant. These elements consist of B, Cl, Cu, Fe, Mn, Mo, and Zn.

Microganisms : A microbe. ***OR*** A living organism which is too small to be seen with the naked eye and can only be observed using a microscope. Includes bacteria and most protists. ***OR*** Microorganisms are microscopic organisms, unicellular except for some fungi, that include bacteria, yeasts and other fungi, and protozoa, all of which can cause human disease.

Micropore : Relatively small soil pore, generally found within structural aggregates and having a diameter < 30 micrometers.

Microsite : Small volume of soil where biological or chemical processes differ from those of the soil as a whole, such as an anaerobic microsite of a soil aggregate or the surface of decaying organic residues.

Microsporum sp. **:** Causes ringworm in humans.

Microtold cells : See M cells.

Microtubules : Small cylinders, about 25 nm in diameter, made of tubulin proteins and present in the cytoplasmic matrix and flagella of eucaryotic cells; they are involved in cell structure and movement.

Microtubules : Thin tubules, made from globular proteins, that serve multiple purposes in eukaryotic cells.

Microtubules : Part of the cytoskeleton consisting of hollow tubes made of globular proteins.

MICU : Mobile intensive care unit.

MID : Minimum infective dose.

Middlebinders : Anti-carbohydrate antibodies can bind either the ends or the middles of polysaccharide chains; the latter antibodies are called middlebinders.

Mid-macron : Used to describe sheep's wool with a thickness of 23-32 microns/ micrometers. This is coarser than merino wool, but finer and softer than many other sheep breeds, such as Romney's. Sheep breeds that are farmed in New Zealand for mid-micron wool include Corriedale, Polwarth, Halfbred and Quarterbred sheep.

MIF : Maximal inspiratory flow.

MIFR : Maximal inspiratory flow rate.

mIg : See membrane immunoglobulin.

MIIC : See MHC class II compartment.

MIK : Methyl isobutyl ketone.

Mildew : 1) Any fungus that attacks plants, paper, cloth and other objects, and forms a downy or powdery whitish growth; 2) The whitish growth formed by some fungi.

Miliary tuberculosis : An acute form of tuberculosis in which small tubercles are formed in a number of organs of the body because of dissemination of *M. tuberculosis* throughout the body by the bloodstream. Also known as reactivation tuberculosis.

Milk : Fluid contained inside a mushroom which often becomes sticky when released.

Miloginite : The name coined by a resident of Milwaukee, Wisconsin for "sewage sludge", so he could scam people into buying sludge instead of paying to dispose of it! When this ingredient is listed in a septic product, you can be assured that the product is fraudulent! You don't use sludge to get rid of sludge.

MIN : Minimal; minor.

Mineralization : The release of inorganic nutrients from organic matter during microbial growth and metabolism. ***OR*** Conversion of an element from an organic form to an inorganic state as a result of microbial decomposition. ***OR*** Metabolism of an organic substrate by bacteria to carbon dioxide and water.

Minimal inhibity concentration (MIC) : The lowest concentration of a drug that will prevent the growth of a particular microorganism.

Minimal lethal concentration (MLC) : The lowest concentration of a drug that will kill a particular microorganism.

Minor elements : See micronutrients.

Minor histocompatibility antigens, min H antigens : Minor histocompatibility antigens (minor H antigens) are peptides of polymorphic cellular proteins bound to MHC molecules that can lead to graft rejection when they are recognized by T cells.

Minor lymphocyte stimulaty (MIs) loci : See mis antigens.

Minus, negative, strand : The virus nucleic acid strand that is complementary in base sequence to the viral mRNA.

Miracidium : The pyriform, ciliated larva of a trematode that developed in and hatched from the egg. The miracidium will penetrate the snail and undergo development to the next stage in the lifecycle.

miRNA : Micro RNA.

Mis antigens, min lymphocyte stimulaty (MIs) loci : Mis antigens are non-MHC antigens that provoke strong primary mixed lymphocyte responses. They are encoded by minor lymphocyte stimulatory (mis) loci, which are endogenous mammary tumor viruses integrated in the mouse genome. MIs antigens are encoded in the 3' long terminal repeat of the integrated virus and act as superantigens. They stimulate a large number of T lymphocytes by binding to the Vb domain of all T-cell receptors bearing the Vb for which the superantigen is specific.

Mismatch repair : The replacement of a base in a heteroduplex structure by one that forms a Watson-Crick base pair.

Missense mutation : A single base substitution in DNA that changes a codon for one amino acid into a codon for another.

Missense mutation : A change in which a codon for one amino acid is replaced by a codon for another amino acid.

Mitochondrial : Cellular structure where energy is generated. Found in all eukaryotic cells.

Mitochondrion : Eukaryotic organelle responsible for processes of respiration and oxidative phosphorylation. ***OR*** The eucaryotic organelle that is the site of electron transport, oxidative phosphorylation, and pathways such as the Krebs cycle; it provides most of a nonphotosynthetic cell's energy under aerobic conditions. It is constructed of an outer membrane and an inner membrane, which contains the electron transport chain. ***OR*** The organelle that is the "engine" of cells where sugar is burned for fuel,the plural is "mitochondria".

Mitogen-activated protein kinases, MAP kinases : Mitogen-activated protein kinases (MAP kinases) are kinases that become phosphorylated and activated on cellular stimulation by a variety of ligands, and lead to new gene expression by phosphorylating key transcription factors.

Mitosis : The type of cell division that makes new body cells. ***OR*** Highly ordered process by which the nucleus divides in eukaryotes. ***OR*** The process whereby replicated chromosomes segregate equally toward opposite poles prior to cell division.

Mitosis : Replication & division of the chromosomes in a cell.

Mixed mesophytic fest : A forest area with a variety of hard wood trees.

Mixed acid fermentation : A type of fermentation carried out by members of the family Enterobacteriaceae in which ethanol and a complex mixture of organic acids are produced.

Mixed lymphocyte reaction : When lymphocytes from two unrelated individuals are cultured together, the T cells proliferate in response to the allogeneic MHC molecules on the cells of the other donor. This mixed lymphocyte reaction is used in testing for histocompatibility.

Mixotroph : Organism able to assimilate organic compounds as carbon sources while using inorganic compounds as electron donors. Compare with autotroph and heterotroph.

Mixotrophic : Refers to microorganisms that combine autotrophic and heterotrophic metabolic processes (they use inorganic electron sources and organic carbon sources).

MK : Marked.

MKV : Measles, killed vaccine.

mL : Milliliter

ML : Mixed Liquor - the combination of raw influent and returned activated sludge. (no, not mixed drinks for human consumption)

ML : Maximum to left; midline.

MLC : Minimum lethal concentration.

MLD : Minimum lethal dose.

MLSS : Mixed Liquor Suspended Solids - the volume of suspended solids (see SS) in the mixed liquor (see ML) of an aeration tank.

MLVSS : Mixed Liquor Volatile Suspended Solids - the volume of organic solids that can evaporate at relatively low temperatures (550 ºC) from the mixed liquor of an aeration tank. This volatile portion is used as a measure or indication of microorganisms present. Volatile substances can also be partially removed by air stripping.

mm : Millimeter.

mM : Millimolar (mmol/l).

MM : Malignant melanoma; mucous membrane; murmur, muscles.

MMF : Maximal midexpiratory flow.

MMFR : Maximal midexpiratory flow rate.

mmol : Millimole.

MMPI : Minnesota multiphasic personality inventory.

MMPII : Matrix metalloproteinase II complex.

MMST : Muscle strength.

MMTV : See mouse mammary tumor virus.

MNT : Molecular nanotechnology.

Mobile genetic element : A segment of the genome that can move as a unit from one location on the genome to another, without any requirement for sequence homology.

MOD : Medical officer of day; moderate.

Modified atmosphere packaging (MAP) : Addition of gases such as nitrogen and carbon dioxide to packaged foods in order to inhibit the growth of spoilage organisms.

Moet : Multiple ovulation and embryo transfer

Moisture content : See water content.

Moisture characteristic curve : See waterretention curve.

Moisture release curve : See waterretention curve.

mol : Mole.

Mold : Any of a large group of fungi that cause mold or moldiness and that exist as multicellular filamentous colonies; also the deposit or growth caused by such fungi. Molds typically do not produce macroscopic fruiting bodies. ***OR*** A filamentous fungus. ***OR*** 1) Any fungus that feeds on live, dead or decaying matter, and forms a slimy, cottony growth; 2) The slimy, cottony growth formed by some fungi.

Molecular biologist : Someone who studies living cells at the molecular level in order to understand the interactions between systems in cells, including the interrelationships between DNA, RNA, and protein synthesis.

Molecular biology : The study of cells at the molecular level in order to understand the interactions between components like DNA, RNA, and proteins. ***OR*** The study of the various molecules and chemical reactions that take place in organisms.

Molecular chaperones : Proteins that aid in the proper folding of unfolded polypeptides or partly denatured proteins and often also help transport proteins across membranes.

Molecular chronometers : Nucleic acid and protein sequences that gradually change over time in a random fashion and at a steady rate, and which therefore can be used to determine phylogenetic relationships.

Molecular epidemiology : The use of nucleotide sequence information to study the diversity and distribution of virus populations.

Molecular mimicry : It has been proposed that infectious agents could provoke autoimmunity by molecular mimicry, the induction of antibodies and T cells that react against the pathogen but also cross-react with self antigens.

Molecular seive chromatography : See Gel filtration chromatography.

Molecular weight : See Gram molecular weight.

Molecularity of a reaction : The number of molecules involved in a specific reaction step.

Molecule : Result of two or more atoms combining by chemical bonding. ***OR*** The smallest division of a compound that still retains or exhibits all the properties of the substance. ***OR*** A combination of atoms of a specific substance forming the smallest unit that can exist alone.

Molecule : Two or more atoms bonded together.

MOM : Milk of magnesia.

Monilia sp. **:** Reported to be allergenic. This fungus produces soft rot of tree fruits. Other members produce a red bread mold. It is infrequently involved in corneal eye infections.

Monocotyledones : The Class (in Division Spermatophyta; Subdivision Angiospermae) of plants which have one nutrient storage area in their seeds.

Monosaccharide : A simple sugar with one carbon backbone, usually consisting of three to six Carbons.

Monoclonal antibody (MAb) : Monoclonal antibodies are produced by a single clone of B lymphocytes. Monoclonal antibodies are usually produced by making hybrid antibody-forming cells from a fusion of nonsecreting myeloma cells with immune spleen cells. ***OR*** Antibody produced from a single clone of cells. This antibody has uniform structure and specificity. ***OR*** An antibody that recognises a specific epitope, or region, of a foreign substance.

Monoculture : Aquaculture in which one species is grown.

Monocyte : Monocytes are white blood cells with a beanshaped nucleus; they are precursors of macrophages. ***Or*** A mononuclear phagocytic leukocyte that circulates briefly in the bloodstream before migrating to the tissues where it becomes a macrophage.

Monocyte-macrophage system : The collection of fixed phagocytic cells (including macrophages, monocytes, and specialized endothelial cells) located in the liver, spleen, lymph nodes, and bone marrow. This system is an important component of the host's general nonspecific defense against pathogens.

Monodictys (Hyphomycetes) : Found on dead wood, stem, tree bark, damp linoleum, and paper. Isolated from soil and air.

Monoecious : Referring to a plant which has both sexes of flowers, both male and female,on the same individual plant (sometimes male and female parts are in the same,perfect, flower and sometimes there are separate, imperfect, male and female flowers).

Monogenic : Only a single gene is involved.

Monokaryon : Fungal hypha in which compartments contain one nucleus.

Monokine : A generic term for a cytokine produced by mononuclear phagocytes (macrophages or monocytes).

Monolayer : A single layer of oriented lipid molecules.

Monomer : One unit of a protein or other structure. ***OR*** A simple organic molecule,such as a monosaccharide or an aminoacid ,that Serves as a subunit from which more complex molecules (polymers) may be Made.

Monomphic : Some antibodies recognize all allelic forms of a polymorphic molecule such as an MHC class I protein; these antibodies are thus said to recognize a monomorphic epitope.

Mononuclear phagocyte system (monocytemacrophage system) : (Formerly known as reticuloendothelial system) collective system of strongly phagocytic cells including macrophages derived from monocytes; histiocytes; microglagia; mononuclear phagocytes (monocytes in blood; macrophages in tissues) and polymorphonuclear leukocytes (e.g., neutrophils, eosinophils)

Monospecificity : An individual lymphocyte carries antigen receptors of a single antigen specificity and thus has the property of monospecificity in response to antigen.

Monotrichous : Having a single flagellum.

Morbidity : Illness.

Morbidity rate (m-bid_i-te) : Measures the number of individuals who become ill as a result of a particular disease within a susceptible population during a specific time period.

Mordant : A substance that helps fix dye on or in a cell.

Morphology : The study of form and structure of organisms (see also phenotype).

Morphometric : Characters are those to do with depth, dimension, geology of shores, sediment distribution, currents, inflow and outflow of water, etc.

Mortality : Death. ***OR*** Death rate or probability of dying at any given age.

Mortality rate : The ratio of the number of deaths from a given disease to the total number of cases of the disease.

Mortar and Pestle : A heavy bowl and thick rod used to grind substances to a powder.

Mosaic : An organism which displays different characteristics in different cells. ***OR*** A pattern of light green and dark green, or green and yellow, patches in a leaf that may be due to virus infection.

Most probable number (MPN) : The statistical estimation of the probable population in a liquid by diluting and determining end points for microbial growth. ***OR*** Method for estimating microbial numbers in soil based on extinction dilutions.

Motile : Motile organisms exhibit or are capable of movement.

Motility : Movement of a cell under its own power.

Mould : A type of fungus that appears as a furry growth, especially in damp or decaying material.

Mouse memmary tumor virus, MMTV : Mouse memmary tumor virus (MMTV) is a retrovirus that encodes a viral superantigen; integrated copies of related viruses encode the endogenous superantigens known as mis antigens.

MP : Menstrual period.

MPC : Mesenchymal precursor cells.

MPC : Maximum permissible concentration.

MPD : Maximum permissible dose.

MPN index : Most probable number of coliform-group organisms per unit volume of sample water. Expressed as a density or population of organisms per 100 ml of sample water.

MR : Maximum to right; mental retardation.

MRI : Magnetic resonance imaging.

mRNA : Messenger RNA. ***OR*** Messenger RNA, translated on ribosomes to produce proteins.

MRSA : A type of antibiotic resistant bacteria, called methicillin-resistant *Staphylococcus* aureus. ***OR*** Abbreviation for either Multiple Antibiotic Resistant Staph. ***OR*** Methicillyn Antibiotic Resistant Staph.

MRU : Minimum recognition unit.

MS : Mass spectrometer or mass spectrometry. ***OR*** Medical student; minor surgery; multiple sclerosis.

MSC : Mesenchymal stem cell.

MSDS : Material Safety Data Sheet - a document that provides pertinent information and a profile of a particular hazardous substance or mixture. An MSDS is normally developed by the manufacturer or formulator of the hazardous substance or mixture. The MSDS is required to be made available to employees and operators whenever there is the likelihood of the hazardous substance or mixture being introduced into the workplace. Some manufacturers prepare MSDS for products that are NOT considered to be hazardous to show that the product or substance is NOT hazardous.

MSTI : Multiple soft tissue injuries.

MSV : Maize streak virus.

MT : Manual traction; maximal therapy; more than; muscles and tendons; muscle test.

MTD : Maximum tolerated dose.

mtDNA : Mitochondrial DNA.

MTR : Masses, tenderness, rebound.

MTRG : Masses, tenderness, rebound, guarding.

Mucigel : Gelatinous material at the surface of roots grown in normal nonsterile soil. It includes natural and modified plant mucilages, bacterial cells, and their metabolic products (e.g., capsules and slimes), and colloidal mineral and organic matter from the soil.

Mucilage : Gelatinous secretions and exudates produced by plant roots and many microorganisms.

Mucilaginous : Sticky, covered in slime.

Mucins : Mucins are highly glycosylated cell-surface proteins. Mucinlike molecules are bound by L-selectin in lymphocyte homing.

Mucociliary blanket : The layer of cilia and mucus that lines certain portions of the respiratory system; it traps microorganisms up to 10 mm in diameter and then transports them by ciliary action away from the lungs.

Mucoid : Sticky or slimy.

Mucor (Zygomycetes) : Often found in soil, plants, hay, stored seeds, and manure. They can be found indoors in house dust, HVAC system dust, and poorly maintained carpets. They are rapid growers and can contaminate many kinds of stored food, including fruits and vegetables. A few species have been recovered from well-documented cases of zygomycosis. In general, infections due to members of this genus are rare.

***Mucor sp.* :** Often found in soil, dead plant material, horse dung, fruits and fruit juice. It is also found in leather, meat, dairy products, animal hair and jute. A Zygomycetes fungus which may be allergenic (skin and bronchial tests). This organism & other Zygomycetes will grow rapidly on most fungal media. May cause mucorosis in immune compromised individuals. The sites of infection are the lung, nasal sinus, brain, eye and skin. Infection may have multiple sites.

Mucosal epithelia : All of the body's internal epithelial organs are lined with epithelium that is coated with mucus, and are therefore called mucosal epithelia. This system is the site of entry for virtually all antigens, and is protected by a unique set of lymphoid organs.

Mucosal-associated lymphoid tissue (MALT) : The defensive immune lymphoid tissue located in the intestinal mucosa. ***OR*** The mucosal-associated lymphoid tissue (MALT) comprises all lymphoid cells in epithelia and in the lamina propria lying below the body's mucosal surfaces. The main sites of mucosal-associated lymphoid tissues are the gut-associated lymphoid tissues (GALT), and the bronchial-associated lymphoid tissues (BALT).

Mulch : (i) Any material such as straw, sawdust, leaves, plastic film, and loose soil, that is spread upon the surface of the soil to protect the soil and plant roots from the effects of raindrops, soil crusting, freezing, or evaporation. (ii) To apply mulch to the soil surface.

Multicellular : Having a body consisting of many cell.

Multi-drug-resistant strains of tuberculosis (MDR-TB) : A multi-drug-resistant strain is defined as *Mycobacterium tuberculosis* resistant to isoniazid and rifampin, with or without resistance to other drugs.

Multiple myeloma : Multiple myeloma is a tumor of plasma cells, almost always first detected as multiple foci in bone marrow. Myeloma cells produce a monoclonal immunoglobulin, called a myeloma protein, that is detectable in the patient's plasma.

Multiple sclerosis : Multiple sclerosis is a neurological disease characterized by focal demyelination in the central nervous system, lymphocytic infiltration in the brain, and a chronic progressive course. It is caused by an autoimmune response to various antigens found in the myelin sheath. ***OR*** An autoimmune disease where a person's immune system attacks and gradually destroys myelin, the substance that forms a protective coat around nerve cells, which disrupts signaling in the brain and spinal cord.

Multiplication : The ability of a microorganism to reproduce and increase in numbers during an infection

Multipolar budding : Blastoconidia developing at different sites on the surface of a parent cell.

Multiseptate : Having several septa.

Mumps : An acute generalized disease that occurs primarily in school-age children and is caused by a paramyxovirus that is transmitted in saliva and respiratory droplets. The principal manifestation is swelling of the parotid salivary glands.

Municipal solid waste : Combined consumer and commercial waste generated within a defined geographic area

MUO : Myocardiopathy of unknown origin.

Murein : See peptidoglycan.

Muriform : A conidium with both longitudinal and transverse septa.

MUS : Muscle.

Mushroom : Large, sometimes edible, fruiting body produced by some fungi.

Must : The juices of fruits, including grapes, that can be fermented for the production of alcohol.

Mutagen : A chemical or physical agent that causes mutations. ***OR*** Substance that causes the mutation of genes. ***OR*** A chemical which causes a base change in DNA strand.

Mutagenesis : A process that leads to a change in the genetic material that is inherited in later generations.

Mutant : Organism, population, gene, or chromosome that differs from the corresponding wild type by one or more base pairs. ***OR*** An organism that carries an altered gene or change in its genome.

Mutarotation : The change in optical rotation of a sugar that is observed immediately after it is dissolved in aqueous solution, as the result of the slow approach of equilibrium of a pyranose or a furanose in its alpha and beta forms.

Mutation : A permanent, heritable change in the genetic material. ***OR*** Heritable change in the base sequence of the DNA of an organism. ***OR*** The genetically inheritable alteration of a gene or group of genes.

Mutualism : Interaction between organisms where both organisms benefit from the association. ***OR*** mutually beneficial symbiotic relationship between two species. ***OR*** Two species living together in a relationship in which both benefit from the association.

Mutualist : An organism associated with another in a relationship that is beneficial to both (and often obligatory).

mV : Millivolt.

MV : Mitral valve.

MVA : Motor vehicle accident.

MVP : Mitral valve prolapsed.

MVV : Maximum volume ventilation; maximum voluntary ventilation.

Myasthenia gravis **:** *Myasthenia gravis* is an autoimmune disease in which autoantibodies against the acetylcholine receptor on skeletal muscle cells cause a block in neuromuscular junctions, leading to progressive weakness and eventually death.

Mycelium : The mass of thread-like filaments making up the 'body' of a fungus. ***OR*** the whole body of a fungus. Colony of hyphae arising from one inoculation. ***OR*** The mass of hyphae making up the thallus of a fungus.

Mycobacterium **:** A genus of aerobic bacteria found in soil and water that are capable of biodegrading multiring compounds such as pahs.

Mycobiont : The fungal partner in a lichen.

Mycolic acids : Complex 60 to 90 carbon fatty acids with a hydroxyl on the b-carbon and an aliphatic chain on the a-carbon; found in the cell walls of mycobacteria.

Mycologist : A person specializing in mycology; a student of mycology.

Mycology : The science and study of fungi.

Mycophage : See mycovirus.

Mycophagous : Organisms that consume fungi, such as mycophagous nematodes.

Mycoplasma : Bacteria that are members of the class Mollicutes and order Mycoplasmatales; they lack cell walls and cannot synthesize peptidoglycan precursors; most require sterols for growth; they are the smallest organisms capable of independent reproduction. ***OR*** Group of bacteria without a cell walls that do not revert to walled forms. Phylogenetically related to clostridia.

Mycoplasmal pneumonia **:** A type of pneumonia caused by *Mycoplasma pneumoniae*. Spread involves airborne droplets and close contact.

Mycorrhiza : Literally "fungus root." The symbiotic association between specific fungi with the fine roots of higher plants.

Mycorrhiza : Literally "fungus root." The symbiotic association between specific fungi with the fine roots of higher plants. ***OR*** a structure formed between a fungus and individual root tips that enables exchange nutrients ***OR*** a mutualistic symbiosis between roots and a fungus.

Mycorrhizosphere : The region around a mycorrhizal fungus in which nutrients released from the fungus increase the microbial population and its activities. ***OR*** Unique microbial community that forms around a mycorrhiza.

Mycosis : Any disease caused by a fungus.

Mycotoxicology : The study of fungal toxins and their effects on various organisms.

Mycotoxin : Secondary metabolites produced by fungi, which are toxic to human and animals in small quantities. Production of a specific mycotoxin tends to be genus-, species-, or even strain-specific

Mycovirus : Virus that infects fungi.

Myeloblast : See myeloid precursor cells.

Myeloid dendritic cells : Dendritic cells can arise from myeloid cells, in which case they are called myeloid dendritic cells, or from lymphoid tissues, in which case they are called lymphoid dendritic cells. Functional differences exist between these two lineages.

Myeloid precurs cells : Give rise to mononuclear phagocyte system (monocytemacrophage system formerly known as reticuloendothelial system).

Myeloid progenits : Myeloid progenitors are cells in bone marrow that give rise to the granulocytes and macrophages of the immune system.

Myeloma cell : A tumor cell that is similar to the cell type found in bone marrow. Also, a malignant, neoplastic plasma cell that produces large quantities of antibodies and can be readily cultivated.

Myeloma proteins : Myeloma proteins are immunoglobulins secreted by myeloma tumors and are found in the patient's plasma.

Myelopoiesis : Myelopoiesis is the production of monocytes and polymorphonuclear leukocytes in bone marrow.

Myiasis : Infestation of organs and tissues of vertebrates by larval dipterans (true flies).

Myocardia lInfarction : Death of a portion of the heart muscle due to lack of blood supply due to a blocked artery.

Myofibril : A unit of thick and thin filaments in a muscle fiber.

Myoglobin : A special protein-type molecule used for oxygen storage in red muscle tissue.

Myopia : Being near sighted,not being able to see distant objects clearly.

Myosin : The main protein of the thick filaments in a muscle myofibril. It is composed of two coiled subunits (Mr about 220,000) that can aggregate to form a thick filament, which is globular at each end.

Myositis : Inflammation of a striated or voluntary muscle.

Myotonia : Increased muscle tension.

Myxamoeba : A free-living amoeboid cell that can aggregate with other myxamoeba to form a plasmodium or pseudoplasmodium. Found in cellular slime molds and the myxomycetes.

Myxobacteria : A group of gram-negative, aerobic soil bacteria characterized by gliding motility, a complex life cycle with the production of fruiting bodies, and the formation of myxospores.

Myxomycetes : Popularly called slime molds. These are not true fungi taxonomically. Some species are found in the soil, in decaying wood, or other organic matter, where they produce structures full of powdery resting spores.

Myxomycota : Division of Kingdom Protista which includes slime molds.

Myxospores : Special dormant spores formed by the myxobacteria.

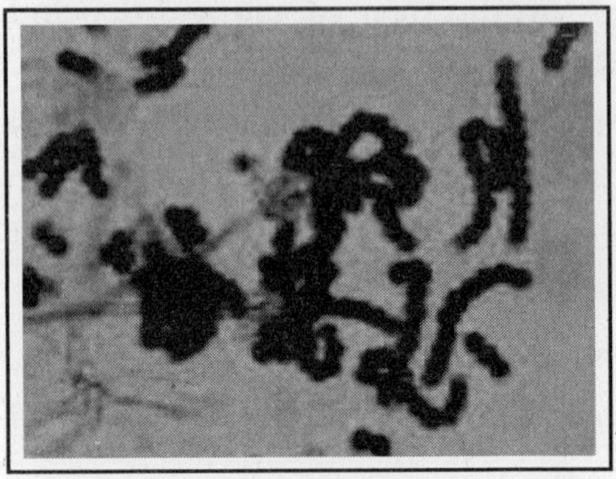

Memnoniella

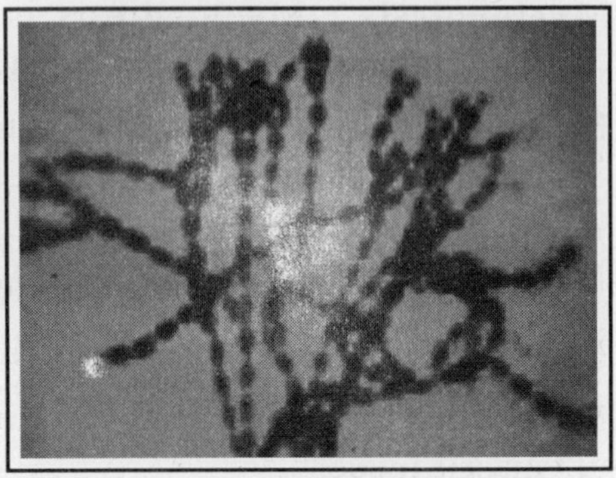

Monilia

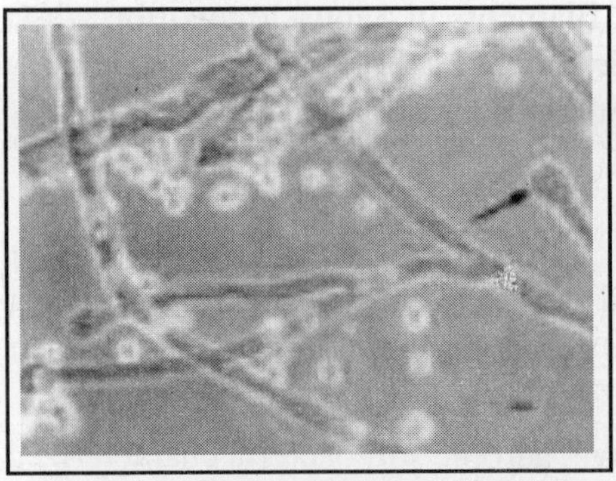

Mucor

Mushroom

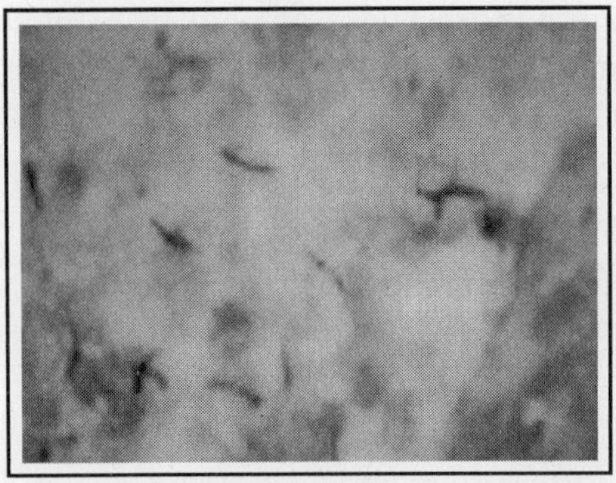

Mycobacteria

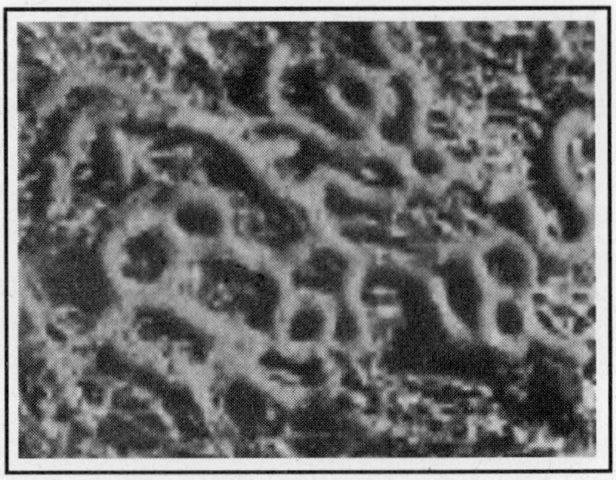

Myxomycetes

N

N : Avogadro's number.

N : Nano (10^{-9}m).

N&M : Nerves and muscles.

N&T : Nose and throat.

N&V : Nausea and vomiting.

NA : Nucleic acid.

NA : No abnormalities; not admitted; not applicable; not available.

NAA : No apparent abnormalities.

NAB : Neutralizing antibody.

N*Acetylglucosamine and *N*Acetylmuramic acid :** Acid sugar derivatives in the peptidoglycan layer of bacterial cell walls. ***OR Sugar derivatives in the peptidoglycan layer of bacterial cell walls.

NAD : Nicotinamide adenine dinucleotide

- NAD+ oxidized form.
- NADH reduced form.

NAD : No active disease; no acute distress; nothing abnormal detected.

NAD^+ : Nicotinamide adenine dinucleotide.

NADH : Reduced nicotinamide adenine dinucleotide.

NADP : Nicotinamide adenine dinucleotide phosphate.

- NADP+ oxidized form .
- NADPH reduced form.

$NADP^+$: Nicotinamide adenine dinucleotide phosphate.

NADPH : Reduced nicotinamide adenine dinucleotide phosphate.

NAG : N-acetylglucosamine.

NAI : No acute infection.

Naive lymphocytes : Naive lymphocytes are lymphocytes that have never encountered their specific antigen and thus have never responded to it, as distinct from memory or effector lymphocytes. All lymphocytes leaving the central lymphoid organs are naive lymphocytes, those from the thymus being naive T cells and those from bone marrow being naive B cells.

NAM : N-acetylmuramic acid.

NANA : N-acetylneuraminic acid.

Nanomaterial : A material used in nanotechnology.

Nanopore : Soil pore having dimensions measured in nanometers. Materials encased in nanopores are beyond the reach of microorganisms and enzymes.

Nanotechnology : Technology developed using molecular scale components; one nanometer (nm) is equal to one millionth of a millimeter (mm).

Napkin (diaper) candidiasis : Typically found in infants whose diapers are not changed frequently and are therefore not kept dry. Caused by *Candida* species of fungi.

NAPL : Nonaqueous phase liquid. This can be lighter than water (LNAPL), or more dense than water (DNAPL).

NAR : Narcotic; no action required.

Narrow-spectrum drugs : Chemotherapeutic agents that are effective only against a limited variety of microorganisms.

Nascent RNA : The initial transcripts of RNA, before any modification or processing.

Natality : Birth rate.

Native : Originating naturally in a particular region; indigenous.

Natural selection : The differential reproductive success of individuals with in a population. ***OR*** Ability of individuals to survive and leave the most reproducing offspring.

Natural antibodies : Undifferentiated serum globulins.

Natural attenuation : The decrease in the level of an enviromental contaminant that results from natural chemical, physical, and biological processes.

Natural classification : A classification system that arranges organisms into groups whose members share many characteristics and reflect as much as possible the biological nature of organisms.

Natural killer (NK) cell : A non-T, non-B lymphocyte present in nonimmunized individuals that exhibits MHC-independent cytolytic activity against tumor cells.

Natural killer cells, NK cells : Natural killer cells (NK cells) are large granular, non-T, non-B lymphocytes, which kill certain tumor cells. NK cells are important in innate immunity to viruses and other intracellular pathogens, as well as in antibody-dependent cell-mediated cytotoxicity (ADCC).

Naturally acquired active immunity : The type of active immunity that develops when an individual's immunologic system comes into contact with an appropriate antigenic stimulus during the course of normal activities; it usually arises as the result of recovering from an infection and lasts a long time.

Naturally acquired passive immunity : The type of temporary immunity that involves the transfer of antibodies from one individual to another.

Nausea : The sensation of being about to vomit. Vomiting, or emesis, is the expelling of undigested food through the mouth. Nausea may be caused by overeating, infection, or irritation of the throat or stomach lining.

NBI : No bone injury.

NBM : Nothing by mouth.

NBS : Normal bowel sounds.

NC : No change; no complaint; non-contributory.

NCD : Not considered disabling .

NCE : New chemical entity.

NCNS : No complications, no sequelae .

NCO : No complaints offered.

ncRNA : Noncoding RNA.

NCS : Nerve conduction studies.

ND : Nondisabling.

NDA : New drug application.

NE : Nerve ending; neurologic examination.

Nearctic realm : The biogeographical realm that includes North America and Greenland.

Necrosis : Damage of living tissues because of infection or injury. ***OR*** localized death of cells, tissue, or organ in response to disease or injury; sum of morphological changes indicative of cell death and caused by the progressive degradative action of cellular components, metabolic byproducts, enzymes and/or toxins. ***OR*** Necrosis is the death of cells or tissues due to chemical or physical injury, as opposed to apoptosis, which is a biologically programmed form of cell death. Necrosis leaves extensive cellular debris that needs to be removed by phagocytes, whereas apoptosis does not.

Necrotic : Dead tissue,usually brown or black or sometimes papery tan.

Necrotic ring : Dead tissue that is in a circular pattern with normal tissue in the middle.

Necrotic spots : Distinct dead spots, for example, brown or dark brown areas of dead cells in a leaf that are brown or dark brown.

Necrotic vascular system : Dead tissue in the vascular system of a plant, seen for example by slicing into a root or stem and looking for brown streaks.

Necrotizing : Causing necrosis.

Necrotizing fasciitis : A disease that results from a severe invasive group A streptococcus infection. Necrotizing fasciitis is an infection of the subcutaneous soft tissues, particularly of fibrous tissue, and is most common on the extremities. It begins with skin reddening, swelling, pain, and cellulitis, and proceeds to skin breakdown and gangrene after 3 to 5 days.

Necrotrophic : Nutritional mechanism by which an organism produces a battery of hydrolytic enzymes to kill and break down host cells and then absorb nutritional compounds from the dead organic matter.

Negative feedbackloop : A control mechanism which works in the opposite direction of the stimulus which triggered it.

Negative control : Repression of biological activity by the presence of a specific molecule.

Negative selection : During intrathymic development, thymocytes that recognize self are deleted from the repertoire, a process known as negative selection. Autoreactive B cells undergo a similar process in bone marrow.

Negative staining : A staining procedure in which a dye is used to make the background dark while the specimen is unstained.

Negri bodies : Masses of viruses or unassembled viral subunits found within the brain neurons of rabies-infected animals.

Nematode : Tiny worms; some are eaten by fungi. ***OR*** Multicellular eukaryote defined as an unsegmented, usually microscopic roundworm. Various species feed on plants, animals, fungi, and bacteria.

NEMS : Nano electro mechanical systems.

Neonate : A new born baby.

Neotropical realm : The biogeographical realm corresponding to South America.

Nephron : Repeating units which process water and wastes with in the kidney.

Neritic zone : The bottom mudin the oceans along the continental shelf.

Nernst equation : An equation that relates the redox potential to the standard redox potential and the concentrations of the oxidized and reduced form of the couple.

Neuritis : Inflammation of nerves or nerve cells (neurons).

Neurodegenerative disease : A disease which affects brain function. Results from the deterioration of neurons. Can be divided into two groups: Conditions causing problems with movements; and conditions affecting memory.

Neuromuscular : The nerve and muscle systems of an animal's body.

Neuron : Nerve cell.

Neurotoxin : A toxin that is poisonous to or destroys nerve tissue; especially the toxins secreted by C. Tetani, *Corynebacterium diphtheriae*, and *Shigella dysenteriae*.

Neurotoxin : Poison which interferes with nerve function, usually by affecting the flow of ions through the cell membrane.

Neustonic : The microorganisms that live at the atmospheric interface of a water body.

Neutralism : Lack of interaction between two organisms in the same habitat.

Neutralization : Blocking of virus infection by antibodies; also, an assay which measures this.

Neutralize : To react another chemical with an acid or a base,causing the pH to become closer to a pH of 7.

Neutralize, neutralizing antibodies, neutralization : Antibodies that can inhibit the infectivity of a virus or the toxicity of a toxin molecule are said to neutralize them. Such antibodies are known as neutralizing antibodies and the process of inactivation as neutralization.

Neutropenia : An abnormal reduction in the number of neutrophils in the blood. ***OR*** Neutropenia describes the situation in which there are fewer neutrophils in the blood than normal.

Neutrophil : A type of white blood cell that can engulf and destroy foreign organisms.

Neutrophil (polymorphonuclear leukocyte) (PMN) : A mature white blood cell in the granulocyte lineage formed in bone marrow. It has a nucleus with three to five lobes and is very phagocytic. ***OR*** granular leukocyte (granulocyte) with multilobed nucleus that is chemotactic, phagocytic, and adherent to immune complexes; primary phagocytic defense against bacterial cells (antibacterial and antifungal); main functions are rapid chemotaxis to and phagocytosis of invading microorganisms; appear first in response to inflammation and produce chemotaxins to attract more PMNs and other mononuclear phagocytes (monocytes; macrophages); present in circulation and can migrate into tissues (diapedesis) with a 23 day lifespan

Neutrophile : Microorganisms that grow best at a neutral pH range between pH 5.5 and 8.0.

Neutrophils, neutrophilic polymphonuclear leukocytes : Neutrophils, also known as neutrophilic polymorphonuclear leukocytes, are the major class of white blood cell in human peripheral blood. They have a multilobed nucleus and neutrophilic granules. Neutrophils are phagocytes and have an important role in engulfing and killing extracellular pathogens.

New organism : Any organism that is coming into New Zealand for the first time, or that has been made by genetic modification; this includes microorganisms, seeds, plants, fish and animals. New organisms in New Zealand are monitored by ERMA and regulated by the Hazardous Substances and New Organisms (HSNO) Act.

NFAT : See nuclear factor of activated T cells.

NFêB : The transcription factor called nfêb is made up of two chains of 50 kda and 65 kda. It is found under normal circumstances in the cytosol, where it is bound to a third chain called iêb, which is an inhibitor of nfêb transcription.

NH_3N : Ammonia nitrogen.

Niacin : Vitamin $B_{3.}$

Niche : The functional role of a species in its environment, including its activities, "job, "relationships, and location. ***OR*** Functional role of a given organism within its habitat. ***OR*** The function of an organism in a complex system, including place of the organism, the resources used in a given location, and the time of use.

Nicolaus Copernicus : Person who lived from 1473 to 1543 and put forth the theory that the planets all revolve around the Sun.

Nicotinamide adenine dinucleotide : An electron acceptor /donor that received electrons from glycolysis and the Krebs cycle and gives them to the electron transport chain.

Nicotinamide adenine dinucleotide (NAD^+) : Important coenzyme, functioning as a hydrogen and electron carrier in a wide range of redox reactions; the oxidized form of the coenzyme is written NAD+, the reduced form as NADH. ***OR*** An electron-carrying coenzyme; it is particularly important in catabolic processes and usually donates its electrons to the electron transport chain under aerobic conditions.

Nicotinamide adenine dinucleotide phosphate ($NADP^+$) : Important coenzyme, functioning as a hydrogen and electron carrier in a wide range of redox reactions; the oxidized form of the coenzyme is written NADP+, the reduced form as NADPH.

Nicotinamide adenine dinucleotide phosphate : An electron-carrying coenzyme that most often participates as an electron carrier in biosynthetic metabolism.

Niger : Black.

Nigrospora (Hyphomycetes) : Some species are common on plants, particularly in the tropics. Occasionally isolated from soil, air, and foodstuffs.

Nigrospora sp. **:** Reported to be allergenic.

NIH : National Institutes of Health.

NIL : Nothing.

NIP : No infection present; no inflammation present.

NIRS : Near-infrared spectroscopy.

Nitrate leaching : The movement of nitrate, a form of nitrogen, from the soil into groundwater. This is an environmental concern because the nitrate can then move from the groundwater into streams, making the water hazardous to drink (if the nitrate levels are very high), and providing good growth conditions for algae.

Nitrate reduction (biological) : Process whereby nitrate is reduced by plants and microorganisms to ammonium for cell synthesis (nitrate assimilation, assimilatory nitrate reduction) or to various lower oxidation states (N_2, N_2O, NO,) by bacteria using nitrate as the terminal electron acceptor in anaerobic respiration.

Nitrification : Biological oxidation of ammonium to nitrite and nitrate, or a biologically induced increase in the oxidation state of nitrogen. ***OR*** An aerobic process in which bacteria change the ammonia and organic nitrogen in wastewater into oxidized nitrogen (usually nitrate). The second-stage BOD is sometimes referred to as the "nitrification stage" (first-stage BOD is called the "carbonaceous stage"). ***OR*** The oxidation of ammonia to nitrate.

Nitrifying bacteria : Bacteria that change the ammonia and organic nitrogen in wastewater into oxidized nitrogen (usually nitrate). ***OR*** Chemolithotrophs capable of carrying out the transformations from NH_3 to NO_2 or NO_2 to NO_3.

Nitrogen cycle : The passage of nitrogen through various valence states, as the result of reactions carried out by a wide variety of different organisms. ***OR*** Sequence of biochemical changes wherein nitrogen is used by a living organism, transformed upon the death and decomposition of the organism, and converted ultimately to its original state of oxidation.

Nitrogen fixation : The metabolic process in which atmospheric molecular nitrogen is reduced to ammonia; carried out by Cyanobacteria, Rhizobium, and other nitrogen-fixing bacteria. ***OR*** Conversion of atmospheric nitrogen into organic nitrogen compounds available to green plants; a process that can be carried out only by certain strains of soil bacteria.

Nitrogen oxygen demand (NOD) : The demand for oxygen in sewage treatment, caused by nitrifying microorganisms.

Nitrogen saturation point : The point at which mineral nitrogen, when added to an ecosystem, can no longer be incorporated into organic matter through biological processes.

Nitrogenase : The enzyme that catalyzes biological nitrogen fixation. ***OR*** Specific enzyme system required for biological N2 fixation.

Nitrogenous base : An aromatic nitrogen-containing molecule with basic properties. Such bases include purines and pyrimidines.

NK : Not known.

NK cells : See natural killer cells.

NK1.1 CD4 T : NK1.1 CD4 T cells are a small subset of T cells that express the NK1.1 marker, a molecule normally found on NK cells. NK1.1 T cells also express a:b T-cell receptors of limited diversity and either the coreceptor molecule CD4 or no co-receptor. They are enriched in the liver, and produce cytokines shortly after stimulation.

NKA : No known allergies.

NKD : No known disease.

NL : Normal; normal limits.

NLM : National Library of Medicine.

NLP : No light perception; normal light perception.

nm : Nanometer.

NM : Neuromuscular; not measured; not mentioned.

NMA : Neurogenic muscular atrophy.

NMJ : Neuromuscular injection.

NMN : Nicotinamide mononudeotide.

NMR : Nuclear magnetic resonance.

nmRNA : Non-messenger RNA.

NN : Neonatal; nerves; nurse notes.

NNR : Neuronal nicotinic receptor.

NNR : Not necessary to return.

N-nucleotides : N-nucleotides are inserted into the junctions between gene segments of T-cell receptor and immunoglobulin heavy-chain V-region genes during gene segment joining. These N-regions are not encoded in either gene segment, but are inserted by the enzyme terminal deoxynucleotidyl transferase (TDT). They markedly increase the diversity of these receptors.

No mold detected : This result indicates that spores, hyphae, or any fungal structures were not observed from the sample.

NOB : Nitrite-oxidizing bacteria.

Nobel prize : A prize awarded every year to people who have outstanding achievements in the fields of physics, chemistry, medicine, literature, economics, or the promotion of peace.

Nocardioforms : Bacteria that resemble members of the genus *Nocardia*; they develop a substrate mycelium that readily breaks up into rods and coccoid elements (a quality sometimes called fugacity).

Nocturnal : Active during night time.

Nodular sclerosis : See Hodgkin's disease.

Nodule : See root nodule.

Nodule bacteria : See rhizobia.

Nodulins : Unique proteins produced in root hairs or nodules in response to rhizobial infection.

NOK : Next of kin.

Nomenclature : System of naming organisms. ***OR*** The branch of taxonomy concerned with the assignment of names to taxonomic groups in agreement with published rules.

Nonamer : Recombination signal sequences (RSS) flanking gene segments consist of a seven-nucleotide heptamer and a nine-nucleotide nonamer of conserved sequence, separated by 12 or 23 nucleotides. RSSs form the target for the site-specific recombinase that joins the gene segments in antigen receptor gene rearrangement.

Noncoding DNA : A piece of DNA that does not contain any genes, or sequences which code for either RNA (e.g., tRNA) or proteins.

Noncompetitive inhibit : An inhibitor of enzyme activity whose effect is not reversed by increasing the concentration of substrate molecule.

Noncyclic photophosphylation : The process in which light energy is used to make ATP when electrons are moved from water to NADP1 during photosynthesis; both photosystem I and photosystem II are involved.

Nongonococcal urethritis (NGU) : Any inflammation of the urethra not caused by *Neisseria gonorrhoeae*.

Nonpolar : Possessing hydrophobic (water repelling) characteristics and not easily dissolved in water.

Nonproductive rearrangements : When T- and B-cell receptor gene segments rearrange, they often form nonproductive rearrangements that cannot encode a protein because the coding sequences are in the wrong translational reading frame.

Nonsense codon : A codon that does not code for an amino acid but is a signal to terminate protein synthesis.

Nonsense mutation : A mutation that converts a sense codon to a nonsense or stop codon. ***OR*** A change in the base sequence that converts a sense codon (one that specifies an amino acid) to one that specifies a stop (a nonsense codon).

Nonseptate : Without septa.

Nonspecific immune response (innate natural immunity) : See nonspecific resistance.

Nonspecific resistance : Refers to those general defense mechanisms that are inherited as part of the innate structure and function of each animal; also known as nonspecific, innate or natural immunity.

Nontarget effect : Impacting organisms other than that intended by a treatment.

NOP : No ocular pain; not otherwise provided for.

Normal flora : (A.k.a., indigenous or resident flora or microbiota) those microbial flora typically occupying a particular niche; given diversity of environmental conditions, organisms tend to segregate; many normal flora perform important functions for the host, including digestive and nutritional functions and competition with pathogenic microorganisms (see also transient flora).

Normal microbiota (also indigenous microbial population, microflora, microbial flora) : The microorganisms normally associated with a particular tissue or structure.

Northern blot : Hybridization of singlestranded nucleic acid (DNA or RNA) to RNA fragments immobilized on a filter.

NOS : Not on staff; not otherwise specified.

Nosocomial : Hospital-acquired infections. Specific species are most well-known for Nosocomial transmission and rarely pass from human to human within their native environments. Since many people are taken to hospitals for physical trauma (car crash, surgery needed, etc.) ***OR*** disease (including cancer, viral or bacterial infections, malfunctioning organs needing transplant), their naturally-protective immune systems may be damaged by the injury. Disease or even by treatment (radiation, chemotherapy, drugs to cancel transplant resistance, etc.), improving the ability of any bacteria present to spread from patient to patient and among hospital staff, themselves. Nosocomial strains spread in hospital environments by developing incresed virulence factors, that enhance spread of infection. Whether acquired by mutatation or "horizontal gene transfer", every Nosocomial infection seeks to add antibiotic resistance genes that enable them to escape treatment with antibiotics that used to rapidly rescue hospital patients by killing them off. At this time, the most common Nosocomial bacteria belong to the following species *Serratia marcescans*, *Acinetobacter baumanii*, *Alcaligenes faecalis*, *Pseudomonas aeruginosa*, *Legionella pneumophila* and the one infective strain of *Escherichia coli*, strain 0157. Although rarely causing serious infections, most health authorities also monitor the import and transfer of *Stenotrophomonas maltophila*, because this species is known for having more genes for antibiotic-resistance than any other known species and is suspected of passing along these genes via horizontal gene transfer, a terrifying prospect. ***OR*** An infection that develops within a hospital (or other type of clinical care facility) and is produced by an infectious organism acquired during the stay of the patient. ***OR*** An infection acquired in the hospital that was not present in the host nor incubating prior to admission and generally occurring within 72 hours after admission

Novum organum **:** Book published by Francis Baconaround 1600.

NP : Nasopharynx; neuropathology; not palpable.

NPC : Neural progenitor cell.

NPC : No previous complaint.

NPDES permit : National Pollutant Discharge Elimination System permit is the regulatory agency document issued by either a federal or state agency which is designated to control all discharges of pollutants from point sources into U.S. waterways. NPDES permits regulate discharges into navigable waters from all point sources of pollution, including industries, municipal wastewater treatment plants, sanitary landfills, large agricultural feed lots and return irrigation flows.

NPH : No previous history.

NPO : Nothing by mouth.

NPO MN : Nothing by mouth after midnight.

NPT : Normal pressure and temperature.

NR : Nerve root; normal range; not remarkable.

N-regions : See N-nucleotides.

NRI : Nerve root involvement or irritation.

NRM : Normal range of motion.

NRN : No return necessary.

NROM : Normal range of motion.

NS : Nerves; neurosurgery; no show, nonspecific.

NSA : No significant abnormalities.

NSAD : No sign of significant disease.

NSAID : Non-steriodal anti-inflammatory drug.

NSC : Neural stem cell.

NSC : No significant change.

NSD : Normal spontaneous delivery; no significant defect.

NSF : No significant findings.

NSG : Nursing.

NSI : No sign of infection, inflammation.

NSR : Normal sinus rhythm.

NSS : Normal saline solution; normal size and shape.

nt : Nucleotide.

NT : Nuclear transfer.

NT : Naso-tracheal; nerve treatment; normal temperature; no test, not tested; not tender.

NTG : Nitroglycerine.

NTP : Normal temperature and pressure.

Nuclear envelope : The complex double-membrane structure forming the outer boundary of the eucaryotic nucleus. It is covered by pores through which substances enter and leave the nucleus.

Nuclear factor of activated T cells, NFAT : The transcription factor called nuclear factor of activated T cells (NFAT) is a complex of a protein called NFATC, as it is held in the cytosol by serine/threonine phosphorylation, and the Fos/Jun dimer known as AP-1. It moves from the cytosol to the nucleus on cleavage of the phosphate residues by calcineurin, a serine/threonine protein phosphatase.

Nuclear membrane : Membrane that encloses the genetic material of eukaryotic cells.

Nuclease : An enzyme that cleaves phosphodiester bonds of nucleic acids.

Nucleic acid : The process of forming a hybrid double-stranded DNA molecule using a heated mixture of single-stranded dnas from two different sources; if the sequences are fairly complementary, stable hybrids will form. ***OR*** An organic acid consisting of joined nuceleotide complexes; the principal tyes are deoxyribonucleic acid (DNA) and ribonucleic acid (RNA). ***OR*** Polymer of nucleotides.

Nucleocapsid : The core of a virus particle consisting of the genome plus a complex of proteins. ***OR*** The nucleic acid and its surrounding protein coat or capsid; the basic unit of virion structure.

Nucleohistone : A complex of DNA and histone.

Nucleoid : An irregularly shaped region in the procaryotic cell that contains its genetic material. ***OR*** Aggregated mass of DNA that makes up the chromosome of prokaryotic cells.

Nucleolus : The organelle, located within the eucaryotic nucleus and not bounded by a membrane, that is the location of ribosomal RNA synthesis and the assembly of ribosomal subunits. ***OR*** A spherical structure visible in the nucleus during interphase. The nucleolus is associated with a site on the chromosome that is involved in ribosomal RNA synthesis.

Nucleophilic compound : Chemical that attracks or is drawn to electrondeficient regions in other chemicals; reducing agents act as nucleophilic compounds.

Nucleophilic group : An electron-rich group that tends to attack an electron-deficient nucleus.

Nucleoside : Nucleotide without the phosphate group. ***OR*** An organic molecule containing a purine or pyrimidine base and a five-carbon sugar (ribose or deoxyribose). ***OR*** A combination of ribose or deoxyribose with a purine or pyrimidine base.

Nucleosome : A complex of histones and DNA found in eucaryotic chromatin; the DNA is wrapped around the surface of the beadlike histone complex. ***OR*** A complex of DNA and an octamer of histone proteins in which a small stretch of the duplex is wrapped around a molecular bead of histone.

Nucleotide : Monomeric unit of nucleic acid, consisting of a sugar (pentose), a phosphate, and a nitrogenous base. ***OR*** The basic structural unit of nucleic acids like DNA and RNA, including: adenine (A), cytosine (C), guanine (G), thymine (T), and uracil (U).

Nucleus : Membrane enclosed structure containing the genetic material (DNA) organized in chromosomes. ***OR*** A organelle in a eukaryotic cells that contains most of the cell's DNA.

Nulliparous : Condition of never having been pregnant.

Numerical aperture : The property of a microscope lens that determines how much light can enter and how great a resolution the lens can provide.

Numerical taxonomy : The grouping by numerical methods of taxonomic units into taxa based on their character states.

Nutraceutical : A pill or capsule that contains the essence of disease-fighting foods e.g. Broccoli extract in the form of a pill taken instead of eating the green vegetable.

Nutrient : Substance taken by a cell from its environment and used in catabolic or anabolic reactions. ***OR*** A substance that supports growth and reproduction.

Nutrient availabilty : The accessibility of a necessary resource, substance or compound providing nourishment to maintain life, i.e. Capable of conversion to energy and structural building blocks.

Nutrients : Substances which are required to support living plants and organisms. Major nutrients are carbon, hydrogen, oxygen, sulfur, nitrogen and phosphorus. Nitrogen and phosphorus are difficult to remove from wastewater by conventional treatment processes because they are water soluble and tend to recycle.

Nutrigenomics : An area of research investigating the interactions between genes and food. This is of interest because our different genotypes, or gene combinations, mean that we all respond differently to different foods. This can have impacts on our health and wellness. Some foods, for example, will be more beneficial (or harmful) for some people than for others.

NV : Neurovascular.

NVD : Nausea, vomiting, diarrhea.

NWB : Non weight bearing.

NYD : Not yet diagnosed.

Nystatin : A polyene antibiotic from *Streptomyces noursei* that is used in the treatment of Candida infections of the skin, vagina, and alimentary tract.

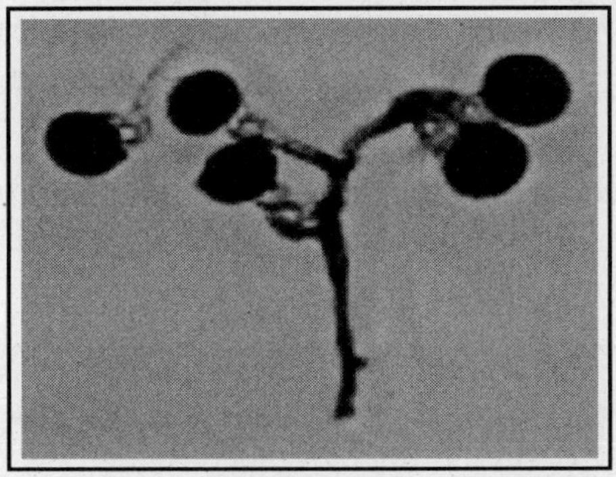

Nigrospora

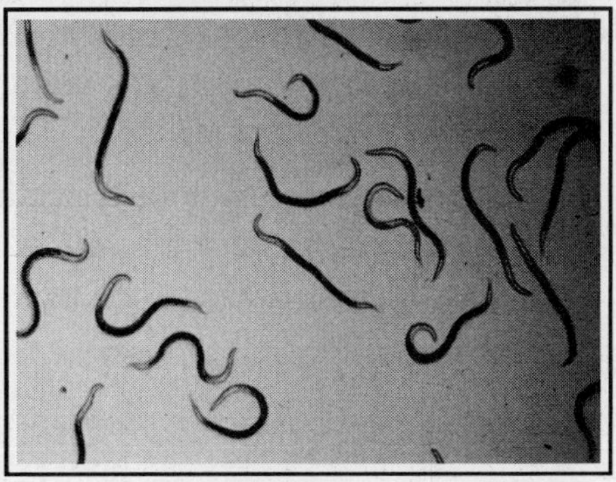

Nematode

O

O : Pint.

O antigen : A polysaccharide antigen extending from the outer membrane of some gram-negative bacterial cell walls; it is part of the lipopolysaccharide.

O&E : Observation and examination.

Oa : Open access.

OA : Osteoarthritis.

OAP : Osteoarthropathy.

OB : Obstetrics.

Obclavate : Club-shaped in reverse; the distal region is smaller.

Obesity : An abnormal accumulation of body fat, usually 20% or more over an individual's ideal body weight. Obesity is associated with increased risk of illness, disability, and death.

OBJ : Object; objective.

OBL : Oblique.

Obligate : (i) Adjective referring to an environmental factor (for example, oxygen) that is always required for growth. (ii) Organism that can grow and reproduce only by obtaining carbon and other nutrients from a living host, such as obligate symbiont.

Obligate aerobe : Bacteria which require the presense of oxygen, such as *Pseudomonas fluorescens*. A few strains of this species are capable of utilizing nitrate to allow anaerobic respiration. *OR* Organisms that grow only in the presence of oxygen..

Obligate anaerobes : Microorganisms that cannot tolerate the presence of oxygen and die when exposed to it.

OBP : Ova, blood and parasites.

Obpyriform : Pear-shaped in reverse; the distal region is larger.

OBS : Organic brain syndrome; observer; obstetrics.

OC : On call.

OCC : Occasionally; occiput; occupation.

Occult infection : An infection with adult worms but no corresponding diagnostic stage. For example in infections with *Dirofilaria immitis* (heartworm) some hosts may not have circulating microfilaria (this may be a result of immune-mediated clearance of the microfilaria, a single-sex infection, chemosterilization of the adult worms by drug treatment, or senescent adults). This would be known as an "occult heartworm infection".

Occupational allergy : When a person's work induces allergy, this is called occupational allergy.

OD : Occupational disease; officer of the day; once daily; overdose; right eye (oculus dexter).

Odmr : Optically detected magnetic resonance.

Odn : Oligo deoxynucleotide.

Odontopathogens : Dental pathogens.

OE : On examination; orthopedic examination; otitis externa.

Oec : Oxygen-evolving complex.

OECD : The Organisation for Economic Co-operation & Development or OECD provides information on trade, economics, education, development, and science. It has 30 member countries including New Zealand, Australia, USA, and Japan, see http://www.oecd.org for a full list of member countries.

OH : Occupational history.

Oil retention boom : A floating baffle used to contain and prevent the spread of floating oil on a water surface.

Okazaki fragment : Small fragments of DNA, which are joined together by ligases during DNA replication. ***OR*** A short segment of single-stranded DNA that is an intermediate in DNA synthesis. In bacteria, Okazaki fragments are 1000-2000 bases in length; in eukaryotes, 100-200 bases in length.

Okazaki fragments : Short stretches of polynucleotides produced during discontinuous DNA replication.

OLA : Oligonucleotide ligation assay.

Oleaceae : Plant family commonly known as the olive family ; the flowers have four petals,but unlike the Cruciferae, these have only two stamens.

Oligochaete : A particular class of segmented worms, including the earthworms, which have few (oligo) body bristles (chaeta).

Oligonucleotide : Short nucleic acid chain, either obtained from an organism or synthesized chemically. *OR* A polynucleotide containing a small number of nucleotides. The linkages are the same as in a polynucleotide; the only distinguishing feature is the small size.

Oligotroph : Microorganism specifically adapted to grow under low nutrient supply. Thought to subsist on the more resistant soil organic matter and be little affected by the addition of fresh organic materials. Sometimes a synonym for autochthonous.

Oligotrophic : Lakes have low nutrient status and relatively low primary productivity.

Oligotrophic environment : An environment containing low levels of nutrients, particularly nutrients that support microbial growth.

Olivaceous : Olive-grey colour.

OM : Occupational medicine; osteopathic manipulation; otitis media.

Omnivore : An animal that eats both plants and animals.

Omnivorous : Feeding on a variety of foods, such as plants, fungi, bacteria, animals and organic detritus.

On : Oligonucleotides.

ONC : Oncology.

Oncogene : A gene whose activity is associated with the conversion of normal cells to cancer cells. *OR* A gene of cellular or viral origin that is responsible for rapid, unruly growth of animal cells. A cancer-causing gene. *OR* Oncogenes are genes involved in regulating cell growth. When these genes are defective in structure or expression, they can cause cells to grow continuously to form a tumor.

Oncology : The study and treatment of cancer'.

Oncosphere : In a cestode life cycle, the oncosphere is the hexacanth (6-hooked) embryo found in the egg.

One-step growth experiment : An experiment used to study the reproduction of lytic phages in which one round of phage reproduction occurs and ends with the lysis of the host bacterial population.

Ontogeny : Developmental history of an individual organism; "Ontogeny recapitulates phylogeny"

Onychomycosis : A fungal infection of the nail plate producing nails that are opaque, white, thickened, friable, and brittle. Also called ringworm of the nails and *Tinea unguium*. caused by *Trichophyton* and other fungi such as *C. albicans*.

OOB : Out of bed.

Oocyst : Cyst formed around a zygote of malaria and related protozoa.

Oogonia : Mitotically dividing female structures that produce primary oocytes and gamete.

Oogonium : Specialized sexual structure formed as a female gametangium by funguslike organisms in the phylum Oomycota.

Oomycetes : A collective name for members of the division Oomycota; also known as the water molds.

OOP : Out of plaster.

Oospore : Thickwalled spore formed within an oogonium by funguslike organisms in the phylum Oomycota.

OP : Operation; out patient.

OPC : Outpatient clinic.

OPD : Outpatient department.

Open reading frame (ORF) : A reading frame sequence not interrupted by a stop codon; it is usually determined by nucleic acid sequencing studies.

Operat : The segment of DNA to which the repressor protein binds; it controls the expression of the genes adjacent to it.

Operon : A group of contiguous genes that are coordinately regulated by two cis-acting elements, a promoter and an operator. Found only in prokaryotic cells. ***OR*** The sequence of bases in DNA that contains one or more structural genes together with the operator controlling their expression.OR Cluster of genes whose expression is controlled by a single operator; typical in prokaryotic cells.

Ophthalmia neonatum : A gonorrheal eye infection in a newborn, which may lead to blindness. Also called conjunctivitis of the newborn.

Opiate : A drug derived from the *Opium poppy*, for example morphine, opium, and heroin.

OPP : Opposite.

Opportunist (opportunistic pathogen) : An organism that is pathogenic only under certain conditions, but nonpathogenic under normal conditions; a usually harmless microorganism that becomes pathogenic under favorable conditions; often a member of the normal microbial flora (see also Pathogen).

Opportunistic infection : Infection caused by a normally harmless microorganism (opportunist) when certain predisposing conditions (disease or conditions that increase host susceptibility) exist.

Opportunistic micro organismm pathogen : A microorganism that is usually free-living or a part of the host's normal microbiota, but which may become pathogenic under certain circumstances, such as when the immune system is compromised. *OR* An opportunistic pathogen is a microorganism that causes disease only in individuals with compromised host defense mechanisms, as occurs in AIDS.

Opsonization : Opsonization is the alteration of the surface of a pathogen or other particle so that it can be ingested by phagocytes. Antibody and complement opsonize extracellular bacteria for destruction by neutrophils and macrophages. *OR* The action of opsonins in making bacteria and other cells more readily phagocytosed. Antibodies, complement (especially C3b), and fibronectin are potent opsonins.

OPT : Outpatient treatment.

Optical activity : The property of a molecule that leads to rotation of the plane of polarization of plane-polarized light when the latter is transmitted through the substance. Chirality is a necessary and sufficient property for optical activity.

Optical tweezer : The use of a focused laser beam to drag and isolate a specific microorganism from a complex microbial mixture.

Optic chiasma : The place where the optic nerves cross to the other side of the brain.

Optimum : The best or maximum that can be achieved.

OPV : Oral polio vaccine.

Oral candidiasis : See thrush.

Oral tolerance : The feeding of foreign antigens leads typically to a state of specific and active unresponsiveness, a phenomenon known as oral tolerance.

Orchitis : Inflammation of the testes.

Order : Taxonomic classification below class and above family.

ORF : Open reading frame.

Organ necrosis : See necrosis.

Organelle : A subcellular membrane-bounded body with a well-defined function. ***OR*** Membrane enclosed body specialized for carrying out certain functions; found only in eukaryotic cells. ***OR*** A structure within or on a cell that performs specific functions and is related to the cell in a way similar to that of an organ to the body. ***OR*** The"body parts"with in a cell.

Organic : Products which haven't been exposed to chemicals, like fertilisers or pesticides.

Organic chemistry : A branch of chemistry specialising in the structure, properties, and reactions of molecules made mostly from carbon and hydrogen atoms.

Organic matter : All of the degradable organics. Living material containing carbon compounds. Used as food by microorganisms.

Organic molecule : A molecule that consists mainly of carbon and hydrogen atoms, and sometimes with oxygen and nitrogen. Includes proteins, carbohydrates, lipids, and nucleic acids.

Organic nitrogen : The nitrogen combined in organic molecules such as proteins, amines, and amino acids.

Organic soil : Soil that contains a high percentage (>200 g kg^{-1}, if saturated with water) of organic carbon.

Organic waste : Waste material which comes mainly from animal or plant sources. Organic waste generally can be consumed by bacteria and other small organisms. Inorganic wastes are chemical substances of mineral origin.

Organism : Any form of animal or plant life. ***OR*** Any living thing including plants, animals or bacteria.

Organotroph : Organism that obtains reducing equivalents (stored electrons) and carbon from organic substrates.

Organ-specific : An autoimmune disease that targets a specific organ is said to be organ-specific.

Orientalrealm : The biogeographical realm consisting of India and South East Asia.

ORIF : Open reduction internal fixation.

Original antigenic sin : Original antigenic sin describes the tendency of humans to make antibody responses to those epitopes shared between the original strain of a virus and subsequent related viruses, while ignoring other highly immunogenic epitopes on the second and subsequent viruses.

Ornithosis : See psittacosis.

ORP : Oxidation reduction potential - the degree of completion of a chemical reaction by detecting the ratio of ions in the reduced form to those in the oxidized form as a variation in electrical potential measured by an ORP electrode assembly.

ORT : Operating room technician.

OS : Left eye (oculus sinister); mouth; oral surgery.

OSHA : The Williams-Steiger Occupational Safety and Health Act of 1970 (OSHA) is a law designed to protect the health and safety of industrial workers and treatment plant operators. It regulates the design, construction, operation and maintenance of industrial plants and wastewater treatment plants. The Act does not apply directly to municipalities, except in those states that have approved plans and have asserted jurisdiction under Section 18 of the OSHA Act. Wastewater treatment plants have come under stricter regulation in all phases of activity as a result of OSHA standards, which also refers to the federal and state agencies which administer OSHA.

Osmophilic micro organisms : Microorganisms that grow best in or on media of high solute concentration.

Osmosis : The movement of water from a region of high concentration to a region of low concentration through a semi-permeable membrane. *OR* Diffusion of water through a membrane from a region of low solute concentration to one of higher concentration.

Osmotic potential : Portion of total soil water potential due to the presence of solutes in soil water.

Osmotic pressure (p) : Expressed in units of atmospheres; reflects the concentration of solute in an aqueous solution. *OR* The pressure generated by the mass flow of water to that side of a membrane-bounded structure that contains the higher concentration of solute molecules. A stable osmotic pressure is seen in systems in which the membrane is not permeable to some of the solute molecules.

Osmotolerent : Organisms that grow over a wide range of salt concentration.

Osteocyte : A bone cell.

Ostiole : An opening or pore in an ascocarp or a pycnidium.

OT : Objective test; occupational therapy.

OTC : Over the counter.

Ouchterlony technique : See double diffusion agar assay.

Outbreak : The sudden, unexpected occurrence of a disease in a given population.

Outer membrane : A special membrane located outside the peptidoglycan layer in the cell walls of Gram-negative bacteria.

Ovary : "Female" plant structure in which female gametophytes are produced and seeds develop or female animal structure in which eggs form.

Ovendry soil : Soil that has been dried at 105°C until it reaches constant mass.

Oviparous : Laying eggs in which the embryos have developed little or not at all. Seen, for example, in the nematodes *Toxocara canis* and *Ancylostoma caninum*.

Oviparous : Laying eggs that continue to develop,and eventually hatch.

Ovoid : Egg-shaped.

Ovoviviparous : Condition in which the organism develops in an egg and hatches within the adult female. Thus a live motile organism emerges from the adult female. Seen, for example, in the nematode *Dirofilaria immitis* and in some sharks and snakes.

Ovulation : The release of gametes in female animals is called ovulation.

Oxaloacetate, Oxaloacetic acid : A 4-carbon chemical used to store CO_2 prior to the Calvin cycle in C4 plants

($HOOCCH_2C{=}OCOOH$).

Oxic : An environment with oxygen. ***OR*** Containing oxygen; aerobic. Usually used in reference to a microbial habitat.

Oxidant : A chemical substance that causes the oxidation (loss of electrons) of another chemical substance.

Oxidase : Oxidoreductase that catalyzes oxidation of substrate with O_2.

Oxidation : Combining elemental compounds with oxygen to form a new compound. A part of the metabolic reaction. *OR* Reaction that causes a molecule to loose electrons (Reacting with oxygen is a common way for this to happen). *OR* Process by which a compound gives up electrons, acting as an electron donor, and becomes oxidized.

Oxidation state : Number of electrons to be added (or subtracted) from an atom in a combined state to convert it to the elemental form.

Oxidation : The loss of electrons from a compound.

Oxidation reduction (redox) reaction : Coupled pair of reactions, in which one compound becomes oxidized, while another becomes reduced and takes up the electrons released in the oxidation reaction. *OR* Reactions involving electron transfers; the reductant donates electrons to an oxidant.

Oxidative damage : Damage to cells caused by reactive oxygen species (ROS); also known as oxidative stress.

Oxidative phosphylation : The synthesis of ATP from ADP using energy made available during electron transport. *OR* Synthesis of ATP involving a membraneassociated electrontransport chain and the creation of a protonmotive force. Also called electrontransport chain phosphorylation. *OR* The formation of ATP as the result of the transfer of electrons to oxygen.

Oxidizing agent oxidant : The electron acceptor in an oxidation-reduction reaction.

Oxidizing bacteria : Any substance such as oxygen (O_2) and chlorine (Cl_2), that can aaccept electrons. When oxygen or chlorine is added to wastewater, organic substances are oxidized. These oxidized organic substances are more stable and less likely to give off odors or to contain disease bacteria.

Oxido-reductase : An enzyme that catalyzes oxidation-reduction reactions.

Oxygen : Element with eight protons.

Oxygenic photosynthesis : Photosynthesis that oxidizes water to form oxygen; the form of photosynthesis characteristic of eucaryotic algae and cyanobacteria.

Oxygenic photosynthesis : Use of light energy to synthesize ATP and NADPH by noncyclic photophos-phorylation with the production of oxygen from water. ***OR*** Use of light energy to synthesize ATP and NADPH by noncyclic photophosphorylation with the production of oxygen from water during carbon dioxide fixation.

Ozonation : The application of ozone to water, wastewater, or air, generally for the purposes of disinfection or odor control.

P

P : Phosphate.

p : Pico (10^{-12}).

P&S : Permanent and stationary.

PO : Per oral; by mouth (see also IV; IP;).

PA : Postero-anterior; pulmonary artery.

PAC : P1-derived artificial chromosome.

Pacemaker enzyme : The enzyme in a metabolic pathway that catalyzes the slowest or rate-limiting reaction; if its rate changes, the pathway's activity changes.

Paecilomyces (Hyphomycetes) : Teleomorph: Byssochlamys (Ascomycetes) A common saprobe found on dead plants and compost. Some species are insect parasites while others cause food spoilage. It is rarely a human pathogen but can cause infection in animals. However, some species, such as *P. variotii, P. marquandii* and *P. lilacinus* are emerging as causative agents of disease in immunocompromised individuals.

***Paecilomyces sp.* :** Commonly found in soil and dust, less frequently in air *P. variotii* can cause paecilomycosis. Linked to wood-trimmers disease and humidifier associated illnesses. They are reported to allergenic. Some members of this genus are reported to cause pneumonia. It may produce arsine gas if growing on arsenic substrate. This can occur on wallpapers covered with paris green.

Paedogenesis : A process in the molluscan intermediate host which involves the production of new individuals by a single larval form in which one trematode egg may eventually develop into hundreds of adults.

PAGE : Polyacrylamide gel electrophoresis.

PAH : Polycyclic Aromatic Hydrocarbons. (rarely, but sometimes used as the abbreviation for polyaluminum hydroxide).

Pakeha : A person of predominantly European descent.

Palearctic realm : The biogeographical realm including Europe and northern Asia.

Paleontology : The study of the remains of formerly-living organisms such as fossils and Dinosaurs.

Paleozoic era : An early geological period 360-340 million years.

Palindrome : A sequence of bases that reads the same in both directions on opposite strands of the DNA duplex (e.g., GAATTC).

Pallid : Off white.

PAM : Percent accepted mutation.

Pandemic : An epidemic that has spread over a large geographic area, throughout an entire country, continent or the whole world.OR a widespread epidemic distributed or occuring widely throughout a region, country, continent, or globally (see also epidemic). ***OR*** An increase in the occurrence of a disease within a large and geographically widespread population (often refers to a worldwide epidemic).

Paneth cell : The granular cell located at the base of glands in the small intestine; it produces the enzyme lysozyme.

Pangea : The name of the super continent which existed 180 million years and contained all current continents.

Pangene : One of the minute body parts thought to have come together to form the Homunculus.

Panning : Lymphocyte subpopulations can be isolated by panning on petri dishes coated with monoclonal antibodies against cell-surface markers, to which the lymphocytes bind.

Pannus : A superficial vascularization of the cornea with infiltration of granulation tissue.

Panthalassa : The one, large ocean that surrounded Pangea.

Pantothenic acid : A B vitamin which is a component of co enzyme A and is needed to convert choline to acetylcholine.

Panzootic : The wide dissemination of a disease in an animal population.

Papaveraceae : Plant family commonly known as the poppy family.

Papilionaceous : Term used to describe some thing which is shaped like a butterfly such as the flowers in family Leguminosae.

Pappus : A modified calyx,for example,the "fuzz" in a dandelion.

Papulospora sp. **:** This fungi is found in soil, textiles, decaying plants, manure, and paper.

Par : Photosynthetically active radiation.

Paractical area, paractex : The paracortical area, or paracortex, is the T-cell area of lymph nodes, lying just below the follicular cortex, which is primarily composed of B cells.

Paralytic shellfish poisoning : Dinoflagellates (*Gonyaulax* spp.) produce a powerful neurotoxin called saxitoxin. Shellfish accumulate saxitoxin and are poisonous when consumed by animals and humans. Saxitoxin paralyzes the striated respiratory muscles by inhibiting sodium transport. Paralytic shellfish poisoning is characterized by numbness of the mouth, lips, face, and extremities.

Paramecium **:** A one-celled organism which is covered with many cilia (paramec=oval;-ium=small).

Parapatric species : Species occupying adjacent areas which meet at a border (para=beside, near; patri=fatherland, habitat).

Paraphysis : A sterile hair like or club-shaped cell found in amongs the asci of Ascomycetes.

Parasexual cycle : Nuclear cycle in which genes of haploid nuclei recombine without meiosis.

Parasite : An heterotrophic organism that needs to obtain its organic nutrients from a living organism. ***OR*** Organisms that live temporarily or permanently on or within other living organisms (plant or animal hosts) for the purpose of obtaining food. ***OR*** an organism benefiting from living in or on another organism. ***OR*** An organism that requires a living host to survive. It lives in or on the host and derives nutrients or other substances from the host. It is generally not able to live saprophytically, and it may often cause extensive damage to the host.

Parasitic phase of life cycle : The development and maturation to adult males and females occurring in the definitive host of a parasite life cycle.

Parasitism : When a smaller organism feeds on a larger, to the detriment of its host. ***OR*** A two species association in which one species, the parasite, lives on or in a second species, the host, for a significant period of its life and obtains nourishment from it. The parasite may or may not cause disease in the host. ***OR*** Feeding by one organism on the cells of a second organism, which is usually larger than the first. The parasite is, to some extent, dependent on the host at whose expense it is maintained.

Parasitoid : A parasite that eats its host's tissues, eventually killing it.

Parasitology : The study of host-parasite relationships. Traditionally this area of study has focused on parasites belonging to the protozoa, helminths and arthropods.

Parasympathetic NS : The functional part of the NS which decreases energy expenditure and stores energy.

Paratenic host (transport host) : A type of intermediate host in which immature helminths may survive for indefinite periods but do not undergo development.

Further parasitic development depends upon infection of the definitive host, which is usually by predation on the paratenic host. Completion of a life cycle may happen with or without a paratenic host, but the presence of a paratenic host may serve to make infection of the definitive host by the parasite more efficient.

Parenteral route : A route of drug administration that is nonoral (e.g., by injection).

Parfocal : A microscope that retains proper focus when the objectives are changed.

Parisitism : One organism living on or in another to obtain nourishment, without providing any benefit to the host organism.

Parkinson's disease : A disease that affects the brain's ability to control muscles, which can cause muscle stiffness, slowness, difficulty starting movements, and tremors.

Paronychia : Inflammation involving the folds of tissue surrounding the nail; usually caused by *Candida albicans*.

Paroxysmal nocturnal hemoglobinuria, PNH : Paroxysmal nocturnal hemoglobinuria (PNH) is a disease in which complement regulatory proteins are defective, so that complement activation leads to episodes of spontaneous hemolysis.

Parthenogenesis : The ability of an unfertilized egg to develop and hatch.

Parthenogenesis : A form of reproduction in which an unfertilized egg develops in to a new individual.

Partial agonist peptides : See altered peptide ligands.

Partial veil : A membrane of sterile tissue stretching between the edge of the cap and the stem to cover the gills in some agarics during development.

Particle density : Density of the soil particles, the dry mass of the particles being divided by the solid (not bulk) volume of the particles, in contrast with bulk density.

Particle size : Effective diameter of a particle measured by sedimentation, sieving or micrometric methods.

Particulate : Free suspended solids.

Passive diffusion : The process in which molecules move from a region of higher concentration to one of lower concentration as a result of random thermal agitation.

Passive hemagglutination : Passive hemagglutination is a technique for detecting antibody in which red blood cells are coated with antigen and the antibody is detected by agglutination of the coated red blood cells.

Passive immunization : The induction of temporary immunity by the transfer of immune products, such as antibodies or sensitized T cells, from an immune vertebrate to a nonimmune one. ***OR*** The injection of or immune serum into a naive recipient is called passive immunization. Cf. Active immunization.

Pasteur effect : The decrease in the rate of sugar catabolism and change to aerobic respiration that occurs when microorganisms are switched from anaerobic to aerobic conditions.

Pasteurization : Process using mild heat to reduce microbial numbers in heatsensitive materials. ***OR*** The process of heating milk and other liquids to destroy microorganisms that can cause spoilage or disease.

PAT : Process analytical technology.

Patella : The knee cap.

Patent : A set of rights given by a federal government to an inventor to protect an invention and prevent others from making, using, or selling it, or claiming it as theirs.

Pathogen : Any virus, bacterium, or other agent that causes disease. ***OR*** Organism able to inflict damage on a host it infects. ***OR*** An infecting agent (microorganism) capable of causing disease

- Opportunistic pathogen a usually harmless microorganism that becomes pathogenic under favorable conditions
- True pathogen any microorganism capable of causing disease; an infecting agent

Pathogen : An organism that can cause a disease on another living organism. ***OR*** A disease-causing organism.

Pathogenic : Harmful; capable of causing disease. ***OR*** Pathogenic micro-organisms, or pathogens, are microorganisms that can cause disease when they infect a host.

Pathogenic potential : The degree that a pathogen causes morbid signs and symptoms.

Pathogenicity : Ability of a parasite to inflict damage on the host. ***OR*** The quality of producing or the ability to produce pathologic changes or disease. ***OR*** The condition or quality of being pathogenic, or the ability to cause disease.

Pathogenicity island : A large segment of DNA in some pathogens that contains the genes responsible for virulence; often it codes for the type III secretion system that allows the pathogen to secrete virulence proteins and damage host cells. A pathogen may have more than one pathogenicity island.

Pathogens : Organisms that cause disease.

Pathogen suppressive soil : Soil where a pathogen does not establish or persist, a pathogen establishes but causes little or no damage, or a pathogen causes disease for a while, but the disease becomes less important even though the pathogen persists in soil.

Pathology : Pathology is the scientific study of disease. The term pathology is also used to describe detectable damage to tissues.

Pathophysiology : The physical characteristics, symptoms, associated with a particular disease.

Pathway architecture : The analysis, design, and modification of biochemical pathways to increase process efficiency.

Pattern recognition receptors, PRRs, pathogen-associated molecular patterns, PAMPs : The term pattern recognition receptors (PRRs) is used to define receptors that bind to pathogen-associated molecular patterns (PAMPs).

Pattern-recognition molecules : Pattern-recognition molecules are receptors of the innate immune system that recognize common molecular patterns on pathogen surfaces.

Payer's patches : Payer's patches are aggregates of lymphocytes along the small intestine, especially the ileum.

PBG : Porphobilinogen.

PC : Phosphatidylcholine. ***OR*** Plastocyanin. ***OR*** Phone call; posterior cervical; present complaint.

PCB : Polychlorinated biphenyls.

PCD : Programmed cell death (apoptosis).

PCN : Penicillin.

PCR : Polymerase chain reaction. A method for amplifying DNA sequences.

PCS : Photon correlation spectroscopy.

PCV : Packed cell volume.

PD : Parkinson's disease. ***OR*** Partial denture; per diem; pulmonary disease.

PDB : Protein data bank.

PDE : Phosphodiesterase.

PDGF : Platelet-derived growth factor.

PDV : Plant-derived vaccine.

PE : Phosphatidylethanolamine. ***OR*** Physical examination; point of entry.

Peat : Unconsolidated soil material consisting largely of undecomposed, or only slightly decomposed, organic matter accumulated under conditions of excessive moisture.

Pecam, CD31 : The cell-adhesion molecule PECAM (CD31) is found both on lymphocytes and at endothelial cell junctions. It is believed that CD31-CD31 interactions enable leukocytes to leave blood vessels and enter tissues.

Pectin : Important component of the plant cell walls containing chains of galacturonic acid that is often esterified with a methyl group.

Pectinate : Like the teeth of a comb.

Pectalis major : The muscles on the upper chest.

Ped : A natural soil aggregate, formed partly through bacterial and fungal growth in the soil. ***OR*** A unit of soil structure such as a block, column, granule, plate, or prism, formed by natural processes.

Pedicel : Stalk usually used to describe the short stem on puff ball spores. ***OR*** A slender stalk. ***OR*** The region of a plant stem below the receptacle(below a flower).

PEG : Pulsed-field gel electrophoresis. ***OR*** Polyethelene glycol.

Pelagic zone : The open water in the ocean.

Pellicle : A film-like or skin-like surface growth. ***OR*** Relatively rigid layer of proteinaceous elements just beneath the cell membrane in many protozoa and algae.

Pelvic inflammaty disease (PID) : A severe infection of the female reproductive organs. The disease that results when gonococci and chlamydiae infect the uterine tubes and surrounding tissue.

Penetration : The stage of viral replication at which the virus genome enters the cell.

Penicillins : A group of antibiotics containing a b-lactam ring, which are active against Gram-positive bacteria.

Penicillium (Hyphomycetes) : Many species are common contaminants on a variety of substrates. May be found indoors in air samples, carpet dust, or on wallpaper. Some species are able to produce mycotoxins, as summarized below. Human pathogenic species are rare, only limited to *P. marneffei*, which causes disease in immunocompromised individuals. Some species are used for commercial production, such as *P. chrysogenum* for the antibiotic penicillin, *P. griseofulvum* for the antibiotic griseofulvin, and *P. roquefortii* for blue cheese.

Penicillium sp. **:** Aw 0.78 - 0.88. A wide number of organisms have placed in this genera. Identification to species is difficult. Often found in aerosol samples. Commonly found in soil, food, cellulose and grains. It is also found in paint and compost piles. It may cause hypersensitivity pneumonitis, allergic alveolitis in susceptible individuals. It is reported to be allergenic (skin). It is commonly found in carpet, wall paper, and in interior fiberglass duct insulation. Some species can produce mycotoxins. Common cause of extrinsic asthma (immeadiate-type hypersensitivity: type I). Acute symptoms include edema and bronchiospasms, chronic cases may develop pulmonary emphysema.

Penicillus (pl. penicilli) : The brush-like conidiophore of Penicillium.

Pentadecacatechol : Pentadecacatechol is the chemical substance in the leaves of the poison ivy plant that causes the cell-mediated immunity associated with hypersensitivity to poison ivy.

Penton pentamer : A capsomer composed of five protomers.

Pentose phosphate pathway : The pathway that oxidizes glucose 6-phosphate to ribulose 5-phosphate and then converts it to a variety of three to seven carbon sugars; it forms several important products (NADPH for biosynthesis, pentoses, and other sugars) and also can be used to degrade glucose to CO_2. ***OR*** The pathway involving the oxidation of glucose-6-phosphate to pentose phosphates and further reactions of pentose phosphates.

Pentose : A sugar with five carbon atoms.

Pentraxins : Pentraxins are a family of acute-phase proteins formed of five identical subunits, to which C-reactive protein and serum amyloid protein belong.

PEP : Phosphoenolpyruvate.

PEPCK : PEP carboxykinase.

Peplomer spike : A protein or protein complex that extends from the virus envelope and often is important in virion attachment to the host cell surface.

Pepsin : An enzyme in the stomach which digests protein in our diets.

Pepsinogen : The precursor molecule,secreted by the stomach lining,from which pepsin is made.

Peptic ulcer disease : A gastritis caused by *Helicobacter pylori*.

Peptide : Two or more amino acids linked together to form a chain. ***OR*** An organic molecule in which a covalent amide bond is formed between the ?-amino group of one amino acid and the ?-carboxyl group of another amino acid, with the elimination of a water molecule. The resulting connection is called a peptide bond.

Peptide interbridge : A short peptide chain that connects the tetrapeptide chains in some peptidoglycans.

Peptide mapping : Same as fingerprinting.

Peptide bond : A bond, formed by dehydration synthesis, between the amino group of one aminoacid and the carboxylic acid group of another aminoacid.

Peptidoglycan : Rigid layer of cell walls of bacteria, a thin sheet composed of Nacetylglucosamine, Nacetylmuramic acid, and a few amino acids. Also called murein. ***OR*** A large polymer composed of long chains of alternating N-acetylglucosamine and N-acetylmuramic acid residues. The polysaccharide chains are linked to each other through connections between tetrapeptide chains attached to the N-acetylmuramic acids. It provides much of the strength and rigidity possessed by bacterial cell walls. ***OR*** The main component of the bacterial cell wall, consisting of a two-dimensional network of heteropolysaccharides running in one direction, cross-linked with polypeptides running in the perpendicular direction. ***OR*** Carbohydrate polymer cross-linked by proteins. It is found in the cell wall of Gram positive bacteria, where it stains with the dye crystal-violet.

Peptidomics : The study of the peptidome: All the peptides synthesised in a particular cell or organism at a particular point in time.

Peptidyl donor site (P site) : The site on the ribosome that contains the peptidyl-tRNA at the beginning of the elongation cycle during protein synthesis.

Peptidyl transferase : The enzyme that catalyzes the transpeptidation reaction in protein synthesis; in this reaction, an amino acid is added to the growing peptide chain.

Peptones : Water-soluble digests or hydrolysates of proteins that are used in the preparation of culture media.

PER OS : By mouth.

Percolation : The movement or flow of water through soil or rocks.

Percurrent : Conidiogenous cell growth where a new axis grows through the previous apex.

Perfect flower : A term referring to a flower which has both male and female flower parts.

Perfate : With holes.

Perfin : Perforin is a protein that can polymerize to form the membrane pores that are an important part of the killing mechanism in cell-mediated cytotoxicity. Perforin is produced by cytotoxic T cells and NK cells and is stored in granules that are released by the cell when it contacts a specific target cell.

Perfin pathway : The cytotoxic pathway that uses perforin protein, which polymerizes to form membrane pores that help destroy cells during cell-mediated cytotoxicity. Perforin is produced by cytotoxic T cells and NK cells and stored in granules that are released when a target cell is contacted.

Periarteriolar lymphoid sheath, PALS : The periarteriolar lymphoid sheath (PALS) is part of the inner region of the white pulp of the spleen, and contains mainly T cells.

Peribacteroid membrane : Plantderived membrane surrounding one to several rhizobia within host cells of legume nodules.

Periconia (Hyphomycetes) : A widespread fungus commonly found on various substrates, including stalks of grasses, herbaceous stems, dead leaves, or leaf spots. The spores of *Periconia* species are often indistinguishable from the spores of smut fungi like *Ustilago* species, when collected on air cassettes. Both genera can have spores that are brown, verruculose, or echinulate, ranging from 10-16 microns in diameter.

Peridinin : Carotenoid pigment found in dinoflagellates.

Peridioles : Largish bodies that contain the spore enclosed within the peridium in the birds nest fungi (Nidulariales). They are dispersed as a unit.

Peridium : The outer wall of an ascocarp. ***OR*** The outer wall of a fungus, especially puffballs.

Perigynous : A plant with an hypanthium surrounding the ovary.

Period of infectivity : Refers to the time during which the source of an infectious disease is infectious or is disseminating the pathogen.

Periodontal disease : A disease located around the teeth or in the periodontium-the tissue investing and supporting the teeth, including the cementum, periodontal ligament, alveolar bone, and gingiva.

Periodontitis : An inflammation of the periodontium.

Periodontium : See periodontal disease.

Periodontosis : A degenerative, noninflammatory condition of the periodontium, which is characterized by destruction of tissue.

Periparturient rise in fecal egg counts (PPR) : An increase in the number of parasite eggs in the feces of animals around parturition. This can be pronounced in ewes, sows, and goats.

Peripheral blood manonuclear cells : Peripheral blood manonuclear cells are lymphocytes and monocytes isolated from peripheral blood, usually by Ficoll-Hypaque™ density centrifugation.

Peripheral tolerance : Peripheral tolerance is tolerance acquired by mature lymphocytes in the peripheral tissues, as opposed to central tolerance, which is acquired by immature lymphocytes during their development.

Peripheral, secondary lymphoid organs : Peripheral or secondary lymphoid organs are the lymph nodes, spleen, and mucosal-associated lymphoid tissues, in which immune responses are induced, as opposed to the central lymphoid organs, in which lymphocytes develop.

Peripheral NS : The distant part of the NS,comprised of the sensory and motor neurons.

Periplasm : The substance that fills the periplasmic space. ***OR*** The region between the inner (cytoplasmic) membrane and the cell wall or outer membrane of a bacterium.

Periplasmic flagella : The flagella that lie under the outer sheath and extend from both ends of the spirochete cell to overlap in the middle and form the axial filament. Also called axial fibrils and endoflagella.

Periplasmic space : Area between the cell membrane and the cell wall in Gramnegative bacteria, containing certain enzymes involved in nutrition.

Periplasmic space periplasm : The space between the plasma membrane and the outer membrane in Gram-negative bacteria, and between the plasma membrane and the cell wall in Gram-positive bacteria.

Peristalsis : The wavelike muscular contraction in the intestines which pushes food along during the process of digestion. ***OR*** Wavelike contractions that move food through the digestive tract. ***OR*** Waves of muscle contractions which push food through the digestive system.

Peristaltic pump : A type of positive displacement pump.

Perithecium : An enclosed ascocarp characterized an apical ostiole and by asci arranged in a basal tuft or hymenium layer. ***OR*** Flaskshaped ascocarp open at the tip; containing asci of fungi in the phylum Ascomycota. ***OR*** the flask-shaped fruiting bodies of certain Ascomycetes. They contain asci. ***OR*** membrane lining the abdominopelvic wall.

Peritonitis : An inflammation of the peritoneum which is the membrane which lines the inside of the abdomen and all of the internal organs. It is most often due to infection by bacteria, but may also be due to some kind of a chemical irritant such as spillage of acid from the stomach, bile from the gall bladder and biliary tract, or enzymes from the pancreas. ***OR*** inflammation of the peritoneum.

Peritrichous : A cell with flagella evenly distributed over its surface. ***OR*** Having flagella attached to many places on the cell surface.

PERLA : Pupils equal, reactive to light & accommodation.

Permanent wilting point : Greatest water content of a soil at which indicator plants, growing in that soil wilt and fail to recover when placed in a humid chamber. Often estimated by the water content at1.5mpa soil matric potential.

Permeable : The property of allowing material to pass through, as a permeable membrane.

Permease : A membrane-bound carrier protein or a system of two or more proteins that transports a substance across the membrane. ***OR*** A protein that catalyzes the transport of a specific small molecule across a membrane.

Peronate : Of stems ensheathed by a volva .

Peroxisomes : Subcellular organelles that contain flavin-requiring oxidases and that regenerate oxidized flavin by reaction with oxygen.

PERRLA : Pupils equal, round, reactive to light & accommodation.

Persistent infection : Infections in which ongoing virus replication occurs, but the virus adjusts its replication and pathogenicity so as to avoid killing host. They differ from chronic infections in that whereas in chronic infections, the virus is usually eventually cleared by the host (unless the infection proves fatal), in persistent infections, the virus may continue to be present and to replicate in the host for its entire lifetime.

Pertussis : An acute, highly contagious infection of the respiratory tract, most frequently affecting young children, usually caused by *Bordetella pertussis* or *B. parapertussis*. Consists of peculiar paroxysms of coughing, ending in a prolonged crowing or whooping respiration; hence the name whooping cough.

PERV : Porcine endogenous retrovirus.

Pesticide : A chemical used to kill pests, often insects or fungi growing on crops. For example, Derris dust is a pesticide used to kill white butterfly caterpillars in vegetable gardens.

Petals : The second layer of flower parts; in insect-pollinated flowers these are frequently enlarged and brightly-colored.

Petechiae (plural) : Round, purple lesions caused by intradermal or submucosal microvascular hemorrhaging, capillary leakage or microhemorrhage; petechia (singular).

Petri dish : A shallow dish consisting of two round, overlapping halves that is used to grow microorganisms on solid culture medium; the top is larger than the bottom of the dish to prevent contamination of the culture.

Peyer's patch : Aggregations of lymphoid tissue concentrated in the ileum.

Peziza (Ascomycetes) : Anamorph: Chromelosporium Popularly referred to as cup fungi. They vary in size and color, but are mostly shades of ocher or brown to gray-violet. Most species are commonly found on old straw, compost, peat, leaf litter, rotting wood, damp soil, and other moist substrates. Can be found indoors in wet basements and wet carpets.

PFGE : Pulsed-field gel electrophoresis.

PFK : Phosphofructokinase.

PG : Polygalacturonase ***OR*** prostaglandin.

PGC : Primordial germ cell.

PGI : Phosphoglucose isomerase.

PGK : Phosphoglycerate kinase.

PGM : Phosphoglycerate mutase.

PGP : Progressive general paralysis.

pH : Negative logarithm of the hydrogen ion activity. The degree of acidity (or alkalinity) of a soil as determined by means of a glass or other suitable electrode or indicator at a specified moisture content or soilwater ratio, and expressed in terms of the pH scale. ***OR*** power of hydrogen; measurement of the amount of hydrogen ion in solution; logarithm of the reciprocal of the hydrogen ion concentration in an aqueous solution used to express its acidity or alkalinity (range from 014). ***OR*** A measure of the acidity or alkalinity of a substance, based on a scale of 0 to 14. A neutral solution has a pH of 7. Acidic solutions have pH values less than 7. Alkaline solutions have pH values greater than 7.

pH : The negative logarithm of the hydrogen ion concentration : $pH = -log[H^+]$.

pH Paper : Test paper which changes color at different ph values. ($pH = -log[H^+]$)

Phaeo : A prefix meaning darkly pigmented.

Phaeophyta : Division with in Kingdom Protista which is commonly known as brown algae.

Phage : Viruses for microorganisms. ***OR*** A bacteriophage.

Phage display library : Antibody-like phage can be produced by cloning immunoglobulin V-region genes in filamentous phage, which thus express antigen-binding domains on their surfaces, forming a phage display library. Antigenbinding phage can be replicated in bacteria and used like antibodies. This technique can be used to develop novel antibodies of any specificity.

Phagocyte : Specialized cell capable of internalizing and usually destroying foreign particulates or microorganisms. ***OR*** A type of white blood cell that can engulf and destroy foreign organisms, like bacteria and get rid of cell debris or particles.

Phagocytic vacuole : A membrane-delimited vacuole produced by cells carrying out phagocytosis. It is formed by the invagination of the plasma membrane and contains solid material.

Phagocytosis : The endocytotic process in which a cell encloses large particles in a membrane-delimited phagocytic vacuole or phagosome and engulfs them. ***OR*** The process of engulfing and digesting extracellular foreign materials. ***OR*** The means by which cells engulf their prey; cellular "eating".

Phagocytosis inhibits : Mechanisms enabling an invading microorganism to resist being engulfed, ingested, and or lysed by phagocytes/ phagolysosomes; patients with a defective/compromised monocytemacrophage system (formerly, RES, reticuloendothelial system) are particularly susceptible to infection.

Phagocytosis, phagocytic cells, phagocytes, phagosome, phagolysosome: Phagocytosis is the internalization of particulate matter by cells. Usually, the phagocytic cells or phagocytes are macrophages or neutrophils, and the particles are bacteria that are taken up and destroyed. The ingested material is contained in a vesicle called a phagosome, which then fuses with one or more lysosomes to form a phagolysosome. The lysosomal enzymes are important in pathogen destruction and degradation to small molecules.

Phagolysosome : A phagosome that has fused with a lysosome within the phagocyte. ***OR*** The vacuole that results from the fusion of a phagosome with a lysosome.

Phagosome : A vacuole within a cell that was formed during phagocytosis.

Phagotrophic : Form of feeding where animals, such as protozoans, engulf particulate nutrients, such as bacterial cells or detritus.

Phagovar : A specific phage type.

Phalanges : The finger and toe bones.

Pharmaceutical : A chemical drug.

Pharyngitis : Inflammation of the pharynx, often due to a *S. pyogenes* infection. ***OR*** An inflammation of the mucus membrane in the throat, a sore throat.

Pharynx : The throat area shared in common by the respiratory and digestive systems.

Phase I trial : The first clinical trial of a new drug or treatment on a small number of people. Phase I trials are used to find out how best to administer a treatment and whether it is safe.

Phase II trial : A clinical trial to test if a new drug or treatment helps patients. Phase II trials are sometimes divided into Phase IIA and Phase IIB. Phase IIA trials are used to work out how much drug should be given and Phase IIB trials are designed to work out how well the drug works at different doses.

Phase III trial : A clinical trial to determine whether a new drug or treatment is better than, worse than, or the same as existing treatments. Phase III trials involve a large number of patients.

Phase-contrast microscope : A microscope that converts slight differences in refractive index and cell density into easily observed differences in light intensity.

PHB : Polyhydroxybutyrate.

Ph.D. : An advanced research degree that is usually completed after doing a first degree (e.g. A Bachelor of Science degree) and an honours or masters degree. In New Zealand, it usually takes three years to complete a bachelor's degree (four years to do an honours' programme) and at least an extra three to complete a Ph.D.

Phe : Phenylalanine.

Phenetic system : A classification system that groups organisms together based on the similarity of their observable charactistics.

Phenol : An organic compound that is an alcohol derivative of benzene.

Phenol coefficient test : A test to measure the effectiveness of disinfectants by comparing their activity against test bacteria with that of phenol.

Phenotype : The specific genetic makeup of an individual. ***OR*** Observable properties of an organism. ***OR*** the appearance and characteristics of an organism that derive from the interaction of genotype (genetic makeup) and environment.

Phenylketon uria : A genetic disorder causing inability to digest phenylalanine. ***OR*** A human disease caused by a genetic deficiency in the enzyme that converts phenylalanine to tyrosine. The immediate cause of the disease is an excess of phenylalanine, which can be alleviated by a diet low in phenylalanine.

Pheo : Pheophytin.

Pheromone : A hormone-like substance that acts as an attractant. ***OR*** A chemical used to pass messages between insects or animals of the same species. It can attract members of the opposite sex, mark territory, or act as a warning. ***OR*** A chemical emitted for communication with other members of the same species.

Phialide : A specialized conidiogenous cell that produces conidia in basipetal succession without increasing in length.

Phialoconidium (pl. phialoconidia) : A conidium produced from a phialide.

Phialophora (Hyphomycetes) : Teleomorph: Ascomycetes occurs in nature as a soft rot fungus on wood often causing a distinct blue stain. Can cause diseases in immunocompromised individuals

Phlegma : One of the four humors,mucus,which people thought was produced by the brain.

Phoma (Coelomycete) : Found in soil and plant materials as saprobes. Will grow on a variety of materials such as butter, paint, cement, and rubber. Occasionally pathogenic to plants and humans, but infection from this fungus is extremely rare.

Phoma sp. **:** A common indoor air allergen. It is similar to the early stages of growth of *Chaetomium* sp. The species are isolated from soil and associated plants (particularly potatoes). Produces pink an purple spots on painted walls. It may have antigens which cross react with those of *Alternaria* sp. It will grow on butter, paint cement and rubber. It may cause phaeohyphomycosis a systematic or subcutaneous disease.

Phosphatase : An enzyme that catalyzes the hydrolytic removal of phosphate from molecules.

Phosphate group : A phosphorus attached to four oxygens (PO_4).

Phosphate group transfer potential : A measure of the ability of a phosphorylated molecule such as ATP to transfer its phosphate to water and other acceptors. It is the negative of the dgo_ for the hydrolytic removal of phosphate.

Phosphate : An ion consisting of a phosphorus atom and four oxygen atoms. Among other things, it is used in the constuction of nucleic acids.

Phosphatidylinositol bisphosphate, PIP : Phosphatidylinositol bisphosphate (PIP2) is a membrane-associated phospholipid that is cleaved by phospholipase C-y to give the signaling molecules diacylglycerol and inositol trisphosphate.

Phosphobacterium : Bacterium that is especially good at solubilizing the insoluble inorganic phosphate in soil.

Phosphodiester : A molecule containing two alcohols esterified to a single molecule of phosphate. For example, the backbone of nucleic acids is connected by 5'-3' phosphodiester linkages between the adjacent individual nucleotide residues.

Phosphodiester bond : Type of covalent bond linking nucleotides together in a polynucleotide.

Phosphogluconate pathway : Another name for the pentose phosphate pathway. This name derives from the fact that 6-phosphogluconate is an intermediate in the formation of pentoses from glucose.

Phospholipase C-g : Phospholipase C-g is a key enzyme in signal transduction. It is activated by protein tyrosine kinases that are themselves activated by receptor ligation, and activated phospholipase C-g cleaves inositol phospholipid into inositol trisphosphate and diacylglycerol.

Phospholipid : Lipids containing a substituted phosphate group and two fatty acid chains on a glycerol backbone. ***OR*** A lipid containing charged hydrophilic phosphate groups; a component of cell membranes.

Phosphorus : The element with 15 protons.

Phosphorus cycle : Sequence of transformations undergone by phosphorus where it is transformed between soluble and insoluble, and organic and inorganic forms.

Phosphylation : The formation of a phosphate derivative of a biomolecule.

Photic zone : Uppermost layer of a body of water or soil that receives enough sunlight to permit the occurrence of photosynthesis.

Photoautotroph : Organism able to use light as its sole source of energy and carbon dioxide as sole carbon source.

Photoautotroph : See photolithotrophic autotrophs.

Photoheterotroph : Organism able to use light as a source of energy and organic materials as carbon source.

Photolithotrophic autotrophs : Organisms that use light energy, an inorganic electron source (e.g., H_2O, H_2, H_2S), and CO_2 as a carbon source.

Photo-microscopy : This technique involves a camera connected to a microscope, and is used to record details of structures inside cells and microorganisms.

Photoorganotrophic heterotrophs : Microorganisms that use light energy and organic electron donors, and also employ simple organic molecules rather than CO_2 as their carbon source.

Photoperiodism : Response to changing amounts of light (daily or seasonally).

Photophosphylation : Synthesis of highenergy phosphate bonds, as ATP, using light energy.

Photeactivation : The process in which blue light is used by a photoreactivating enzyme to repair thymine dimers in DNA by splitting them apart. ***OR*** DNA repair in which the damaged region is repaired with the help of light and an enzyme. The lesion (break) is repaired without excision (cutting out) from the DNA.

Photoreceptor : Body part or sensory neurons which are sensitive to light.

Photosynthesis : Process of using light energy to synthesize carbohydrates from carbon dioxide. ***OR*** A process in green plants and some microorganisms that changes carbon dioxide and water into carbohydrates using the energy in sunlight. The process often releases oxygen into the atmosphere as a by-product. ***OR*** The trapping of light energy and its conversion to chemical energy, which is then used to reduce CO_2 and incorporate it into organic form.

Photosynthesis : The process of capturing light energy, converting it to chemical energy, and storing it by manufacturing sugar.

Photosystem I : The photosystem in eucaryotic cells that absorbs longer wavelength light, usually greater than about 680 nm, and transfers the energy to chlorophyll P700 during photosynthesis; it is involved in both cyclic photophosphorylation and noncyclic photophosphorylation.

Photosystem II : The photosystem in eucaryotic cells that absorbs shorter wavelength light, usually less than 680 nm, and transfers the energy to chlorophyll P680 during photosynthesis; it participates in noncyclic photophosphorylation.

Phototaxis : Movement toward light. ***OR*** Moving in response to light.

Phototroph : Organism that uses light as the energy source to drive the electron flow from the electron donors, such as water, hydrogen, or sulfide. ***OR*** A microorganism which gains energy from sunlight (radiant energy).

Phototrophs : Organisms that use light as their energy source.

Phototropism : Growing or leaning in response to light.

Phragmoconidium (pl. Phragmoconidia) : A conidium having two or more transverse septa.

Phycobilin : Watersoluble pigment that occurs in cyanobacteria and functions as the light harvesting pigments for Photosystem II.

Phycobiliproteins : Photosynthetic pigments that are composed of proteins with attached tetrapyrroles; they are often found in cyanobacteria and red algae.

Phycobilisomes : Special particles on the membranes of cyanobacteria that contain photosynthetic pigments and electron transport chains.

Phycobiont : The algal or cyanobacterial partner in a lichen.

Phycocyanin : A blue phycobiliprotein pigment used to trap light energy during photosynthesis. ***OR*** Blue, water-soluble pigment found in the cyanobacteria and the red algae.

Phycoerythrin : Red, water-soluble pigment found in the cyanobacteria and red algae. ***OR*** A red photosynthetic phycobiliprotein pigment used to trap light energy.

Phycology : The study of algae; algology.

Phyllosphere : The surface of plant leaves.

Phylogenetic phyletic classification system : A classification system based on evolutionary relationships rather than the general similarity of contemporary characteristics.

Phylogenetic tree : A graph made of nodes and branches, much like a tree in shape, that shows phylogenetic relationships between groups of organisms and sometimes also indicates the evolutionary development of groups.

Phylogeny : Evolutionary history of a species or other taxonomic group. ***OR*** Ordering of species into higher taxa and the construction of evolutionary trees based on evolutionary (genetic) relationships.

Physicochemical : Characters are those such as temperature profiles, pH, inorganic ion content and distribution, CO_2, oxygen profiles with depth, etc.

Physiology : Biochemical reactions. ***OR*** The study of organic processes or functions of living organisms or parts of organisms. ***OR*** The study of the functioning of the various body parts.

Phytoaccumulation : See phytoextraction. A process in which plants are able to degrade (break down) organic pollutants through their metabolic processes.

Phytochemical : A chemical found in plants, and which may promote health or reduce the risk of disease.

Phytoextraction : Use of plants to extract contaminants (such as metals) from the environment (especially soil). When the plants are saturated with contaminants they are harvested.

Phytomining : Use of plants to extract inorganic substances of economic value (precious metals, etc.).

Phytoplankton : A community of floating photosynthetic organisms, largely composed of algae and cyanobacteria.

Phytoremediation : The use of plants and their associated microorganisms to remove, contain, or degrade environmental contaminants.

Phytostabilization : Use of soil amendments and plants to reduce bioavailability and offsite migration of contaminants.

Phytovolatilization : Use of plants to volatilize contaminants (solvents, etc.) from soil or water.

P_i : Inorganic phosphate (also called ortho-phosphate ion).

Pi : Phosphatidyl-inositol

PI : Present illness; personal injury.

Pib : Product Information Bulletin. General information on a product.

PID : Pelvic inflammatory disease; prolapsed intervertebral disc.

Piedra : A fungal disease of the hair in which white or black nodules of fungi form on the shafts.

Pigment : Any colorful compound, used by living things to absorb or block sunlight, and in sexual displays.

Pileus : A cap especially in Agarics.

Pili (plural) : Former term for fimbriae (still used as alternative term); term more properly applied to those organelles (Fpilus) responsible for bacterial conjugation (transfer of nucleic acids between closely related strains or species = "bacterial sex"); F pilus (sex pili) are hollow organelles that allow for the transfer of portions of a bacterial chromosome from a donor cell to a recipient cell (conjugation); pilus (singular).

Pilose : With long hairs.

Pilot scale : The size of a system that is between the laboratory bench-size and a full-size system.

Pilot trial : A first test run of a new product, usually on a small scale.

Pilus (plural pili) : Fimbria like structure that is present on fertile cells and is involved in DNA transfer during conjugation. Sometimes called sex pilus.

Pin floc : Excessive solids carryover. May occur from time to time as small suspended sludge particles in the supernatant. There are two kinds grey - ashlike, inert, has low BOD - indicates old sludge; and brown, but a portion neither settles nor rises, has high BOD - indicates young sludge.

Pinaceae : Gymnosperm family in which the leaves are needle-shaped.

Pinocytosis : The endocytotic process in which a cell encloses a small amount of the surrounding liquid and its solutes in tiny pinocytotic vesicles or pinosomes.

PIP_2 : Phosphatidylinositol-4,5-bisphosphate.

PIR : Protein information resource.

Pistil : A "female" plant part consisting of ovary, style, and stigma.

Pitch length (pitch) : The number of base pairs per turn of a duplex helix.

Pitched : Pertaining to inoculation of a nutrient medium with yeast, for example, in beer brewing.

Pithomyces (Hyphomycetes) : This genus is common in soil and on dead or decaying plant materials. Requires high moisture level for spore germination. Can potentially produce cyclodepsipeptides, sporidesmolides, and sporidesmin.

Pithomyces sp. **:** Grow on dead grass in pastures. Causes facial eczema in ruminants.

PK : Pyruvate kinase.

PKA : Protein kinase A.

PKC : Protein kinase C.

PKU : Phenylketonuria.

Placebo : An inactive substance or treatment. In clinical trials, the effects of a new drug or treatment are often compared with the effects of a placebo.

Placental mammal : A mammal whose young develop with in the uterus and gain nutrition via a placenta and umbilical cord.

Plage : A smooth area just below the apiculus found on some rough spores.

Plague : An infectious and usually deadly epidemic disease. ***OR*** An acute febrile, infectious disease, caused by the bacillus *Yersinia pestis*, which has a high mortality rate; the two major types are bubonic plague and pneumonic plague.

Plane : Flat especially of caps in agarics.

Planet : A relatively large heavenly body which shines by reflected sunlight and revolves around the sun.

Plankton : Free-floating, mostly microscopic microorganisms that can be found in almost all waters; a collective name. ***OR*** The small organisms that float around near the surface of the ocean.

Planktonic : See plankton.

Plant geography : The study of the geographic distribution of plant species.

Plant growth promoting rhizobacteria (PGPR) : (PGPR) Broad group of soil bacteria that exert beneficial effects on plant growth usually as root colonizers.

Plant virus : A parasite that is subcellular, composed of nucleic acid and protein, and is transmissible. Plant viruses replicate only within host cells. Plant viruses are typically named after the plant they are discovered in and the major symptom that results, for example tobacco mosaic virus (TMV).

Plaque (plak) : Localized area of lysis or cell inhibition caused by virus infection on a lawn of cells. ***OR*** 1. A clear area in a lawn of bacteria or a localized area of cell destruction in a layer of animal cells that results from the lysis of the bacteria by bacteriophages or the destruction of the animal cells by animal viruses. 2. The term also refers to dental plaque, a film of food debris, polysaccharides, and dead cells that cover the teeth. It provides a medium for the growth of bacteria (which may be considered a part of the plaque), leading to a microbial plaque ecosystem that can produce dental decay. ***OR*** A circular clearing on a lawn (continuous layer) of bacterial or culture cells, resulting from cell Iysis and production of phage or animal virus progeny.

Plasma : Plasma is the fluid component of blood containing water, electrolytes, and the plasma proteins. ***OR*** Fluid portion of blood or lymph containing suspended particulate components (see also serum).

Plasma cell : A mature, differentiated B lymphocyte chiefly occupied with antibody synthesis and secretion; a plasma cell lives for only 5 to 7 days. ***OR*** Plasma cells are terminally differentiated B lymphocytes and are the main antibody-secreting cells of the body. They are found in the medulla of the lymph nodes, in splenic red pulp, and in bone marrow.

Plasma membrane : The membrane that surrounds the cytoplasm. ***OR*** The selectively permeable membrane surrounding the cell's cytoplasm; also called the cell membrane, plasmalemma, or cytoplasmic membrane.

Plasmablast : A plasmablast is a B cell in a lymph node that already shows some features of a plasma cell.

Plasmid : An extrachromosomal genetic element found in bacteria, not essential for growth. Usually contains genetic information for resistance to an antimicrobial agent or for degradation of additional substrates. ***OR*** Covalently closed, circular piece of DNA which, as an extrachromosomal genetic element, is not essential for growth. ***OR*** A small ring of DNA found in bacterial cells. It contains genes that are additional to the genes carried on the bacterial chromosome. Some eukaryotic cells, like yeasts may also have plasmids.

Plasmodial (acellular) slime mold : A member of the division Myxomycota that exists as a thin, streaming, multinucleate mass of protoplasm, which creeps along in an amoeboid fashion.

Plasmodium : A stage in the life cycle of myxomycetes (plasmodial slime molds); a multinucleate mass of protoplasm surrounded by a membrane. Also, a parasite of the genus *Plasmodium*.

Plasmodium vivax **:** The parasitic member of Division Apicomplexa which causes malaria.

Plasmogamy : Fusion of the contents of two cells, including cytoplasm and nuclei.

Plasmolysis : The process in which water osmotically leaves a cell, which causes the cytoplasm to shrivel up and pull the plasma membrane away from the cell wall.

Plastid : Specialized cell organelles containing pigments or protein materials. ***OR*** A cytoplasmic organelle of algae and higher plants that contains pigments such as chlorophyll, stores food reserves, and often carries out processes such as photosynthesis.

Plate count : Number of colonies formed on a solid culture medium when uniformly inoculated with a known amount of soil, generally as a dilute soil suspension. The technique estimates the number of certain organisms present in the soil sample.

Platelet : A blood cell that causes blood clots to form and stop bleeding.

Platelet activating factor, PAF : Platelet activating factor (PAF) is a lipid mediator that activates the blood clotting cascade and several other components of the innate immune system.

Platelets : Platelets are small cell fragments found in the blood that are crucial for blood clotting. They are formed from megakaryocytes.

Plate tectonics : The study of the interactions among the plates which make up the earth's crust.

Plato : Person who lived from 427 to 343 BC and developed the idea of the two worlds ; the perfect, invisible one and the imperfect, visible one.

Pleiotropy : The ability of one gene to have multiple effects.

Pleomorphic : Having more than one form. ***OR*** Taking a variety of shapes and forms; multiple morphologies. ***OR*** Refers to bacteria that are variable in shape and lack a single, characteristic form.

Pleura : Serous membrane investing the lungs and lining thoracic cavity

- visceral pleura, pleura surrounding lobes of lungs and lining the lung tissues
- parietal pleura, pleura lining walls of thorax
- pleurisy chest pain deriving from inflammation of pleura

Pleurisy : An infection in the pleural membranes lining the chest cavity.

Pleurocercoid : Second larval type in Pseudophyllidian tapeworms in which after the crustacean is eaten by freshwater fish and the procercoid is liberated, the pleurocercoid is develops in the muscles of the new host and possesses the characteristic scolex and is the only stage which is infective to final host.

Pleurocystidium : Cystidium occurring on the face of a gill.

Pleurogenous : Born on the sides of a conidiophore or hyphae.

Plicate : Folded like a fan; especially of caps in agarics.

Plp : Pyridoxal-5-phosphate.

Pluripotent : The ability of a cell to differentiate into many different cell types.

Pluripotent (pleuripotent) stem cell : See hematopoietic stem cell.

Plus strand positive strand : The virus nucleic-acid strand that is equivalent in base sequence to the viral MRNa.

Pm : Micrometer.

PMA : Positive mental attitude; progressive muscular atrophy.

PMB : Plant made biologics.

PMD : Private medical doctor.

PMF : Plant molecular farming.

PMH : Past medical history.

PMI : Past medical illness; point of maximum impact.

PMP : Plant made pharmaceutical.

PN : Pain; post-nasal.

PNA : Peptide nucleic acid.

Pneumocystis pneumonia, Pneumocystis carinii pneumonia (PCP) : A type of pneumonia caused by the protist *Pneumocystis carinii*.

Pneumonia : Inflammation of the lungs with consolidation as the lung becomes firm (solid) as alveoli are filled with exudate. ***OR*** An infection of the lungs by the bacterium *Streptococcus pneumonia.*

Pneumonic plague : See plague.

PNI : Peripheral nerve injury.

P-nucleotides : P-nucleotides are nucleotides found in junctions between gene segments of the rearranged V-region genes of antigen receptors. They are an inverse repeat of the sequence at the end of the adjacent gene segment, being generated from a hairpin intermediate during recombination, and hence are called palindromic or P-nucleotides.

PO : By mouth; post-operative.

POC : Point of care.

Podzol : Ash-like,gray layer of soil over red, acidic humus.

Poikilothermic : Term used to describe an organism with varying body temperature,usually because the organism is also exothermic.

Point mutation : A mutation that affects only a single base pair in a specific location.

Point source outbreak : Large numbers of people acquiring infection and disease from a common source and within a limited period of time (see also sporadic cases).

Poison ivy : Poison ivy is a plant whose leaves contain pentadecacatechol, a potent contact sensitizing agent and a frequent cause of contact hypersensitivity.

Pol : Polymerase.

Polar : Possessing hydrophilic characteristics and generally water soluble.

Polar flagellation : Condition of having flagella attached at one end or both ends of the cell.

Polar flagellum : A flagellum located at one end of an elongated cell.

Polar group : A hydrophilic (water-loving) group.

Polar molecule : A molecule that has a slightly positive charge at one end and a slightly negative charge at the other. Water is a polar molecule.

Polar mutation : A mutation in one gene that reduces the expression of a gene or genes distal to the promoter in the same operon.

Polarimeter : An instrument for determining the rotation of polarization of light as the light passes through a solution containing an optically-active substance.

Poliomyelitis : An acute, contagious viral disease that attacks the central nervous system, injuring or destroying the nerve cells that control the muscles and sometimes causing paralysis; also called polio or infantile paralysis.

Pollen : Another name for the male gametophyte in gymnosperms and angiosperms.

Pollinat : An organism,often an insect,which aids in the transfer of pollen.

Pollution : The impairment (reduction) of water quality by agriculture, domestic or industrial wastes (including thermal and radioactive wastes) to such a degree as to hinder any beneficial use of the water or render it offensive to the senses of sight, taste, or smell or when sufficient amounts of waste creates or poses a potential threat to human health or the environment.

Polyamine : A hydrocarbon containing more than two amino groups.

Polybetahydroxybutyrate (PHB) : Common storage material of prokaryotic cells consisting of betahydroxybutyrate or other betaalkanoic acids.

Poly-b-hydroxybutyrate : A linear polymer of b-hydroxybutyrate used as a reserve of carbon and energy by many bacteria.

Polycistronic messenger RNA : In prokaryotes, an RNA that contains two or more cistrons; note that only in prokaryotic mRNAs can more than one cistron be utilized by the translation system to generate individual proteins.

Polyclonal antibody : A mixture of antibodies that together recognise different epitopes, or regions, of a foreign substance.

Polyclonal antiserum : Mixture of antibodies to a variety of antigens or to a variety of determinants on a single antigen.

Polyculture : Fish farming in which two or more compatible or symbiotic species of fish are grown together. Also known as Multiculture.

Polygenic : The major histocompatibility complex is both polygenic, containing several loci encoding proteins of identical function, and polymorphic, having multiple alleles at each locus.

Polygonaceae : From genus *Polygonum*, plant family characterized by swellings and a ring. Of bristles at each joint (node) of the stem ; the flowers lack petals but the Sepals often are colored.

Polygonal activation, polygonal mitogens : Antigen activates specific lymphocytes, whereas all mitogens, by definition, activate most or all lymphocytes, a process known as polygonal activation because it involves multiple clones of diverse specificity. Such mitogens are known as polygonal mitogens.

Poly-Ig : The poly-Ig receptor binds polymeric immunoglobulins, especially IgA, at the basolateral membrane of epithelia and transports them across the cell, where they are released from the apical surface. This transcytosis transfers IgA from its site of synthesis to its site of action at epithelial surfaces.

Polymer : Large molecule formed by polymerization of monomeric units. ***OR*** A chemical formed by the union of many monomers (a molecule of low molecular weight). Polymers are used with other chemical coagulants to aid in binding small suspended particles to form larger chemical flocs for easier removal from water. All polyelectrolytes are polymers, but not all polymers are polyelectrolytes. ***OR*** A large molecule constructed from many smaller identical units. These include proteins, nucleic acids, and starches.

Polymer : A large molecule made of many, smaller, repeating, identical or similar subunits.

Polymerase : An enzyme that catalyzes the synthesis of a polymer from monomers.

Polymerase chain reaction (PCR) : The polymerase chain reaction or PCR uses high temperature and unique thermostable enzymes to replicate DNA. It has revolutionized molecular biology. ***OR*** An *in vitro* technique used to synthesize large quantities of specific nucleotide sequences from small amounts of DNA. It employs oligonucleotide primers complementary to specific sequences in the target gene and special heat-stable DNA polymerases. ***OR*** Method for amplifying DNA *in vitro*, involving the use of oligonucleotide primers complementary to nucleotide sequences in target genes and the copying of the target sequences by the action of DNA polymerase.

Polymorphism : Polymorphism literally means existing in a variety of different shapes. Genetic polymorphism is variability at a gene locus in which the variants occur at a frequency of greater than 1%. The major histocompatibility complex is the most polymorphic gene cluster known in humans. ***OR*** Occurrence of several distinct forms of a species in the same habitat at the same time.

Polymorphonuclear leukocyte (PMN) : A leukocyte that has a variety of nuclear forms. ***OR*** Polymorphonuclear leukocytes are white blood cells with multilobed nuclei and cytoplasmic granules. There are three types of polymorpho-nuclear leukocyte: the neutrophils with granules that stain with neutral dyes, the eosinophils with granules that stain with eosin, and the basophils with granules that stain with basic dyes. ***OR*** granular leukocytes with single multilobed nucleus; differentiated into classes on basis of dye uptake; constitute 50-70% of circulating leukocytes; 60-70% of all leukocytes are granulocytes and 90% of these are neutrophils; stain with neutral dyes (see also neutrophil).

Polynucleotide : A chain structure containing nucleotides linked together by phosphodiester (5'-3') bonds. The polynucleotide chain has a directional sense with a 5' and a 3' end.

Polynucleotide phosphylase : An enzyme that polymerizes ribonucleotide diphosphates. No template is required.

Polypeptide : A linear polymer of amino acids held together by peptide linkages. The polypeptide has a directional sense, with an amino- and a carboxy-terminal end. ***OR*** A chain of aminoacids bonded together.

Polypetalous : Term used to refer to a plant with an united flower parts attached to the receptacle.

Polyphasic taxonomy : An approach in which taxonomic schemes are developed using a wide range of phenotypic and genotypic information.

Polypodiaceae : Plant family to which many of the ferns belong.

Polyprotein : A long polypeptide encoding several mature proteins which are subsequently released by protease cleavage.

Polyribosome (polysome) : A complex of an mRNA and two or more ribosomes actively engaged in protein synthesis. ***OR*** A complex of several ribosomes with a messenger RNA; each ribosome is translating the same message.

Polysaccharide : A linear or branched chain structure containing many sugar molecules linked by glycosidic bonds. ***OR*** Long chain of monosaccharides (sugars) linked by glycosidic bonds. ***OR*** A biopolymer made up of many individual sugar molecules linked together. Examples include cellulose, starch or glycogen. ***OR*** A long chain of monosaccharides joined by dehydration synthesis.

Polysome : Strings of ribosomes attached by strands of mRNA.

Polyspecificity, polyreactivity : Some antibodies show polyspecificity, the ability to bind to many different antigens. This is also known as polyreactivity.

Polyurethane : A type of material with multiple uses as a plastic, glue, paint, foam, or as a water-resistant coating.

Pons : The closest,larger bulge at the top of the spinal cord.

Pontiac fever : A bacterial disease caused by *Legionella pneumophila* that resembles an allergic disease more than an infection. First described from Pontiac, Michigan. See Legionnaires' disease..

Popres : The orifices of the tubes of Polyporaceae.

Population : An assemblage of organisms of the same type. ***OR*** A group of the same species of organism in the same area at the same time.

Pore space : Portion of soil bulk volume occupied by soil pores.

Pin : Major proteins in Gramnegative cell outer membranes grouped as dimers or trimers to form transmembrane channels that allow transport of certain molecules into the cells; "cellular pores". ***OR*** A protein channel in the lipopolysaccharide layer of Gramnegative bacteria.

Pin proteins : Proteins that form channels across the outer membrane of Gram-negative bacterial cell walls. Small molecules are transported through these channels.

Poconidium (pl. Poconidia) : A conidium produced through a small pore in a conidiogenous cell.

Porosity : Volume of pores in a soil sample (nonsolid volume) divided by the bulk volume of the sample.

Porphyria : A dominant genetic in ability to make porphyrin.

Porphyrin : A complex planar structure containing four substituted pyrroles covalently joined in a ring and frequently containing a central metal atom. For example, heme is a porphyrin with a central iron atom.

Porphyrin ring : A multi-ring, carbon-based molecule with nitrogens at its central corners,found in heme groups and chlorophyll.

Portal (route) of entry exit : Means by which an infectious agent enters the host; three portals of entry i) inhalation, ii) ingestion, iii) direct penetration.

Positive control : A system that is turned on by the presence of a regulatory protein.

Positive selection : Only those developing T cells whose receptors can recognize antigens presented by self MHC molecules can mature in the thymus, a process known as positive selection. All other developing T cells die before reaching maturity.

Positive feed backloop : A control mechanism that acts in the same direction as the stimulus that triggered it.

Post mortem : After death.

Post-doctoral : Research of further studies after gaining a Doctor of Philosophy.

Post-graduate study : Study at a level above a Bachelor's degree.

Postherpetic neuralgia : The severe pain after a herpes infection.

Postpartum : Adjective pertaining to the time just after giving birth to a baby.

Posttranscriptional modification : The processing of the initial RNA transcript, heterogeneous nuclear RNA, to form MRNa. ***OR*** The covalent bond changes that occur in a polypeptide chain after it leaves the ribosome and before it becomes a mature protein.

Potable : Refers to water suitable for drinking.

Potable water : Water that does not contain objectionable pollution, contamination, minerals, or infective agents and is considered satisfactory for drinking.

POTW : Publicly owned treatment works, as opposed to an industrially owned facility or pipe system.

Pour plate : A Petri dish of solid culture medium with isolated microbial colonies growing both on its surface and within the medium, which has been prepared by mixing micro-organisms with cooled, still liquid medium and then allowing the medium to harden. ***OR*** Method for performing a plate count of microorganisms. A known amount of a serial dilution is placed in a sterile Petri dish and then a melted agar medium is added and the inoculum mixed well by gently swirling. After growth the number of colony forming units is counted.

POX : Point of exit.

PP : Positive pressure; presenting problem; private physician.

PP&A : Palpation, percussion and auscultation.

PPD : Packs per day; permanent partial disability.

PPF : Photosynthetic photon flux.

PPH : Past pertinent history; post-partum hemorrhage.

PP_i : Pyrophosphate ion.

ppm : Parts Per Million - the unit commonly used to designate the concentration of a substance in a wastewater in terms of weight ie. One pound per million pounds, etc. ppm is synonymous with the more commonly used term mg/l (milligrams per liter).

PQ : Plastoquinone.

PR : Partial remission; progress record; pulse rate.

Pre-B cells : During B-cell development, pre-B cells are cells that have rearranged their heavy-chain genes but not their light-chain genes.

Pre-B-cell receptor, pre-B-cell receptor complex : Expression of the pre-B-cell receptor, or pre-B-cell receptor complex, is a critical event in B-cell development. Expression of this receptor, which is a complex of at least five proteins, causes the pre-B cell to enter the call cycle, to turn off the RAG genes, to degrade the RAG proteins, and to expand by several cell divisions. Then the signal ceases, and the pre-B cell is ready to rearrange its light chains.

Prebiotic : Nutrients often added to food that favour the growth of beneficial bacteria in the gut over that of harmful ones. ***OR*** A substance that feeds good bacteria in the gut.

Precipitation (precipitin) reaction : The reaction of an antibody with a soluble antigen to form an insoluble precipitate.

Precipitin : The antibody responsible for a precipitation reaction.

Precipitin reaction : The precipitin reaction was the first quantitative technique for measuring antibody production. The amount of antibody is determined from the amount of precipitate obtained with a fixed amount of antigen. The precipitin reaction also can be used to define antigen valence and zones of antibody or antigen excess in mixtures of antigen and antibody.

Pre-clinical : Extensive research of a new health-care product conducted before a clinical trial. Testing in the pre-clinical stage is done on laboratory models and animals.

Predation : One species benefits at the expense of another. ***OR*** Relationship between two organisms whereby one organism (predator) engulfs or captures and digests the second organism (prey).

Predation : When a larger animal eats other (smaller) animals.

Predilection site : The site within a host where the parasite is normally found.

Prednisone : Prednisone is a synthetic steroid with potent anti-inflammatory and immunosuppressive activity used in treating acute graft rejection, autoimmune disease, and lymphoid tumors.

Preferential flow : The process whereby free water and its constituents move by preferred pathways through a porous medium.

Preimplantation genetic diagnosis : When a cell is taken from a human egg fertilised *in vitro* and its DNA is tested for inherited diseases.

Preliminary treatment : The removal of metal, rocks, rags, sand, eggshells, and similar materials which may hinder the operation of a treatment plant. Preliminary treatment is accomplished by using equipment such as racks, bar screens, comminutors, and grit removal systems.

Pre-parasitic phase of life cycle : The development to the infective stage occurring outside the definitive host in the environment or in a second (intermediate) host of the parasite life cycle.

Prepatent period (ppp) : The period of time from infection until mature adult parasites are producing eggs or larvae.

Presbyopia : Being farsighted,not being able to see close objects clearly.

Pre-T-cell receptor : In T-cell development, TCR â chains expressed by CD44lowCD25+ thymocytes pair with a surrogate á chain called ptá (pre-T-cell á) to form a pre-T-cell receptor that exits the endoplasmic reticulum via the Golgi as a complex with the CD3 molecules.

Pretreatment facility : Industrial wastewater treatment plant consisting of one or more treatment devices designed to remove sufficient pollutants from wastewaters to allow an industry to comply with effluent limits established by the US EPA General and Categorical Pretreatment Regulations or locally derived prohibited discharge requirements and local effluent limits. Compliance with effluent limits allows for a legal discharge to a POTW.

Prevalence : Number of cases of disease occurring within a population at any one given point in time.

Prevalence rate : Refers to the total number of individuals infected at any one time in a given population regardless of when the disease began.

Pribnow box : A special base sequence in the promoter that is recognized by the RNA polymerase and is the site of initial polymerase binding.

Primary (azurophilic) granules (a.k.a., lysosomes) : Contain acid hydrolases, myeloperoxidase and lysozyme.

Primary (frank) pathogen : Any organism that causes a disease in the host by direct interaction with or infection of the host.

Primary amoebic meningoencephalitis : An infection of the meninges of the brain by the free-living amoebae *Naegleria* or *Acanthamoeba*.

Primary focus : During T-dependent antibody responses, a primary focus of B-cell activation forms in the vicinity of the margin between T and B cell areas of lymphoid tissue. Here, the T and B cells interact and B cells can differentiate directly into antibody-forming cells or migrate to lymphoid follicles for further proliferation and differentiation.

Primary follicles : Lymphoid tissues contain lymphoid follicles made up of follicular dendritic cells and B lymphocytes. The primary follicles contain resting B lymphocytes and are the site at which germinal centers form when they are entered by activated B cells, forming secondary follicles.

Primary immune response, Primary immunization : The primary immune response is th adaptive immune response to an initial exposure to antigen. Primary immunization, also known as priming, generates both the primary immune response and immunological memory.

Primary interaction : The binding of antibody molecules to antigen is called a primary interaction, as distinct from secondary interactions in which binding is detected by some associated change such as the precipitation of soluble antigen or agglutination of particulate antigen.

Primary lymphoid organ : See central lymphoid organ.

Primary metabolites : Microbial metabolites produced during the growth phase of an organism.

Primary producer : Photoautotrophic and chemoautotrophic organisms that incorporate carbon dioxide into organic carbon and thus provide new biomass for the ecosystem. ***OR*** Organism that adds biomass to the ecosystem by synthesizing organic molecules from carbon dioxide and simple inorganic nutrients.

Primary production : The incorporation of carbon dioxide into organic matter by photosynthetic organisms and chemoautotrophic bacteria. ***OR*** The production of natural products, for example meat, wool, crops, and wood, that are not turned into value-added products prior to selling, is referred to as primary production.

Primary structure : In a polymer, the sequence of monomers and the covalent bonds. In proteins, it refers to the amino acid sequence.

Primary treatment : The first step of sewage treatment, in which physical settling and screening are used to remove particulate materials. ***OR*** A wastewater treatment process that takes place in a rectangular or circular tank and allows those substances in wastewater that readily settle or float to be separated from the water being treated.

Primate : An animal group that includes human, monkeys, and apes.

Primer : Molecule (usually a polynucleotide) to which DNA polymerase can attach the first nucleotide during DNA replication. ***OR*** A short nucleotide strand that is required for DNA replication to begin. The primer binds by complementary base pairing to the DNA which will be copied. Primers are also needed to start the polymerase chain reaction (PCR).

Priming : Priming of antigen-specific naive lymphocytes occurs when antigen is presented to them in an immunogenic form; the primed cells will differentiate either into armed effector cells or into memory cells that can respond in second and subsequent immune responses.

Primosome : A multiprotein complex that catalyzes synthesis of RNA primer at various points along the DNA template.

Prion : An infectious particle that is the cause of slow diseases like scrapie in sheep and goats; it has a protein component, but no nucleic acid has yet been detected.

PRN : As needed.

Pro : Proline.

Pro-B cells : During B-cell development, pro-B cells are cells that have displayed B-cell surface marker proteins but have not yet completed heavychain gene rearrangement. They are divided into early pro-B cells and late pro-B cells.

Probe : A short, labeled nucleic acid segment complementary in base sequence to part of another nucleic acid, which is used to identify or isolate the particular nucleic acid from a mixture through its ability to bind specifically with the target nucleic acid.

Probiotic : (1) The oral administration of either living microorganisms or substances to promote the health and growth of an animal or human. (2) A living organism that may provide health benefits beyond its nutritional value when it is ingested. ***OR*** A commercial product containing selected strains of natural commensal species of bacteria, known to thrive in the acid pH of the digestive system, while producing valuable food-digesting enzymes, without causing any harm to the host organism. Such products often contain "prebiotic" nutrients, compounds which selectively encourage growth of the included beneficial bacteria, enhancing their ability to "out-compete" undesirable organisms, including pathogenic (disease-causing) species. ***OR*** Microorganisms, usually bacteria, which have health benefits, and can be used to inhibit the presence of other, harmful microorganisms).

Procaryote : Microorganisms which do not have an organized nucleus surrounded by a nuclear membrane. Bacteria and Archaea stand alone in this primitive kingdom. Plants, fungi, animals, birds, fish and all other organisms fit into the kingdom of "Eucaryotes".

Procaryotic cells : Cells that lack a true, membrane-enclosed nucleus; bacteria are procaryotic and have their genetic material located in a nucleoid.

Procaryotic species : A collection of strains that share many stable properties and differ significantly from other groups of strains.

Procercoid : The larvae of Pseudophyllidian tapeworms in which after a crustacean ingests a ciliated coracidium, the procercoid develops in the body cavity of the crustacean.

Prochiral molecule : A nonchiral molecule that lacks handedness and is optically inactive, but would become chiral by a change in one of the substituents at the chiral center. A prochiral molecule may react with an enzyme so that two groups that have a mirror-image relationship to each other are treated differently.

Prodromal stage : The period during the course of a disease in which there is the appearance of signs and symptoms, but they are not yet distinctive and characteristic enough to make an accurate diagnosis.

Prodrome (prodromal period) : (Period of) precursor symptom(s) indicative of the imminent onset of morbidity or disease.

Prodrug : An inactive form of a drug; the drug is converted to its active form by processes in the body of the person that has taken it.

Productive rearrangements : Any lymphocyte receptor chain can be rearranged in either of two ways, productive and nonproductive. Productive rearrangements are in the correct reading frame for the receptor chain in question.

Profile : See soil profile.

Progametangium : The cell that gives rise to a gametangium and a proximal suspensor during the early stages of sexual reproduction in zygomycetous fungi.

Progenits : Progenitors are the more differentiated progeny of stem cells that give rise to distinct subsets of mature blood cells and lack the capacity for self-renewal possed by true stem cells.

Proglottid : One segment of the body (strobila) of a tapeworm. Proglottids may be immature, mature (male and/or female sex organs present) or gravid (full of eggs).

Programmed cell death : See apoptosis.

Prokaryote : Single celled organism which does not contain any internal membranes, such as a nuclear membrane. Includes all bacteria. ***OR*** Organism lacking a unit membrane bound nucleus and other organelles, usually having its DNA in a single circular molecule. ***OR*** A unicellular organism that contains a single chromosome, no nucleus, no membrane-bound organelles, and has characteristic ribosomes and biochemistry. ***OR*** Simple, one-celled organisms that do not have their DNA enclosed in a true nucleus and lack many other organelles such as mitochondria and chloroplasts.

Proliferative kidney disease : A protozoan disease caused by an unclassified myxozoan in salmonids throughout the world.

PROM : Passive range of motion.

Promoter : Site on DNA where the RNA polymerase binds and begins transcription. ***OR*** The region on DNA at the start of a gene that the RNA polymerase binds to before beginning transcription. ***OR*** A DNA sequence where RNA polymerase binds to begin transcription.

Propagated epidemic : An epidemic that is characterized by a relatively slow and prolonged rise and then a gradual decline in the number of individuals infected. It usually results from the introduction of an infected individual into a susceptible population, and the pathogen is transmitted from person to person.

Propagule : Cell unit capable of developing into a complete organism.

Prophage : The latent form of a temperate phage that remains within the lysogen, usually integrated into the host chromosome. ***OR*** State of the genome of a temperate virus when it is replicating in synchrony with that of the host, typically integrated into the host genome or the silent phage genome. Some prophages integrate into the host genome; others replicate autonomously. The prophage state is maintained by a phage-encoded repressor.

Prophase : First stage of mitosis or meiosis during which the chromosomes shorten, thicken, and become distinct (and in meiosis they pair up).

Propolis : A wax-like substance collected by honey bees from tree buds. It is used to seal gaps in the hive.

Proprotein : A protein that is made in an active form, so that it requires processing to become functional.

Prostaglandin : Prostaglandins, like leukotrienes, are lipid products of the metabolism of arachidonic acid that have a variety of effects on a variety of tissues, including activities as inflammatory mediators. ***OR*** An oxygenated eicosanoid that has a hormonal function. Prostaglandins are unusual hormones in that they usually have effects only in that region of the organism where they are synthesized.

Prostate gland : The largest of the male accessory glands which put sits secretions directly in to the urethra.

Prostheca : An extension of a bacterial cell, including the plasma membrane and cell wall, that is narrower than the mature cell.

Prosthetic group : Synonymous with coenzyme except that a prosthetic group is usually more firmly attached to the enzyme it serves. ***OR*** Tightly bound, nonprotein portion of an enzyme; not the same as coenzyme. ***OR*** A tightly bound cofactor that remains at the active site of an enzyme during its catalytic activity.

Protamines : Highly basic, arginine-rich proteins found complexed to DNA in the sperm of many invertebrates and fish.

Protandrous : When an organism changes sexes,exhibiting male traits first, then female.

Protease : An enzyme that hydrolyzes proteins to their constituent amino acids. Also called a proteinase.

Proteasome : A large, cylindrical protein complex that degrades ubiquitin-labeled proteins to peptides in an ATP-dependent process. ***OR*** Cytosolic proteins are degraded by a large catalytic multisubunit protease called a proteasome. It is thought that peptides that are presented by MHC class I molecules are generated by the action of proteasomes, and two interferon-inducible subunits of some proteasomes are encoded in the MHC.

Protectin : Protectin (CD59) is a cell-surface protein that protects host cells from being damaged by complement. It inhibits the formation of the membrane-attack complex by preventing the binding of C8 and C9 to the C5b,6,7 complex.

Protective immunity : Protective immunity is the resistance to specific infection that follows infection or vaccination.

Protein : An organic molecule that consists of amino acids joined together in a specific sequence. Proteins are made inside cells by gene expression. They can have a variety of functions inside and outside cells, including giving cells their shape and structure. Some proteins are enzymes and speed up chemical reactions. Some are responsible for the movement of muscles. Others are hormones. Others are found in cell membranes and help transport atoms and molecules in and out of cells. They are a main constituent of many foods and an important part of a balanced diet (the amino acids released by the break down of proteins we have eaten are used by the body's cells to make new proteins). ***OR*** Class of biochemical compounds constructed from amino acids. Proteins may be structural, such as those that make up hair and cartilage, or they may be reactive, such as the enzymes. ***OR*** A macro molecule made from aminoacids; some consist of one, long polypeptide chain while others are made from several intert wined polypeptides.

Protein A : Protein A is a membrane component of *Staphylococcus aureus* that binds to the Fc region of IgG and is thought to protect the bacteria from igg antibodies by inhibiting their interactions with complement and Fc receptors. It is useful for purifying IgG antibodies.

Protein engineering : The rational design of proteins by constructing specific amino acid sequences through molecular techniques, with the objective of modifying protein characteristics.

Protein kinase C, PKC : Protein kinase C (PKC) is a family of serine/threonine kinases that are activated by diacylglycerol and calcium as a result of signaling via many different receptors.

Protein kinases, protein phosphatases, protein tyrosine kinases, protein tyrosine phosphatases : Protein kinases add phosphate groups to proteins, and protein phosphatases remove these phosphate groups. Enzymes that add phosphate groups to tyrosine residues are called protein tyrosine kinases. These enzymes have crucial roles in signal transduction and regulation of cell growth. Their activity is regulated by a second set of molecules called protein tyrosine phosphatases that remove the phosphate from the tyrosine residues.

Protein splicing : The post-translational process in which part of a precursor polypeptide is removed before the mature polypeptide folds into its final shape; it is carried out by self-splicing proteins that remove inteins and join the remaining exteins.

Protein subunit : One of the components or monomers of a multicomponent protein.

Protein synthesis : The formation of a specific protein. Individual amino acids are joined together in a specific sequence, determined by an mRNA molecule (translation). The sequence of the mRNA is determined by the sequence in a gene (transcription).

Proteinaceous : Describes any structure which is composed of protein.

Proteobacteria : A large group of bacteria, primarily Gram-negative, that 16S rRNA sequence comparisons show to be phylogenetically related; proteobacteria contain the purple photosynthetic bacteria and their relatives and are composed of the a, b, g, d, and e subgroups.

Proteoglycan : A protein-linked heteropolysaccharide in which the heteropolysaccharide is usually the major component.

Proteome : The complete collection of proteins that an organism produces. ***OR*** All of the proteins synthesised by an individual cell or organism at one particular point in time.

Proteomics : The study of the proteome (all the proteins synthesised in a particular cell or organism at a particular point in time).

Protist : A relatively undifferentiated organism that can survive as a single cell. ***OR*** A large and diverse kingdom of organisms that are mostly unicellular (such as diatoms and protozoans like the amoeba) although some can be colonial and others are multicellular (like kelp).

Protista : Old taxonomic term referring to algae, fungi, and protozoa (collectively, the eukaryotic protists), and the prokaryotes.

Protists : Eucaryotes with unicellular organization, either in the form of solitary cells or colonies of cells lacking true tissues.

Protocooperation : A positive, but not obligatory, interaction between two different organisms in which both parties benefit.

Protogynous : When an organism changes sexes,exhibiting female traits first,then male.

Protomer : An individual subunit of a viral capsid; a capsomer is made of protomers.

Proton acceptor : A functional group capable of accepting a proton from a proton donor molecule.

Proton motive force (PMF) : Energized state of a membrane created by expulsion of protons through action of an electron transport chain. ***OR*** The force arising from a gradient of protons and a membrane potential that is thought to power ATP synthesis and other processes. ***OR*** The thermodynamic driving force for proton translocation.

Proto-oncogene : A cellular gene that can undergo modification to a cancer-causing gene (oncogene). ***OR*** Proto-oncogenes are cellular genes that regulate growth control. When mutated or aberrantly expressed, they can contribute to the malignant transformation of cells, leading to cancer. Cf. Oncogenes.

Protoplasm : Complete cellular contents, cytoplasmic membrane, cytoplasm, and nucleus; usually considered the living portion of the cell, thus excluding those layers peripheral to the cell membrane.

Protoplast : Cell from which the wall has been removed. ***OR*** A bacterial or fungal cell with its cell wall completely removed. It is spherical in shape and osmotically sensitive.

Protoplast fusion : The joining of cells that have had their walls weakened or completely removed.

Protothecosis : A disease of humans and animals produced by the green alga *Prototheca moriformis*.

Prototroph : A microorganism that requires the same nutrients as the majority of naturally occurring members of its species.

Prototype : A model use to test how well something works.

Protozoa : A unicellular protist that is considered to be animal-like because it gets its energy from ingested food. ***OR*** A group of motile microscopic animals (usually single-celled and aerobic) that sometimes cluster into colonies and often consume bacteria as an energy source. ***OR*** Tiny animals; most feed on microorganisms. ***OR*** Protests that are animal-like in their nutrition.

Protozoan protozoon : Unicellular eukaryotic microorganisms that move by either protoplasmic flow (amoebae), flagella (flagellates) or, cilia (ciliates). Most species feed on bacteria, fungi, or detrital particles. ***OR*** A microorganism belonging to the Protozoa subkingdom. A unicellular or acellular eucaryotic protist whose organelles have the functional role of organs and tissues in more complex forms. Protozoa vary greatly in size, morphology, nutrition, and life cycle.

Proviral DNA : Viral DNA that has been integrated into host cell DNA. In retroviruses it is the double-stranded DNA copy of the RNA genome.

Provirus : A provirus is the DNA form of a retrovirus when it is integrated into the host cell genome, where it can remain transcriptionally inactive for long periods of time.

Prp : Prion protein.

Prpp : 5-phosphoribosyl-1-pyrophosphate.

PRRE : Pupils round, regular and equal.

PRT : Patient refused test.

Pruinose : Covered with a bloom rather like chalk dust; finely powdered.

Ps : Phosphatidylserine. ***OR*** Photosystem.

PS : Paraspinous; physical status; present symptoms.

P-selectin : See selecting.

Pseudo stratified ciliated columnar cells : Epithelial cells which bear cilia, are columnar shaped,and have one layer of cells, but appear to be stratified due to staggered heights (found lining the trachea).

Pseudocycle : A sequence of reactions that can be arranged in a cycle but that usually do not function simultaneously in both directions. Also called a futile cycle, since the net result of simultaneous functioning in both directions would be the expenditure of energy without accomplishing any useful work.

Pseudohyphae : A string of elongated blastoconidia formed in some yeasts that resemble a hypha-like filament.

Pseudomonad : Member of the genus *Pseudomonas*, a large group of Gramnegative, obligately respiratory (never fermentative) bacteria.

Pseudomurein : A modified peptidoglycan lacking D-amino acids & containing N-acetyltalosaminuronic acid instead of N-acetylmuramic acid; found in methanogenic archaea. ***OR*** A modified peptidoglycan found in methanogenic archaea.

Pseudoplasmodium : A sausage-shaped amoeboid structure consisting of many myxamoebae and behaving as a unit; the result of myxamoebal aggregation in the cellular slime molds; also called a slug.

Pseudopodium : Extension of the cytoplasm in some types of cells,used for movement or engulfing prey.

Pseudopodium pseudopod : Protrusion of an amoeboid cell formed by the extrusion or streaming of the cytoplasm (but still enclosed in the membrane) for the purpose of movement or feeding. ***OR*** A nonpermanent cytoplasmic extension of the cell body by which amoebae and amoeboid organisms move and feed.

PSI : Proteomics standards initiative.

Psittacosis (nithosis) : A disease due to a strain of *Chlamydia psittaci*, first seen in parrots and later found in other birds and domestic fowl (in which it is called ornithosis). It is transmissible to humans.

PSP : Posterior spinous process.

Pssa : Phosphorylation site specific antibody.

Pstv : Potato spindle tuber virus.

Psychrophile : Organism able to grow at low temperatures and showing a growth temperature optimum < 15 ºC. ***OR*** Organism able to grow at low temperatures (ºC) and showing a growth temperature optimum < 15 ºC. Not able to grow above 20 ºC. ***OR*** A micro-organism that grows well at 0°C and has an optimum growth temperature of 15°C or lower and a temperature maximum around 20 °C. ***OR*** Liking cold temperatures; optimal growth at 15 to 20 ºC.

Psychrophilic bacteria : Bacteria whose optimum temperature range is between 0 and 20 ºC.

Psychrotroph : A microorganism that grows at 0 °C, but has a growth optimum between 20 and 30 °C, and a maximum of about 35 °C. ***OR*** Organism able to grow at 0 °C and above 20 °C.

PT : Patient; permanent and total; physical therapy; pneumothorax.

PTA : Prior to admission.

PTD : Permanent total disability.

PTGS : Post-transcriptional gene silencing.

PTH : Phenylthiohydantoin.

PTHRP : Parathyroid hormone related protein.

PTR : Patellar tendon reflex.

PTSD : Post-traumatic stress disorder.

PTA : See pre-T-cell receptor.

Pubescent : Covered with fine short hairs.

Pubis : The lower front hip bone.

PUD : Peptic ulcer disease; pulmonary disease.

Puerperal fever : An acute, febrile condition following childbirth; it is characterized by infection of the uterus and/or adjacent regions and is caused by *Streptococci*.

Pulmonary circuit : The blood flow to and from the lungs to pick up oxygen.

Pulmonary anthrax : A form of anthrax involving the lungs. Also known as woolsorter's disease.

Pulmonary surfactant A : The pulmonary surfactant A and D proteins are members of the pentraxin family that play various roles in the acute-phase response.

Pulmonary syndrome hantavirus : See hantavirus pulmonary syndrome.

Pulse-chase : An experiment in which a short labeling period is followed by the addition of an excess of the same, unlabeled compound to dilute out the labeled material. Useful for observing time-dependent behavior of compounds.

Punctate : Marked with minute dots points scales or hollows.

Punctated equilibrium : The theory that there are "short" periods of much evolutionary change followed by long periods of species stability.

Pupil : The opening in the center of there is for the light to go in to the interior of the eye.

Pure culture : A population of cells that are identical because they arise from a single cell. ***OR*** Population of microorganisms composed of a single strain. Such cultures are obtained through selective laboratory procedures and are rarely found in a natural environment.

Purine : A basic, heterocyclic, nitrogen-containing molecule with two joined rings that occurs in nucleic acids and other cell constituents; most purines are oxy or amino derivatives of the purine skeleton. The most important purines are adenine and guanine. ***OR*** A heterocyclic ring structure with varying functional groups. The purines adenine and guanine are found in both DNA and RNA.

Purine nucleotide phosphylase deficiency, PNP : Purine nucleotide phosphorylase deficiency (PNP) is an enzyme defect that results in severe combined immunodeficiency. This enzyme is important in purine metabolism, and its deficiency causes the accumulation of purine nucleosides, which are toxic for developing T cells, causing the immune deficiency.

Puromycin : An antibiotic that inhibits polypeptide synthesis by competing with aminoacyl-trna for the ribosomal binding site A.

Purple membrane : An area of the plasma membrane of Halobacterium that contains bacteriorhodopsin and is active in photosynthetic light energy trapping.

Purpura : Purple patches on skin due to microhemmorhaging in skin, mucous membranes or serosal surfaces.

Purulent : Consisting of or containing pus.

Pus : Pus is the mixture of cell debris and dead neutrophils that is present in wounds and abscesses infected with extracellular encapsulated bacteria. ***OR*** Viscous yellowish fluid product of inflammatory response that contains dead leukocytes in a thin liquid (liquor puris).

Putrefaction : Biological decomposition of organic matter with the production of ill-smelling products associated with anaerobic conditions. ***OR*** The microbial decomposition of organic matter, especially the anaerobic breakdown of proteins, with the production of foul-smelling compounds such as hydrogen sulfide and amines.

PV : Paravertebral.

PX : Physical examination; prescription; prognosis.

Pycnidium (pl. pycnidia) : An asexual fruiting body containing conidia.

Pyoderma : Any purulent skin disease.

Pyogenic : Pusproducing; "pyo" = pus; "genic" = generation.

Pyogenic bacteria : Bacteria with large capsules are difficult for phagocytes to ingest. Such encapsulated bacteria often produce pus at the site of infection, and are thus called pyogenic bacteria. Pyogenic organisms used to kill many young people. Now, pyogenic infections are largely limited to the elderly.

Pyranose : A simple sugar containing the six-membered pyran ring.

Pyrenoid : The differentiated region of the chloroplast that is a center of starch formation in green algae and stoneworts. ***OR*** Starch-storage organelles in the chloroplasts of various algae.

Pyridoxine : Vitamin B_6.

Pyriform : Pear-shaped.

Pyrimidine : A basic, heterocyclic, nitrogen-containing molecule with one ring that occurs in nucleic acids and other cell constituents; pyrimidines are oxy or amino derivatives of the pyrimidine skeleton. The most important pyrimidines are cytosine, thymine, and uracil. ***OR*** A heterocyclic six-membered ring structure. Cytosine and uracil are the main pyrimidines found in RNA, and cytosine and thymine are the main pyrimidines found in DNA.

Pyrogenic : Causing fever; "pyro" = fire.

Pyrophosphate : A molecule formed by two phosphates in anhydride linkage.

Pyruvic acid : A 3-carbon molecule, $CH_3COCOOH$ which results from the splitting of glucose in glycolysis.

Pyuria : Pus in the urine.

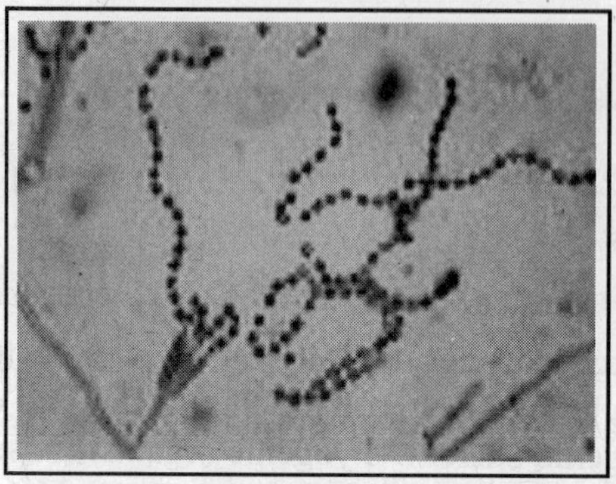

Paecilomyces

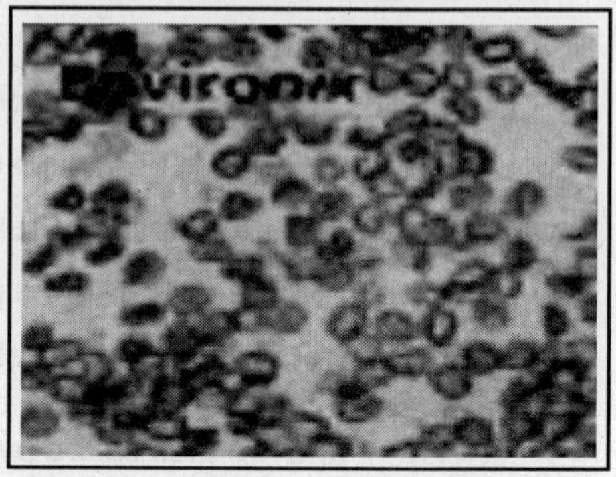

Periconia

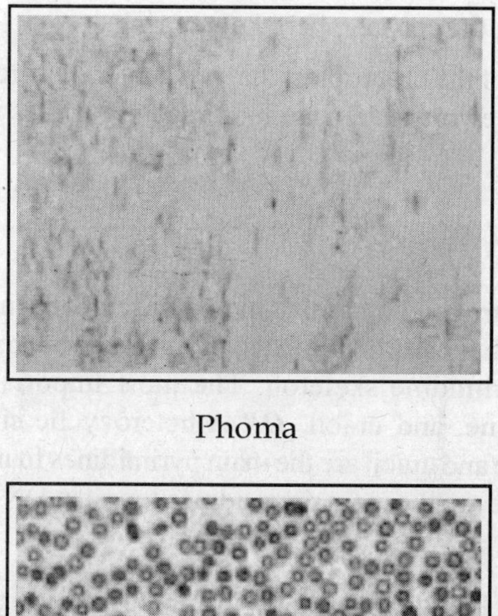

Phoma

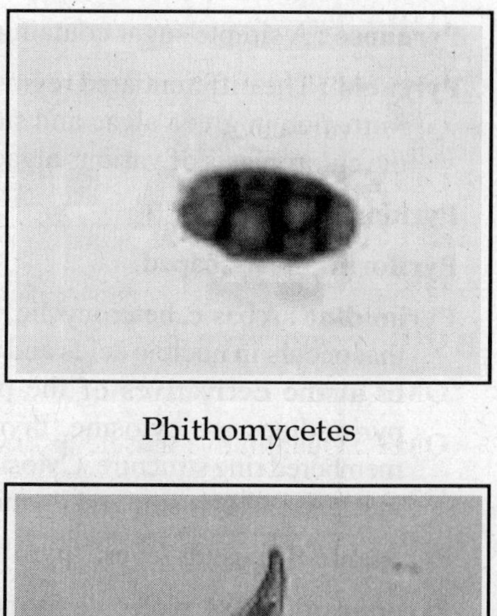

Phithomycetes

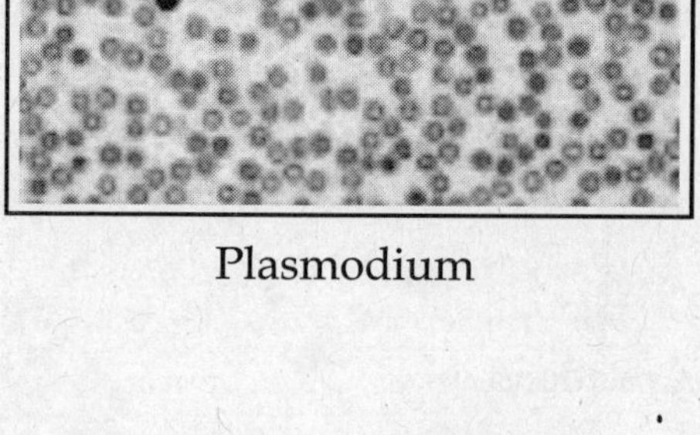

Plasmodium

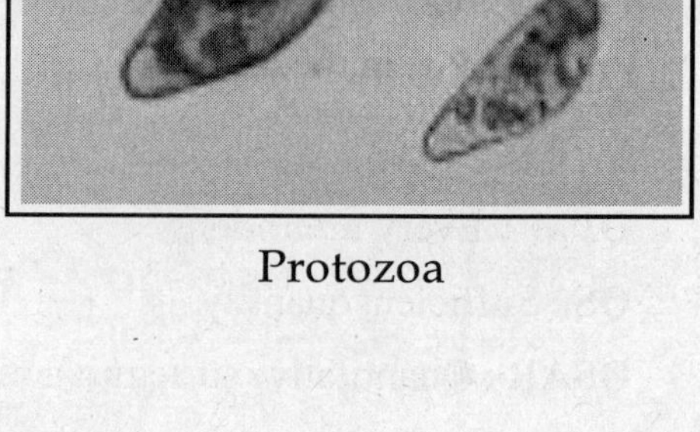

Protozoa

Q

Q : Ubiquinone (coq).

Q : Every.

Q fever : An acute zoonotic disease caused by the rickettsia *Coxiella burnetii*.

q.d. : Every day; Once per day (Latin, quaque die) (see also b.i.d; q.i.d.; t.i.d.).

q.i.d. : Four times per day (Latin, quater in die) (see also b.i.d; q.d.; t.i.d.).

Q_{10} : Relative increase in a reaction rate with temperature. It is expressed as the increase over a 10 °C interval.

Q2H : Every two hours.

QCP : Quantitative chromatin profiling .

QD : Every day.

QELS : Quasi-elastic light scattering.

QH : Every hour.

QH_2 : Reduced coenzyme Q (ubiquinol).

QID : Four times a day.

QM;QAM : Every morning.

QMT : Quantitative muscle testing.

QN : Every night.

QNS : Quantity not sufficient.

QOD : Every other day.

QOH : Every other hour.

QON : Every other night.

QP : As much as desired; at will.

QPCR : Quantitative PCR.

QPM : Every afternoon.

QS : Sufficient quantity.

QSAR : Quantitative structure-activity relationship.

QSTR : Quantitative structure toxicology relationship.

QT : Quantity; quart; quiet.

QTL : Quantitative trait locus.

Quadriceps femis : The muscle that is the front of the thigh.

Quaternary structure : In a protein, the way in which the different folded subunits interact to form the multisubunit protein.

Quellung reaction : The increase in visibility or the swelling of the capsule of a microorganism in the presence of antibodies against capsular antigens.

Quum sensing : The process in which bacteria monitor their own population density by sensing the levels of signal molecules that are released by the microorganisms. When these signal molecules reach a threshold concentration, the population density has attained a critical level or quorum, and quorum-dependent genes are expressed.

QV : As much as desired.

R

R : Gas constant.

R factors R plasmids : Plasmids bearing one or more drug resistant genes.

R group : Shorthand for the side chain of an amino acid.

R loop : A triple-stranded structure in which RNA displaces a DNA strand by DNA-RNA hybrid formation in a region of the DNA.

R/O unit : Reverse osmosis unit for water purification in small aquariums and miniature yard-ponds, utilizes a membrane under pressure to filter dissolved solids and pollutants from the water. Two different filter membranes can be used the CTA (cellulose triacetate) membrane is less expensive, but only works with chlorinated water and removes 50-70% of nitrates, and the TFC membrane, which is more expensive, removes 95% of nitrates, but is ruined by chlorine.

R5P : Ribose-5-phosphate.

RA : Rheumatoid arthritis; right arm.

Rabies : An acute infectious disease of the central nervous system, which affects all warm-blooded animals (including humans). It is caused by an ssRNA virus belonging to the genus *Lyssavirus* in the family *Rhabdoviridae.*

Rac : See small G proteins.

Rachis : An extension of a sympodial proliferating conidiogenous cell bearing conidia.

Rack : Evenly spaced parallel metal bars or rods located in the influent channel to remove rags, rocks, and cans from wastewater.

Racking : The removal of sediments from wine bottles.

Racquet hyphae : A hypha composed of a number of cells swollen at one end resembling a tennis racquet.

RAD : Radiate.

Radappertization : The use of Gamma rays from a cobalt source for control of microorganisms in foods.

Radiation bone marrow chimeras : Radiation bone marrow chimeras are mice that have been heavily irradiated and then reconstituted with bone marrow cells of a different strain of mouse, so that the lymphocytes differ genetically from the environment in which they develop. Such chimeric mice have been important in studying lymphocyte development.

Radioactive : A substance that has an unstable nucleus, which can fall apart and release energy.

Radioimmunoassay, RIA : An immunological assay employing radioactive antibody or antigen for the detection of certain substances in body fluids. ***OR*** A very sensitive assay technique that uses a purified radioisotope-labeled antigen or antibody to compete for antibody or antigen with unlabeled standard and samples to determine the concentration of a substance in the samples. ***OR*** Antigen-antibody interaction can be studied by radioimmunoassay (RIA) in which antigen or antibody is labeled with radioactivity. An unlabeled antigen or antibody is attached to a solid support such as a plastic surface, and the fraction of the labeled antibody or antigen retained on the surface is determined in order to measure binding.

Radioisotope : An isotope of an element that undergoes spontaneous decay with the release of radioactive particles.

Radionuclide : An isotope of artificial or natural origin that exhibits radioactivity.

Radius : The bone on the thumb side of the lower arm; the Radius Rotates around.

Raf : See small G proteins.

Ram : Male sheep.

Ranunculaceae : Plant family, commonly known as the buttercup family, whose flowers have multiple superior ovaries and numerous stamens (**rana, ranunculi**=frog, little frog, medicinal plant ;**-aceae**=ending for plant families).

Rapamycin : Rapamycin, or sirolimus, is an immuno-suppressive drug that blocks cytokine action.

RAPD : Random amplified polymorphic DNA.

Rapid-start complex : The complex that RNA polymerase forms at the promoter site just before initiation.

Ras : See small G proteins.

RAS : Return activated sludge - settled activated sludge that is collected in the secondary clarifier and returned to the aeration basin to mix with incoming raw settled waste-water.

RASVSS : Return activated sludge volatile suspended solids.

RBC : Rotating biological contactor - an attached culture wastewater treatment system.

rbcL : A gene which is located in the chloroplast of photosynthetic organisms. It codes for the large subunit of the protein rubisco, and its sequence has been useful in plant phylogenies.

RDD : Rational drug design.

RDM : Readmission.

rDNA : Recombinant DNA.

Reaction center : A photosynthetic complex containing chlorophyll (or bacteriochlorophyll) and other components, within which occurs the initial electron transfer reactions of photophosphorylation.

Reactivation tuberculosis : See miliary tuberculosis.

Reactive oxygen species (ROS) : Chemicals like hydrogen peroxide (H2O2), superoxide (O_2^-), and the hydroxyl radical (OH.). Produced naturally by the human body, they contribute to the anti-microbial activity of white blood cells and control some cell signalling processes. At higher levels they can cause DNA and RNA damage.

Reading frame : The way in which nucleotides in DNA and mRNA are grouped into codons or groups of three for reading the message contained in the nucleotide sequence.

Reagent : A pure chemical substance that is used to make new products or is used in chemical tests to measure, detect, or examine other substances. ***OR*** A substance used in a chemical reaction to detect, analyse, or produce another substance.

Reagin : Antibody that mediates immediate hypersensitivity reactions. IgE is the major reagin in humans.

Reannealing : Process where two complementary single strands of DNA automatically hybridize back into a single, doublestranded molecule upon cooling.

Rearrangement : Antigen receptor expression requires gene segment rearrangement in developing lymphocytes. The expressed V-region sequences are compose of rearranged gene segments.

REC : Rear end collision; receive; recommend; recurrent.

Recalcitrance : The inability of a nutrient to be degraded or transformed by a cell. Possible reasons for recalcitrance include the lack of necessary membrane transport mechanisms for the chemical or lack of enzymes that facilitate its transformation or degradation. Recalcitrance should not be confused with bioavailability, although distinguishing the two in the environment can be very difficult.

Recalcitrant : Resistant to microbial attack.

Receptacle : The "end" of a plant stem immediately under a flower—occasionally,absinthe Rosaceae,this may been larged and extended in to an hypanthium surrounding the ovary.

Receptor : A specific molecule on the surface of a cell which is used by a virus for attachment.

Receptor editing : The replacement of a light chain of a self-reactive antigen receptor on immature B cells with a light chain that does not confer autoreactivity is known as receptor editing. This has also been shown with heavy chains.

Receptor-associated tyrosine kinases : The antigen receptors of lymphocytes are associated with receptor-associated tyrosine kinases, mainly of the Src family, which bind to receptor tails via their SH2 domains.

Receptor-mediated endocytosis : Receptor-mediated endocytosis is the internalization into endosomes of molecules bound to cell-surface receptors. Antigens bound to B-cell receptors are internalized by this process.

Recessive : Will not be expressed if a dominant allele is present.

Recessive lethal gene : A recessive lethal gene encodes a protein that is needed for the human or animal to develop to adulthood; when both copies are defective, the human or animal dies in utero or early after birth.

Recipient : In any situation in which cells or tissues are transplanted, they come from a donor and are placed in a recipient or host. ***OR*** A person that receives something, for example, a new organ or blood transfusion.

Recognition site : A specific sequence of DNA that is recognised by a restriction enzyme.

Recombinant DNA : DNA molecule containing DNA originating from two or more sources.

Recombinant DNA technology : The techniques used in carrying out genetic engineering; they involve the identification and isolation of a specific gene, the insertion of the gene into a vector such as a plasmid to form a recombinant molecule, and the production of large quantities of the gene and its products.

Recombinant-vector vaccine : The type of vaccine that is produced by the introduction of one or more of a pathogen's genes into attenuated viruses or bacteria. The attenuated virus or bacterium serves as a vector, replicating within the vertebrate host and expressing the gene(s) of the pathogen. The pathogen's antigens induce an immune response.

Recombination : Process by which genetic elements in two separate genomes are brought together in one unit. ***OR*** The process in which a new recombinant chromosome is formed by combining genetic material from two organisms.

Recombination activating genes : See RAG-1 and RAG-2.

Recombination repair : A DNA repair process that repairs damaged DNA when there is no remaining template; a piece of DNA from a sister molecule is used.

Recombination signal sequences, RSSs : Recombination signal sequences (RSSs) are short stretches of DNA that flank the gene segments that are rearranged to generate a V-region exon. They always consist of a conserved heptamer and nonamer separated by 12 or 23 base pairs. Gene segments are only joined if one is flanked by an RSS containing a 12 base pair spacer and the other is flanked by an RSS containing a 23 base pair spacer-the 12/23 rule of gene segment joining.

Recombination : The transfer to offspring of genes not found together in either of the parents.

Rectum : The short terminal (distal) portion of the large intestine extending from the terminal colon to the anus.

Rectus abdominis : Segmented muscles on the front of the abdomen.

Recycle : The use of water or wastewater within (internally) a facility before it is discharged to a treatment system.

RED : Reduce, reduction.

Red pulp : The nonlymphoid area of the spleen in which red blood cells are broken down is called the red pulp.

Red tides : Red tides occur frequently in coastal areas and often are associated with population blooms of dinoflagellates. Dinoflagellate pigments are responsible for the red color of the water. Under these conditions, the dinoflagellates often produce saxitoxin, which can lead to paralytic shellfish poisoning.

Redfield ratio : The carbon-nitrogen-phosphorus ratio of aquatic micro-organisms. This ratio is important for predicting limiting factors for microbial growth.

Redia : In the trematode life cycle, the redia is the larval form (possessing an oral sucker) that develops from the sporocyst and the redia will give rise to the cercariae in the snail (intermediate host).

Redox : Biological reductions/oxidations. These reactions usually require enzymes to mediate the electron transfer. The sediment in the bottom of a lake, sludge in a sewerage works or septic tank will have a very low redox potential and will likely be devoid of any oxygen. This sludge or waste water will have a very high concentration of reductive anaerobic bacteria, indeed the bulk of the organic matter may in fact be bacteria. As the concentration of oxygen increases the oxidation potential of the water will increase. A low redox potential or small amount of oxygen is toxic to anaerobic bacteria, therefore as the concentration of oxygen and redox potential increases the bacterial population changes from reductive anaerobic bacteria to oxidative aerobic bacteria. Measurement of redox potential is also referred to as ORP.

Redox couple : An electron donor and its corresponding oxidized form.

Redox potential (E°) : The relative tendency of a pair of molecules to release or accept an electron. The standard redox potential (E°) is the redox potential of a solution containing the oxidant and reductant of the couple at standard concentrations.

Reducing agent reductant : Any substance, such as the base metal (iron) or the sulfide ion that will readily donate (give up) electrons. The opposite of an oxidizing agent. ***OR*** The electron donor in an oxidation-reduction reaction.

Reducing equivalent (power) : Electrons stored in reduced electron carriers such as NADH, NADPH and $FADH_2$.

Reduction : Process by which a compound accepts electrons.

Reduction potential : Inherent tendency of a compound to act as an electron donor or an electron acceptor. measured in millivolts.

Reductive dechlination : Removal of Cl as Cl from an organic compound by reducing the carbon atom from CCl to CH. ***OR*** The cleavage of carbon-halogen bonds by anaerobic bacteria that creates a strong electron-donating environment.

Reed-Sternberg cells : Reed-Sternberg cells are large malignant B cells that are found in Hodgkin's disease.

Refraction : The deflection of a light ray from a straight path as it passes from one medium (e.g., glass) to another (e.g., air).

Refractive index : The ratio of the velocity of light in the first of two media to that in the second as it passes from the first to the second.

Refracty materials : Material difficult to remove entirely from wastewater such as nutrients, color, taste, and odor-producing substances and some toxic materials.

Regnier de Graaf : Person who lived in the 1600s and thought he saw the homunculus in an egg cell.

Regular flower : A term referring to a flower which has radial symmetry (note that this applies to individual flowers, not to flower clusters or in florescences).

Regulation : Control over the speed at which something happens.

Regulator T cell : Regulator T cells control the development of effector T cells. Two types exist: T-helper cells (CD41 cells) and T-suppressor cells. There are three subsets of T-helper cells: TH1, TH2, and TH0. TH1 cells produce IL-2, IFN-g, and TNF-b. They effect cell-mediated immunity and are responsible for delayed-type hypersensitivity reactions and macrophage activation. TH2 cells produce IL-4, IL-5, IL-6, IL-10, IL-13. They are helpers for B-cell antibody responses and humoral immunity; they also support IgE responses and eosinophilia. TH0 cells exhibit an unrestricted cytokine profile.

Regulatory enzyme : An enzyme in which the active site is subject to regulation by factors other than the enzyme substrate. The enzyme frequently contains a nonoverlapping site for binding the regulatory factor that affects the activity of the active site.

Regulatory gene : A gene whose principal product is a protein designed to regulate the synthesis of other genes.

Regulatory mutants : Mutant organisms that have lost the ability to limit synthesis of a product, which normally occurs by regulation of activity of an earlier step in the biosynthetic pathway.

Regulatory, suppress T cells : Regulatory or suppressor T cells can inhibit T-cell responses.

Regulon : A collection of genes or operons that is controlled by a common regulatory protein.

Release : The stage of viral replication at which virus particles escape the infected cell.

REM : Rapid eye movements; removal.

Renal artery : The main artery providing blood to the kidney.

Renaturation : The process of returning a denatured structure to its original native structure, as when two single strands of DNA are reunited to form a regular duplex, or an unfolded polypeptide chain is returned to its normal folded three-dimensional structure.

Renewable : Resources able to be sustained or renewed indefinitely, either because of inexhaustible supplies or because of new growth.

Repair synthesis : DNA synthesis following excision (cutting out) of damaged DNA.

Repand : Upturned especially of caps in Agarics.

Repetitive DNA : A DNA sequence that is present in many copies per genome.

Replica plating : A technique for isolating mutants from a population by plating cells from each colony growing on a nonselective agar medium onto plates with selective media or environmental conditions, such as the lack of a nutrient or the presence of an antibiotic or a phage; the location of mutants on the original plate can be determined from growth patterns on the replica plates. ***OR*** A technique in which an impression of a culture is taken from a master plate and transferred to a fresh plate. The impression can be of bacterial clones or phage plaques.

Replication : Conversion of one doublestranded DNA molecule into two identical doublestranded DNA molecules. ***OR*** The process in which an exact copy of parental DNA or RNA is made with the parental molecule serving as a template.

Replication fork : The Y-shaped structure where DNA is replicated. The arms of the Y contain template strand and a newly synthesized DNA copy. ***OR*** The Y-shaped region of DNA at the site of DNA synthesis; also called a growth fork.

Replicative form : A double-stranded form of nucleic acid that is formed from a single-stranded virus genome and used to synthesize new copies of the genome.

Replicon : A genetic element that behaves as an autonomous replicating unit. It can be a plasmid, phage, or bacterial chromosome. ***OR*** A unit of the genome that contains an origin for the initiation of replication and in which DNA is replicated.

Reporter gene : A gene that produces a protein that can be easily measured is called a reporter gene. Often used in genetic modification, the reporter gene is inserted along with other DNA of interest. The presence of the protein coded for by the reporter gene is used to indicate whether the foreign DNA has been taken up by the cell or not.

Repressible enzyme : An enzyme whose level drops in the presence of a small molecule, usually an end product of its metabolic pathway.

Repression : Process by which the synthesis of an enzyme is inhibited by the presence of an external substance (the repressor).

Repress : A regulatory protein that inhibits transcription from one or more genes. It can combine with an inducer (resulting in specific enzyme induction) or with an operator element (resulting in repression).

Repress protein : A protein coded for by a regulator gene that can bind to the operator and inhibit transcription; it may be active by itself or only when the corepressor is bound to it.

RER : Rough endoplasmic reticulum.

Reservoir : In the context of human pathogens, a place where disease causing organisms reside when they are not living in a human host. ***OR*** A site, alternate host, or carrier that normally harbors pathogenic organisms and serves as a source from which other individuals can be infected.

Reservoir host : An organism other than a human that is infected with a pathogen that can also infect humans.

Residue : Substance left behind after a process or treatment. For example, after washing dishes a residue of dishwashing liquid may be left behind if they are not rinsed.

Residue sphere : The region surrounding organic matter such as a seed or plant part in which microbial growth is stimulated by increased organic matter availability.

Resistance : The ability to withstand adverse (harmful, unfavourable) conditions. For example, someone who has high resistance to a particular disease might not get it even if lots of people around them are sick.

Resistant : Possessing qualities that hinder development of a given pathogen.

Resolution : The ability of a microscope to separate or distinguish between small objects that are close together.

Resonance hybrid : A molecular structure that is a hybrid of two structures that differ in the locations of some of the electrons. For example, the benzene ring can be drawn in two ways, with double bonds in different positions. The actual structure of benzene is in-between these two equivalent structures.

Respiration : An energy-yielding process in which the energy substrate is oxidized using an exogenous or externally derived electron acceptor. ***OR*** The energy producing process of breathing, by which an organism supplies its cells with oxygen and relieves itself of carbon dioxide. A type of heterotrophic metabolism that uses oxygen in which 38 moles of ATP are derived from the oxidation of 1 mole of glucose, yielding 380,000 cal. (An additional 308,000 cal is lost as heat.) ***OR*** Catabolic reactions producing ATP in which either organic or inorganic compounds are primary electron donors and exogenous compounds are the ultimate electron acceptors.

Respiratory burst : The respiratory burst occurs when an activated phagocytic cell increases its oxygen consumption to support the increased metabolic activity of phagocytosis. The burst generates highly toxic oxygen products such as singlet oxygen, superoxide radical, hydrogen peroxide, hydroxyl radical, and hypochlorite.

Respiratory burst : When neutrophils and macrophages take up opsonized particles, this triggers a metabolic change in the cell called the respiratory burst. It leads to the production of a number of mediators.

Respiratory syncytial : The virus known as respiratory syncytial virus (RSV) is a human pathogen that is a common cause of severe chest infection in young children, often associated with wheezing, as well as in immuno-compromised patients and patients with AIDS.

Respiratory syncytial virus (RSV) : A member of the family Paramyxoviridae and genus Pneumovirus; it is a negative-sense ssRNA virus that causes respiratory infections in children.

Restricted transduction : A transduction process in which only a specific set of bacterial genes are carried to another bacterium by a temperate phage; the bacterial genes are acquired because of a mistake in the excision of a prophage during the lysogenic life cycle.

Restriction endonuclease (restriction enzyme) : Enzyme that recognizes and cleaves specific DNA sequence, generating either blunt or singlestranded (sticky) ends. ***OR*** Enzyme that recognizes and cleaves specific DNA sequence, generating either blunt or single stranded (sticky) ends. ***OR*** An enzyme that causes a break, or cut, in both strands of DNA at a specific sequence, usually about 4-8 base pairs long. Also called a restriction endonuclease.

Restriction fragment length polymorphism : (RFLP) Method to identify differences between similar genes from different organisms. Digestion of genes with restriction endonucleases followed by separation of the resulting fragments by gel electrophoresis yields banding patterns that are characteristic of the individual gene. ***OR*** Method to identify differences between similar genes from different organisms. Digestion of genes with restriction endonucleases followed by separation of the resulting fragments by gel electrophoresis yields banding patterns that are characteristic of the individual gene.

Restriction-modification system : A pair of enzymes found in most bacteria (but not eukaryotic cells). The restriction enzyme recognizes a certain sequence in duplex DNA and makes one cut in each unmodified DNA strand at or near the recognition sequence. The modification enzyme methylates (or modifies) the same sequence, thus protecting it from the action of the restriction enzyme.

Resupinate : Fruiting structure that lies flat on the substratum with the hymenium covering the upper surface.

RET : Retention; retired; returned.

Reticulate : Net-like, marks made by lines veins or ridges which cross one another.

Reticulate body (RB) : The form in the chlamydial life cycle whose role is growth and reproduction within the host cell.

Reticuloendothelial system : See mononuclear phagocyte system.

Retina : The lining of the back of the eye which contains the actual photo receptors.

Retinoblastoma : A genetic condition in which the retinas develop fatal tumors just after birth—the person will survive and remain healthy if his /her eyes are removed.

Retrogressive conidial development : The conidiogenesis cell becomes shorter during the successive development of conidia.

Retrovirus : Virus containing single stranded RNA as its genetic material and producing a complementary DNA by action of the enzyme reverse transcriptase. ***OR*** A group of viruses with RNA genomes that carry the enzyme reverse transcriptase and form a DNA copy of their genome during their reproductive cycle.

Rev : The Rev protein is the product of the rev gene of the human immunodeficiency virus (HIV). The Rev protein promotes the passage of viral RNA from nucleus to cytoplasm during HIV replication.

REV : Review.

Reverse transcriptase : The enzyme reverse transcriptase is an essential component of retroviruses, as it translates the RNA genome into DNA before integration into host cell DNA. Reverse transcriptase also enables RNA sequences to be converted into complementary DNA (cDNA), and so to be cloned, and thus is an essential reagent in molecular biology.

Reverse transcriptase (RT) : An RNA-dependent DNA polymerase that uses a viral RNA genome as a template to form a DNA copy; this is a reverse of the normal flow of genetic information, which proceeds from DNA to RNA. ***OR*** An enzyme that synthesizes DNA from an RNA template, using deoxyribonucleotide triphosphates. ***OR*** Process of copying information found in RNA into DNA.

Reverse transcriptase-polymerase chain reaction, RT-PCR : The reverse transcriptase-polymerase chain reaction (RT-PCR) is used to amplify RNA sequences. The enzyme reverse transcriptase is used to convert an RNA sequence into a cDNA sequence, which is then amplified by PCR.

Reversible covalent modification : A mechanism of enzyme regulation in which the enzyme's activity is either increased or decreased by the reversible covalent addition of a group such as phosphate or AMP to the protein.

Reye's syndrome : An acute, potentially fatal disease of childhood that is characterized by severe edema of the brain and increased intracranial pressure, vomiting, hypoglycemia, and liver dysfunction. The cause is unknown but is almost always associated with a previous viral infection (e.g., influenza or varicella-zoster virus infections).

RF : Replicative form. ***OR*** Radio frequence or release factor ***OR*** replicative form. ***OR*** Renal failure.

RFA : Right forearm.

RFLP : Restriction fragment length polymorphism.

RFT : Renal function test.

RH : Recombinant human ***OR*** Right hand.

RHA : Rheumatoid arthritis.

Rheotaxis : Moving in response to a current.

Rheotropism : Leaning or growing in response to a current.

Rhesus blood group antigen : The Rhesus blood group antigen (Rh) is a red cell membrane antigen that is also detectable on the red blood cells of rhesus monkeys. Anti-Rh antibodies do not agglutinate human red blood calls, so antibody to Rh antigen must be detected by using a Coombs test.

Rhesus factor : A specific molecule that may or may not be present on the surfaces of red blood cells; first identified in the blood of a Rhesus monkey. A person who is Rhesus negative should not be given a blood transfusion from someone who is Rhesus positive.

Rheumatic fever : Rheumatic fever is caused by antibodies elicited by infection with some *Streptococcus* species. These antibodies cross-react with kidney, joint, and heart antigens. ***OR*** A disease that causes high temperature, painful swelling of the joints, and damage to heart valves. ***OR*** An autoimmune disease characterized by inflammatory lesions involving the heart valves, joints, subcutaneous tissues, and central nervous system. The disease is associated with hemolytic *Streptococci* in the body. It is called rheumatic fever because two common symptoms are fever and pain in the joints similar to that of rheumatism.

Rheumatic heart disease : Permanent damage to heart valves, caused by rheumatic fever.

Rheumatoid arthritis, rheumatoid factor : Rheumatoid arthritis is a common inflammatory joint disease that is probably due to an autoimmune response. The disease is accompanied by the production of rheumatoid factor, an IgM anti-IgG antibody that can also be produced in normal immune responses.

Rhinitis : An inflammation of the mucus membrane in the nose,usually due to a cold or allergies.

Rhinocladiella (Hyphomycetes) : Teleomorph: capronia (ascomycetes).

Rhizobacteria : Bacteria that aggressively colonize roots.

Rhizobia : Bacteria capable of living symbiotically in roots of leguminous plants, from which they receive energy and often fix molecular dinitrogen. Collective common name for *Rhizobium* and closely related genera.

Rhizobium **:** Genus of nitrogen-fixing bacteria found in clover root nodules.

Rhizoid : Root like structure that helps to hold an organism to a substrate. ***OR*** A short branching root-like hyphae seen in some Zygomycetes.

Rhizomorph : Mass of fungal hyphae organized into long, thick strands usually with a darkly pigmented outer rind and containing specialized tissues for absorption and water transport. ***OR*** A mycelial strand where hyphae aggregate together in parallel. Commonly found attached to the base of fruiting bodies.

Rhizomucor sp. **:** The Zygomycetous fungus is reported to be allergenic. It may cause mucorosis in immune compromised individuals. It occupies a biological niche similar to *Mucor* sp. It is often linked to occupational allergy. May cause mucorosis in immune compromised individuals. The sites of infection are the lung, nasal sinus, brain, eye and skin. Infection may have multiple sites.

Rhizoplane : Plant root surfaces and usually strongly adhering soil particles.

Rhizopoda : Division with in Kingdom Protista that includes organisms such as amoeba that move by means of pseudopodia.

Rhizopus (Zygomycete) : Frequently isolated from soil and agricultural products, such as cereals and vegetables. Can cause infection in immuno-compromised, malnourished or severely burned people.

Rhizopus sp. **:** The Zygomycetous fungus is reported to be allergenic. It may cause mucorosis in immune compromised individuals. It occupies a biological niche similar to *Mucor* sp. It is often linked to occupational allergy. May cause mucorosis in immune compromised individuals. The sites of infection are the lung, nasal sinus, brain, eye and skin. Infection may have multiple sites.

Rhizosphere : Soil surrounding plant roots. ***OR*** Zone of soil immediately adjacent to plant roots in which the kinds, numbers, or activities of microorganisms differ from that of the bulk soil. ***OR*** A region around the plant root where materials released from the root increase the microbial population and its activities.

Rhizosphere competence : Ability of an organism to colonize the rhizosphere.

Rho factor : A protein involved in the termination of transcription of some messenger RNAs.

Rho factor : The protein that helps RNA polymerase dissociate from the terminator after it has stopped transcription.

Rhodophyta : Division with in Kingdom Protista commonly known as red algae, many of which (such as nori) are used as food and are good sources of dietary iodine.

Rhodotorula (Yeast) : A reddish yeast frequently isolated from air, soil, water, fruit juice, dairy products, and other substrates. Typically found as a saprobe in moist environments indoor such as carpeting, cooling coils, water tanks, humidifiers, and drain pans. Reported to be allergenic. Has been found to colonize terminally ill patients.

Rhodotula sp. **:** A reddish yeast typically found in moist environments such as carpeting, cooling coils and drain pans. In some countries it is the most common yeast genus identified in indoor air. This yeast has been reported to be allergenic. Positive skin tests have been reported. It has colonized terminally ill patients.

Rhoptry : Saclike, electron dense structure in the anterior portion of a zoite of a member of the phylum Apicomplexa; perhaps involved in the penetration of host cells.

RHS : Right hand side.

RI : Respiratory illness, infection.

RIA : Radioimmunoassay.

Rib : Bones protecting the chest cavity.

Riboflavin : Vitamin B-2 (also formerly known as vitamin G).

Ribonucleic acid (RNA) : Polymer of nucleotides connected via a phosphateribose backbone, involved in protein synthesis. ***OR*** A polynucleotide composed of ribonucleotides joined by phosphodiester bridges.

Ribose : The five-carbon sugar found in RNA.

Ribosomal RNA (rRNA) : Types of RNA found in the ribosome; some participate actively in the process of protein synthesis. ***OR*** The RNA present in ribosomes; ribosomes contain several sizes of single-stranded rRNA that contribute to ribosome structure and are also directly involved in the mechanism of protein synthesis. ***OR*** The RNA parts of the ribosome. ***OR*** Types of RNA found in the ribosome; some participate actively in the process of protein synthesis.

Ribosome : The organelle where protein synthesis occurs; the message encoded in mRNA is translated here.

Ribosomes : Small cellular particles made up of ribosomal RNA and protein. They are the site, together with mRNA, of protein synthesis.

Ribotyping : Ribotyping is the use of *E. coli* rRNA to probe chromosomal DNA in southern blots for typing bacterial strains. This method is based on the fact that rRNA genes are scattered throughout the chromosome of most bacteria and therefore polymorphic restriction endonuclease patterns result when chromosomes are digested and probed with rRNA.

Ribulose-1,5-bisphosphate carboxylase : The enzyme that catalyzes the incorporation of CO_2 in the Calvin cycle.

RIC : Right iliac crest.

RID : Ruptured intervertebral disc.

Rights : Things that should nearly always be allowed. So, most people believe that humans have such rights as the right to life, the right to freedom of speech, the right to a fair trial and so on. Under certain circumstances a person may lose a right - I don't have the right to freedom of speech if that means shouting 'fire' in a crowded public place where there is no fire. If someone has a right to something, it usually means that one or more people have duties to that person.

Rig mortis : A stiffening of the muscles after death because no new ATP is formed so calcium ions can not be pumped out of the muscle cells.

Rimose : Splitting radially.

RIMS : Research information management system.

Ring : The membranous remnant of the partial veil

Ringspot : A circle of tissue that is yellow, brown, or darker in color from the center tissue. Often ringspots are a circle of brown on a leaf with green in the center.

Ringworm : The common name for a fungal infection of the skin, even though it is not caused by a worm and is not always ring-shaped in appearance.

RIP : Ribosome-inactivating protein.

RISC : RNA-induced silencing complex .

Rise period burst : The period during the one-step growth experiment when host cells lyse and release phage particles.

RIVD : Ruptured intervertebral disc.

RK : HMG-coa reductase kinase.

RKK : Reductase kinase kinase.

RLB : Right lateral bending.

RLE : Right lower extremity. ***OR*** Right lateral flexion.

RLL : Right lower lateral; right lower leg.

RLQ : Right lower quadrant.

RLR : Right lumbar region.

RLX : Right lower extremity.

RNA : Ribonucleic acid. ***OR*** ribonucleic acid; generated in cells by transcription and required for the synthesis of proteins. ***OR*** a polynucleotide in which the sugar is ribose. ***OR*** "Ribonucleic acid". The nucleic acid which carries the DNA message into parts of the cell where it is interpreted and used. The 18s ribosomal RNA sequence has been used in many groups of organisms to reconstruct phylogeny.

RNA interference (RNAi) : The formation of a double-stranded RNA molecule that is subsequently broken down by natural processes within a cell. This prevents the original single-stranded RNA molecule from being translated into a protein. Also called RNA silencing or RNA inactivation.

RNA polymerase : The enzyme that catalyzes the synthesis of mRNA under the direction of a DNA template. ***OR*** The enzyme which catalyses the formation of RNA during transcription. ***OR*** An enzyme that catalyzes the formation of RNA from ribonucleotide triphosphates, using DNA as a template.

RNA splicing : The excision of a segment of RNA, followed by a rejoining of the remaining fragments.

RNAi : RNA interference.

RNP : Ribonucleoprotein.

RO : Reverse osmosis. ***OR*** Routine order; rule out.

Robert Hooke : Person who, in 1665,was the first to see and name cells—actually,what he first saw was the cell walls that were the remains of formerly-living cork cells.

Rocky mountain spotted fever : A disease caused by *Rickettsia rickettsii*.

ROH : Alcohol.

Rolling circle replication : A mechanism for the replication of circular DNA. A nick in one strand allows the 3' end to be extended, displacing the strand with the 5' end, which is also replicated, to generate a double- stranded tail that can become larger than the unit size of the circular DNA.

Rolling-circle mechanism : A mode of DNA replication in which the replication fork moves around a circular DNA molecule, displacing a strand to give a tail that is also copied to produce a new double-stranded DNA.

ROM : Range of motion, movement; rupture of membranes; Romberg test.

ROMCP : Range of motion complete and pain free.

Romney : Breed of sleep.

Rongoa : Remedy, preservative against sickness or death.

Root nodule : Specialized structure occurring on roots, especially of leguminous plants, in which bacteria fix dinitrogen and make it available for the plant. ***OR*** Gall-like structures on roots that contain endosymbiotic nitrogen-fixing bacteria (e.g., *Rhizobium* or *Bradyrhizobium* is present in legume nodules).

ROQ : Right outer quadrant.

Rosaceae : Plant family commonly known as the rose family whose flowers have numerous stamens and the ovary surrounded by an hypanthium.

Roseola infantum : A skin eruption that produces a rose-colored rash in infants. Caused by the human herpesvirus 6. The disease is short-lived and characterized by a high fever of 3 to 4 days' duration.

ROT : Rotation.

Royal jelly : A substance made by honey bees to feed the queen bee.

RPM (Revolutions per minute) : A measure of the speed with which a centrifuge is spinning. ***OR*** revolutions per minute.

RR : Respiration rate - the weight of oxygen utilized by the total weight of MLSS in a given time. ***OR*** Recovery room; regular rhythm; respiratory rate.

RRCT : Regular rate, clear tones.

rRNA : Ribosomal RNA.

RRR : Regular rate and rhythm.

RS : Respiratory system; right side.

RSR : Regular sinus rhythm.

R strategy : Ecological strategy where organisms rely on high reproductive rates for continued survival within the community. Populations of Rstrategists are subject to extreme fluctuations.

RSV : Rous sarcoma virus.

RTC : Return to clinic.

RTW : Return to work.

Ru1,5P : Ribulose-1,5-bisphosphate.

Ru5P : Ribulose-5-phosphate.

RUA : Right upper arm.

Rubella : A moderately contagious skin disease that occurs primarily in children 5 to 9 years of age that is caused by the rubella virus, which is acquired by droplet inhalation into the respiratory system; German measles.

Rubeola : See measles.

Rubisco : Protein which fixes carbon in photosynthetic organisms. It binds molecules of carbon dioxide to a five-carbon molcule. Rubisco is the most common protein on earth.

RuBP : Ribulose-1,5-bisphosphate

RUE : Right upper extremity.

Rugose : Coarsely wrinkled.

Rugulose : Finely wrinkled.

Rumen : The expanded upper portion or first compartment of the stomach of ruminants. ***OR*** The first stomach in the digestive system of a ruminant animal. Most food collects here immediately after being swallowed. It is later returned to the mouth as cud for thorough chewing.

Ruminant : An herbivorous animal that has a stomach divided into four compartments and chews a cud consisting of regurgitated, partially digested food.

Ruminants : Hoofed mammal with an even number of toes (eg. Sheep, cattle, deer, and camels) that chew the cud and have a complex 3- or 4-chambered stomach.

Run : The straight line movement of a bacterium.

Runoff : Water running down slopes rather than sinking in (again, result of poor humus content) ex. erosion due to deforestation.

RUQ : Right upper quadrant.

Rusts : Obligate parasitic fungi, which belong to Teliomycetes – Uredinales that cause plant diseases.

Rut : A periodic sexual excitement in the male of some animals that is analogous to estrous in the female.

RX : Prescription; treatment.

RXN : Reaction.

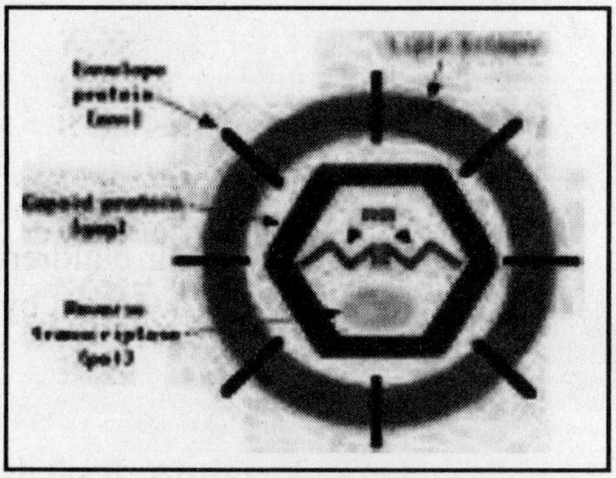

Retrovirus

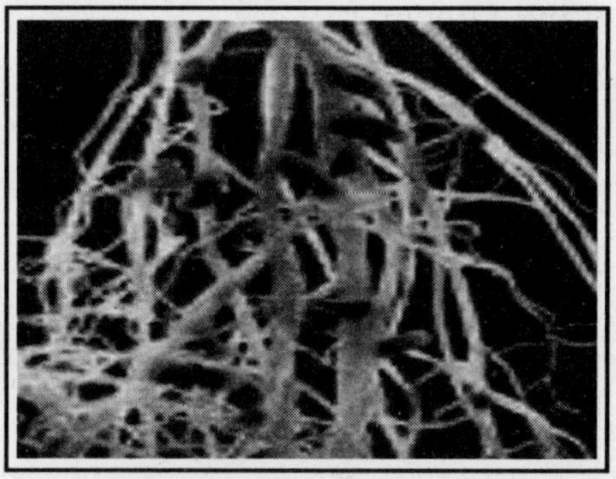

Rhizoboum nodules

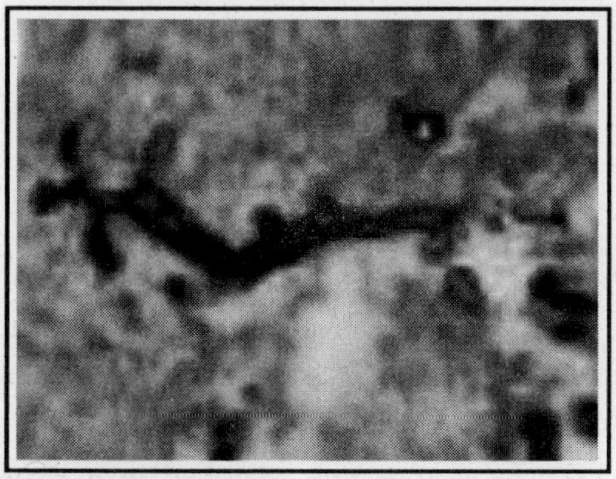

Rhizomucor

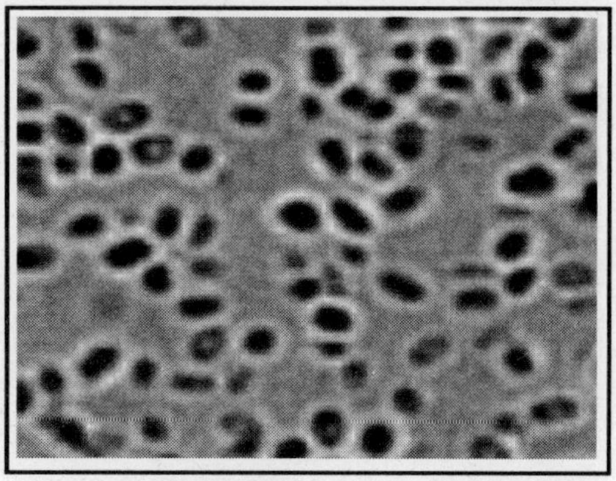

Rodotorula

S

S : Second.

S* :** Sedimentation coefficient. ***OR Svedberg unit.

S phase : The period during the cell cycle when the chromosome is replicated.

S7P : Sedoheptulose-7-phosphate.

SAB : Spontaneous abortion.

SAC : Sacral; shortarm cast.

Saccate : Of the volva cup or bag-shaped

Saccharomyces sp. **:** Reported to be allergenic. Baker's yeast.

Sacrum : Five fused vertebrae which are joined to the pelvis

SAE : Serious adverse event.

SAGE : Serial analysis of gene expression.

Saline soil : Soil containing sufficient soluble salt to adversely affect the growth of most crop plants.

Salmonellosis : An infection with certain species of the genus *Salmonella*, usually caused by ingestion of food containing salmonellae or their products. Also known as Salmonella gastroenteritis or Salmonella food poisoning.

Salting in : The increase in solubility that is displayed by typical globular proteins upon the addition of small amounts of certain salts, such as ammonium sulfate.

Salting out : The decrease in protein solubility that occurs when salts such as ammonium sulfate are present at high concentrations.

Salvage pathway : A family of reactions that permits, for instance, nucleosides as well as purine and pyrimidine bases resulting from the partial breakdown of nucleic acids to be re-utilized in nucleic acid synthesis.

SAM : S-adenosylmethionine.

Sand : Soil particle between 0.05 and 2.0 mm in diameter.

Sandwich ELISA : The technique of sandwich ELISA uses antibody bound to a surface to trap a protein by binding to one of its epitopes. The trapped protein is then detected by an enzyme-linked antibody specific for a different epitope on the protein's surface. This gives the assay a high degree of specificity.

Sanguis : One of the four humors,blood, which people used to think was produced by the heart.

Sanitization : Elimination of pathogenic or deleterious organisms, insect larvae, intestinal parasites, and weed seeds. ***OR*** Reduction of the microbial population on an inanimate object to levels judged safe by public health standards; usually, the object is cleaned.

SAP : Stress-activated protein.

Saprobe : A fungus that feeds by external digestion of dead organic matter, and usually has the ability to adapt rapidly on the substrate.

Saprophyte : Nonparasitic nutritional mechanism by which an organism obtains its food exclusively from the degradation of nonliving organic material. ***OR*** An organism (bacteria, fungi) that utilizes decaying material as a primary nutritional source. ***OR*** Obtaining organic nutrients from dead material. ***OR*** An organism, usually a fungus, that feeds on and helps to decompose decaying organic matter such as dung, corpses, etc.

Saprophytic : Bacteria that breakdown bodies of dead plants and animals (non-living organic material), returning organic materials to the food chain. Saprophytic bacteria are usually non-pathogenic, too.

Saprozoic nutrition : Having the type of nutrition in which organic nutrients are taken up in dissolved form; normally refers to animals or animal-like organisms.

SAR : Sodium Adsorption Ratio - this ratio expresses the relative activity of sodium ions in the exchange reactions with the soil. ***OR*** structure activity relationship.

Sartius : A muscle that extends from the ilium, across the front of the thigh, to the inside of the knee so that it can rotate the leg outward.

SAS : Short arm splint.

SAT : Satisfactory.

Satellite DNA : A DNA fraction whose base composition differs from that of the main component of DNA, as revealed by the fact that it bands at a different density in a CSCI gradient. Usually repetitive DNA or organelle DNA.

Satellite phenomenon : See syntrophism.

SAW : Suface acoustic wave.

Saxitoxin : Neurotoxin found in a variety of dinoflagellates. If ingested, it may cause respiratory failure and cardiac arrest.

SB : Small bowel; sternal border.

SBD : Senile brain disease.

SBE : Single-base extension. ***OR*** Short below elbow; subacute bacterial endocarditis.

SBO : Small bowel obstruction.

SBR : Strict bed rest.

SC : Synaptonemal complex. ***OR*** Sacrococcygeal; scapula; self care; subcutaneous.

SC&M : Sensation, circulation and motion.

SCA : Specific combining ability.

Scabrous : Rough with short projections in the form of granules or scales.

Scaffolding proteins : Special proteins that are used to aid procapsid construction during the assembly of a bacteriophage capsid and are removed after the completion of the procapsid.

Scalanaturae : Aristotle's arrangement of all organisms from least to most complex.

Scale : A platelike organic structure found on the surface of some cells (chrysophytes).

Scaling-up : Changing a process, like food manu-facturing or fermentation, to produce much larger amounts of the final product.

Scanning electron microscope (SEM) : An electron microscope that scans a beam of electrons over the surface of a specimen and forms an image of the surface from the electrons that are emitted by it.

Scanning probe microscope : A microscope used to study surface features by moving a sharp probe over the object's surface (e.g., the scanning tunneling microscope).

Scanning tunneling microscope : A type of scanning probe microscope used to image a surface by moving a fine probe over it at a constant height, which is maintained by keeping a constant electron flow (tunneling current) between the tip and surface.

Scapula : The shoulder blade.

SCAR : Sequence characterized amplified region.

Scarlatina : See scarlet fever.

Scarlet fever : A disease that results from infection with a strain of *Streptococcus pyogenes* that carries a lysogenic phage with the gene for erythrogenic (rash-inducing) toxin. The toxin causes shedding of the skin. This is a communicable disease spread by respiratory droplets.

Scatchard analysis : Scatchard analysis is a mathematical analysis of equilibrium binding that allows the affinity and valence of a receptor-ligand interaction to be determined.

Scavenger receptors : Scavenger receptors on macrophages and other cells bind to numerous ligands and remove them from the blood. The Kupffer cells in the liver are particularly rich in scavenger receptors.

SCD : Service connected disability.

SCE : Sister chromatid exchange.

SCFM : Cubic feet of air per minute at standard conditions of temperature, pressure and humidity (0, 14.7 psi and 50% relative humidity).

Schizogony : Multiple asexual fission. ***OR*** In schizogony the nucleus of the original invading protozoan undergoes repeated divisions to form many nuclei within the cell. Then each nucleus becomes surrounded by cytoplasm and a cell membrane forms around each new organism. These new stages within the schizont are known as merozoites.

Schizont (meront) : An asexual reproducing stage of some protozoa. It reproduces by schizogony.

Schizophrenia : A mental disorder caused by brain dysfunction, resulting in hallucinations, delusions and distorted thinking.

SCI : Spinal cord injury.

SCID, severe combined immunodeficiency : See severe combined immunodeficiency.

SCL : Scleroderma; sclerosis.

Sclera : The white covering of the eyeball.

Sclerotium(sclerotia) **:** A mass of thick-walled cells formed by the vegetative hyphae that function as an organ of perennation. ***OR*** Modified fungal hyphae that form a compact, hard vegetative resting structure with a thick pigmented outer rind. ***OR*** A (often spherical) aggregate of hyphae with a hard dark-coloured rind, generally thought to be a resting stage with better capacity for survival than mycelium. It can germinate to produce hyphae, or fruit bodies.

SCNT : Somatic cell nuclear transfer.

Scolex : The anterior end ("head") of the adult cestode, it is the hold-fast organ that anchors the worm in the GI tract.

Scopulariopsis (Hyphomycetes) : Teleomorph: Microascus (Ascomycetes) Mainly soil-borne, but also frequently isolated from wood, grain, fruit, paper, and food such as meat and dairy products. Also isolated from indoor environments. Most species can liberate arsenic gaseous compounds that can lead to arsenic poisoning. Has recently been associated with invasive human infections.

Scopulariopsis sp. **:** It may produce arsine gas if growing on arsenic substrate.. This can occur on wallpapers covered with paris green. It has been found growing on a wide variety of materials including house dust. It is associated with type III allergy.

SCP : Single-cell protein.

Scrophulariaceae : Plant family to which snap dragons belong,from genus *Scrophularia*.

Scytalidium (Hyphomycetes) : Isolated from wood and soil.

Scytonemin : A pigment produced by cyanobacteria that absorbs UV radiation, and thereby acts as a passive sunscreen.

SD : Sterile dressing; subdural.

SDD : Sterile dry dressing.

SDS : Sodium dodecylsulfate.

SDS : Sudden death syndrome.

SDS-PAGE : SDS-PAGE is the common abbreviation for polyacrylamide gel electrophoresis (PAGE) of proteins dissolved in the detergent sodium dodecyl sulfate (SDS). This technique is widely used to characterize proteins, especially after labeling and immuno-precipitation. Sodium dodecyl sulphate PAGE.

Sebacious gland : Holocrine gland in the skin that secretes sebum into hair follicles.

Sebum : Fatty secretion of the sebacious glands; thick greasy semifluid lubricating substance composed of fat and epithelial debris.

SEC : Size exclusion chromatography.

Second law of thermodynamics : Physical and chemical processes proceed in such a way that the entropy of the universe (the system and its surroundings) increases to the maximum possible.

Second messenger : A diffusible small molecule, such as camp, that is formed at the inner surface of the plasma membrane in response to a hormonal signal.

Second set rejection : When the recipient of a first tissue or organ graft has rejected that graft, a second graft from the same donor is rejected more rapidly and vigorously in what is called a second set rejection.

Second signal : The co-stimulatory signal required for lymphocyte activation is often called a second signal, with the first signal coming from the binding of antigen by the antigen receptor. Both signals are required for the activation of most lymphocytes.

Secondary antibody : A secondary antibody response is the antibody response induced by a second or booster injection of antigen-a secondary immunization. The secondary response starts sooner after antigen injection, reaches higher levels, is of higher affinity than the primary response, and is dominated by IgG antibodies. Therefore, the response to each immunization is increasingly intense, so the secondary, tertiary, and subsequent responses are of increasing magnitude.

Secondary granules : (Specific) granules contain lactoferrin and lysozyme; oxygendependent (respiratory burst) and oxygenindependent (cationic proteins, lysozyme, lactoferin) killing.

Secondary interactions : See primary interactions.

Secondary metabolite : Product of intermediary metabolism released from a cell, such as an antibiotic.

Secondary structure : In a protein or a nucleic acid, any repetitive folded pattern that results from the interaction of the corresponding polymeric chains. In proteins, the most common are ?-strands (sheets) and ?-helices.

Secondary treatment : The biological degradation of dissolved organic matter in the process of sewage treatment; the organic material is either mineralized or changed to settleable solids. ***OR*** A wastewater treatment process used to convert dissolved or suspended materials into a form more readily separated from the water being treated. Usually the process follows primary treatment by sedimentation. The process commonly is a type of biological treatment process followed by secondary clarifiers that allow the solids to settle out from the water being treated.

Secretory component : The secretory component attached to IgA antibodies in body secretions is a fragment of the poly-Ig receptor left attached to the IgA after transport across epithelial cells.

Secretory IgA (sIgA) : The primary immunoglobulin of the secretory immune system. See IgA. ***OR*** mucosal IgA; antigenbinding site attaches to target bacteria and Fc portion binds mucin, trapping/clearing bacteria.

Secretory vacuole : In protists and some animals, these organelles usually contain specific enzymes that perform various functions such as excystation. Their contents are released to the cell exterior during exocytosis.

Sedimentation : The process of subsidence and deposition of suspended matter from a wastewater by gravity.

Seeding : Introduction of microorganisms into a biological oxidation unit to minimize the time required to build a biological sludge. Also referred to as inoculation with cultured organisms.

Segmented genome : A virus genome that is divided into several parts or fragments, each probably coding for the synthesis of a single polypeptide; segmented genomes are very common among the RNA viruses.

Seine net : A net designed to collect aquatic organisms inhabiting natural waters from the shoreline to 3' depths is called a seine net. Most often a plankton seine.

Seizures : A sudden attack, convulsion, spasm or sensation.

Selected : A cell is said to be selected by antigen when its receptors bind that antigen. If the cell starts to proliferate as a result, this is called clonal selection, and the cell founds a clone; if the cell is killed as a result of binding antigen, this is called negative selection or clonal deletion.

Selectins : Selectins are a family of cell-surface adhesion molecules of leukocytes and endothelial cells that bind to sugar moieties on specific glycoproteins with mucinlike features. ***OR*** A family of cell adhesion molecules that are displayed on activated endothelial cells; examples include P-selectin and E-selectin. Selectins mediate leukocyte binding to the vascular endothelium.

Selection pressure : An external feature that changes the traits an organism needs to survive and reproduce.

Selective breeding : A human-controlled breeding programme in which specific plants or animals are chosen to breed (produce offspring) because they have particularly useful or valuable features. The aim is to produce offspring with specific characteristics.

Selective media : Culture media that favor the growth of specific microorganisms; this may be accomplished by inhibiting the growth of undesired microorganisms.

Selective medium : Medium that allows the growth of certain types of microorganisms in preference to others. For example, an antibiotic containing medium allows the growth of only those microorganisms resistant to the antibiotic.

Selective toxicity : The ability of a chemotherapeutic agent to kill or inhibit a microbial pathogen while damaging the host as little as possible.

Self antigens : Antigens in the body of an individual are by convention called self antigens. Lymphocytes are screened during their immature stages for reactivity with self antigens, and those that do respond undergo apoptosis.

Self tolerance, tolerance : Tolerance is the failure to respond to an antigen; when that antigen is borne by self tissues, tolerance is called self tolerance. See also tolerance.

Self-assembly : The spontaneous formation of a complex structure from its component molecules without the aid of special enzymes or factors.

SEM : Scanning electron microscope; semen, seminal.

Semiconservative replication : Duplication of DNA in which the daughter duplex carries one old strand and one new strand.

Semilunar valve : The valve between a ventricle and its major, out going artery.

Semimacronematous : Having a conidiophore that is only slightly morphologically different from the vegetative hyphae.

Semipermeable : The characteristic of allowing only some molecules, usually smaller or uncharged ones, to pass through.

Semitendinosus : The inner muscle on the back of the thigh.

SENS : Sensory, sensation.

Sensitization, sensitized : Allergic reactions require prior immuni-zation, called sensitization, by the allergen that elicits the acute response. Allergic reactions occur only in sensitized individuals.

Sensy testing : Testing a new product for it's sensory effects, such as smell, taste, and texture.

SEP : Separate, separation.

Sepals : The outer most layer of flower parts which surround the other three layers of parts.

Sepsis, septic : An infection in the blood. ***OR*** Sepsis is infection of the bloodstream. This is a very serious and frequently fatal condition. Infection of the blood with Gram-negative bacteria triggers septic shock through the release of the cytokine TNF-a. ***OR*** A condition produced by anaerobic bacteria. If severe, the wastewater turns black, gives off foul odors, contains little or no dissolved oxygen and creates a high oxygen demand.

Septate : Divided by cross walls. ***OR*** Divided by a septum or cross wall; also with more or less regular occurring cross walls.

Septic shock : Sepsis associated with severe hypotension despite adequate fluid resuscitation, along with the presence of perfusion abnormalities that may include, but are not limited to, lactic acidosis, oliguria, or an acute alteration in mental status. Gram-positive bacteria, fungi, and endotoxin-containing Gram-negative bacteria can initiate the pathogenic cascade of sepsis leading to septic shock.

Septic tank : Untreated liquid household wastes (sewage) will quickly clog your absorption field if not properly treated. The septic tank is a holding tank in which this treatment can take place. When sewage enters the septic tank, the heavy solids settle to the bottom of the tank; the lighter solids, fats and greases partially decompose and rise to the surface and form a layer of scum. The solids that have settled to the bottom are attacked by bacteria and form sludge. ***OR*** A tank used to process small quantities of domestic sewage. Solid material settles out and is partially degraded by anaerobic bacteria as sewage slowly flows through the tank. The outflow is further treated or dispersed in aerobic soil.

Septicemia : A disease associated with the presence in the blood of pathogens or bacterial toxins.

Septicemia (sepsis) : Systemic disease associated with persistent presence of bacterial cells, bacterial toxins or other bacterial products in the bloodstream.

Septicity : Septicity is the condition in which organic matter decomposes to form foul-smelling products associated with the absence of free oxygen. If severe, the wastewater turns black, gives off foul-odors, contains little or no dissolved oxygen and creates a heavy oxygen demand.

Septum (pl. septa) : A cross wall in a hypha. ***OR*** Crosswall (partition) dividing a parent cell into two daughter cells during binary fission or occurring between adjacent cells in hyphae. ***OR*** A partition or cross-wall that occurs between two cells in a bacterial (e.g., actinomycete) or fungal filament, or which partitions off fungal structures such as spores. Septa also divide parent cells into two daughter cells during bacterial binary fission.

SEQ : Sequela.

Sequela (singular) : A followon (sequel) to a disease.

Sequelae : Plural of sequela.

Sequence motif : A sequence motif is a pattern of nucleotides or amino acids shared by different genes or proteins that often have related functions. Sequence motifs observed in peptides that bind a particular MHC glycoprotein are based on the requirements for particular amino acids to achieve binding to that MHC molecule.

Ser : Serine.

Serial dilution : Series of stepwise dilutions (usually in sterile water) performed to reduce the populations of microorganisms in a sample to manageable numbers.

Serial dilution : A series of step wise dilutions in which each solutions serve as the source of the aliquot for then extdilution in the series.

Seroconversion : Seroconversion is the phase of an infection when antibodies against the infecting agent are first detectable in the blood.

Serological tests : Diagnostic methods that are based on specific antibody-antigen reactions to determine if a substance or pathogen is present. One example is ELISA.

Serology, serological assays : Study of antigenantibody reactions *in vitro*. ***OR*** The branch of immunology that is concerned with *in vitro* reactions involving one or more serum constituents (e.g., antibodies and complement). ***OR*** Serology is the use of antibodies to detect and measure antigens using serological assays, so called because these assays were originally carried out with serum, the fluid component of clotted blood, from immunized individuals.

Serotonin : Mast cells in mice store the small soluble mediator serotonin in their granules.

Serotype : Different strains of bacteria and other pathogens can sometimes be distinguished by their serotype, the ability of an immune serum to agglutinate or lyse some strains of bacteria and not others.

Serotyping : A technique or serological procedure that is used to differentiate between strains (serovars or serotypes) of microorganisms that have differences in the antigenic composition of a structure or product.

Serpins : Serpins are a large family of protease inhibitors.

Serpula (Basidiomycete) : Wood-attacking fungi. *Serpula lacrymans* is popularly referred to as the dry rot fungus or house fungus.

Serpula lacrymans **:** Common cause of extrinsic asthma (immeadiate-type hypersensitivity: type I). Acute symptoms include edema and bronchiospasms, chronic cases may develop pulmonary emphysem.

Serrate : Saw-toothed,used to describe the margin of gills.

Serrulate : Finely toothed.

Serum : Serum is the fluid component of clotted blood. ***OR*** Fluid portion of blood minus cells and clotting factors; inherently fatal to many microorganisms (serum killing; serum sensitivity) (see also plasma).

Serum resistance : The type of resistance that occurs with bacteria such as *Neisseria gonorrhoeae* because the pathogen interferes with membrane attack complex formation during the complement cascade.

Serum sickness : Serum sickness occurs when foreign serum or serum proteins are injected into a person. It is caused by the formation of immune complexes between the injected protein and the antibodies formed against it. It is characterized by fever, arthraigias, and nephritis.

Serving size : The portion (or amount) of food used as a reference on food labels. It tends to be based on the average amount eaten in one sitting.

Sessile : Without a stalk.

Settleable solids : Those solids in suspension which will pass through a 2000 micron sieve and settle in one hour under the influence of gravity.

Settling basin : A basin used during water purification to chemically precipitate out fine particles, microorganisms, and organic material by coagulation or flocculation.

Severe combined immune deficiency, SCID : Severe combined immune deficiency (SCID) is an immune deficiency disease in which neither antibody nor T-cell responses are made. It is usually the result of T-cell deficiencies. The SCID mutation causes severe combined immune deficiency in mice.

Severe combined immunodeficiency synrome : A disorder of the immune system, which makes patients more likely to get infections.

Sewage : The used water and water-carried solids from homes that flow in sewers to a wastewater treatment plant. The preferred term is wastewater.

Sex pilus : A thin protein appendage required for bacterial mating or conjugation. The cell with sex pili donates DNA to recipient cells.

SF : Spinal fluid.

SFC : Supercritical fluid chromatography.

SFP : Spinal fluid pressure.

SGD : Saccharomyces genome database.

SH : Self help; serum hepatitis; shoulder; surgical history.

SH2 domain, Src-family tyrosine kinases : See Src-family tyrosine kinases.

Sheath : Tubular structure formed around a chain of cells or around a bundle of filaments. ***OR*** A hollow tubelike structure surrounding a chain of cells and present in several genera of bacteria.

Shigellosis : The diarrheal disease that arises from an infection with a member of the genus Shigella. Often called bacillary dysentery.

SHIME system (simulated human intestinal microbial ecosystem) : A set of connected chemostat-like reactors that provide a sequence of environments similar to the human digestive system.

Shine-Dalgarno sequence : A segment in the leader of procaryotic mRNA that binds to a special sequence on the 16S rRNA of the small ribosomal subunit. This helps properly orient the mRNA on the ribosome.

Shingles (zoster) : A reactivated form of chickenpox caused by the varicella-zoster virus.

Shock load : The arrival at a plant of a waste which is toxic to organisms in sufficient quantity or strength to cause operating problems. Possible problems include odors and sloughing off of the growth or slime on a trickling-filter media. Oganic or hydraulic overloads also can cause a shock load.

Short tandem repeat : A short DNA sequence (from 2-5 nucleotides) that is repeated multiple times. Short tandem repeats can be used to get a DNA profile.

ShRNA : Short hairpin *RNA*.

SI : Sacroiliac; self-inflicted.

SID : Once per day.

Siderochromes : Compounds produced by microorganisms that are involved with the uptake of iron by those microorganisms (see siderchromes).

Siderophore : A small molecule that complexes with ferric iron and supplies it to a cell by aiding in its transport across the plasma membrane. ***OR*** Nonporphyrin metabolite secreted by certain microorganisms that forms a highly stable coordination compound (chelate) with iron; a highaffinity iron binding compound. There are two major types: catecholates and hydroxamates. ***OR*** extracellular iron chelating compounds ("sidero" = iron; "phore" = a carrier).

SIDS : Sudden infant death syndrome.

Sigma factor : A protein that helps the RNA polymerase core enzyme recognize the promoter at the start of a gene. ***OR*** A subunit of RNA polymerase that recognizes specific sites on DNA for initiation of RNA synthesis.

Sign : An objective change in a diseased body that can be directly observed (e.g., a fever or rash).

Signal joint : A signal joint is formed by the precise joining of recognition signal sequences in the process of somatic recombination that generates T-cell receptor and immunoglobulin genes.

Signal sequence : A (usually N-terminal) sequence of a protein that directs its processing or localization within the cell.

Signal transducers and activats of transcription, STATs, Janus kinases : See Janus kinases.

Signal transduction : Signal transduction describes the general process by which cells perceive changes in their environment. In lymphocytes, the most important changes are those occurring during infection that generate antigens that stimulate the cells of the immune system to bring about an adaptive immune response.

Signs (Clinical) : Objective evidence of a disease as observed by the physician.

SIJ : Sacroiliac joint.

Silent mutation : A mutation that does not result in a change in the organism's proteins or phenotype even though the DNA base sequence has been changed.

Silica : Amorphous silicon dioxide (glass). It is a structural component in many organisms, such as diatoms and horsetails.

Silt : Soil particle with a diameter between 0.002 and 0.05 mm.

Simple matching coefficient (SSM) : An association coefficient used in numerical taxonomy; the proportion of characters that match regardless of whether or not the attribute is present.

siNA : Small interfering *NA*.

SINE : Short interspersed nuclear element.

Single nucleotide polymphism (SNP) : One nucleotide base in a DNA sequence is changed to a different nucleotide base.

Single radial immunodiffusion (RID) assay : An immunodiffusion technique that quantitates antigens by following their diffusion through a gel containing antibodies directed against the test antigens.

Single-chain Fv : A single-chain Fv fragment, comprising a V region of a heavy chain linked by a stretch of synthetic peptide to a V region of a light chain, can be made by genetic engineering.

Single-copy DNA : A region of the genome whose sequence is present only once per haploid complement.

Single-positive thymocytes : During T-cell maturation in the thymus, mature T cells are detected by the expression of either the CD4 or the CD8 co-receptor and are therefore called single-positive thymocytes.

Sinuate : Notched used to describe gills which have a notch before attaching to the stem.

siRNA : Small interfering *RNA*.

Sirolimus : Sirolimus is the drug name that has been adopted for the chemical rapamycin; the two terms are used interchangeably in the literature.

SIT : Situational.

Site : (i) In ecology, area described or defined by its biotic, climatic, and soil conditions as related to its capacity to produce vegetation. (ii) Area sufficiently uniform in biotic, climatic, and soil conditions to produce a particular climax vegetation.

Site directed mutagenesis : Insertion of a different nucleotide at a specific site in a molecule using recombinant DNA methodology. ***OR*** An intentional alteration in a DNA sequence that can be genetically inherited.

Site-specific recombination : Recombination of nonhomologous genetic material with a chromosome at a specific site.

SIW : Self inflicted wound.

SK : Skeletal.

Skin-associated lymphoid tissue (SALT) : The lymphoid tissue in the skin that forms a first-line defense as a part of nonspecific immunity.

Slage : Fermented plant material with increased palatability and nutritional value for animals, which can be stored for extended periods.

Slash-and-burn agriculture : The cutting down and burning of tropical vegetation to make mineral nutrients available for use by introduced agricultural crops.

S-layer : A regularly structured layer composed of protein or glycoprotein that lies on the surface of many bacteria. It may protect the bacterium and help give it shape and rigidity.

SLB : Short leg brace.

SLC : The chemokine known as SLC is produced by lymphatic vessels and attracts dendritic cells. ***OR*** Short leg cast.

Slime : The viscous extracellular glycoproteins or glycolipids produced by *Staphylococci* and *Pseudomonas aeruginosa* bacteria that allows them to adhere to smooth surfaces such as prosthetic medical devices and catheters. More generally, the term often refers to an easily removed, diffuse, unorganized layer of extracellular material that surrounds a bacterial cell.

Slime layer : A diffuse or loosely fitting capsule; also known as glycocalyxor A layer of diffuse, unorganized, easily removed material lying outside the bacterial cell wall. ***OR*** Diffuse layer of polysaccharide exterior to the cell wall in some bacteria.

Slime mold : A common term for members of the divisions Acrasiomycota and Myxomycota. ***OR*** Nonphototrophic eukaryotic micro-organism lacking cell walls, which aggregate to form fruiting structures (cellular slime molds) or simply masses of protoplasm (acellular slime molds).

Slow sand filter : A bed of sand through which water slowly flows; the gelatinous microbial layer on the sand grain surface removes waterborne micro-organisms, particularly Giardia, by adhesion to the gel. This type of filter is used in some water purification plants.

Slow virus disease : A progressive, pathological process caused by a transmissible agent (virus or prion) that remains clinically silent during a prolonged incubation period of months to years after which progressive clinical disease becomes apparent.

SLR : Straight leg rasing.

SLS : Short leg splint.

SLT : Slight.

Sludge (sluj) : Intermittent releases or discharges of industrial wastes. ***OR*** The settleable solids separated from liquids during processing; the deposits of foreign materials on the bottoms of streams or other bodies of water. ***OR*** A general term for the precipitated solid matter produced during water and sewage treatment; solid particles composed of organic matter and microorganisms that are involved in aerobic sewage treatment (activated sludge).

Sludge age : A measure of the length of time a particle of suspended solids has been retained in the activated sludge process.

SLWC : Short leg walking cast.

Small G proteins : Small G proteins are monomeric G proteins such as *Ras*, that act as intracellular signaling molecules downstream of many transmembrane signaling events. They bind GTP in their active form, and hydrolyze it to GDP to become inactive.

Small intestine : The portion of the digestive tract between the stomach and the ileocecal junction at the beginning of the large intestine; composed of the duodenum, jejunum, and ileum; approximately twenty feet in length.

Small pre-B cells : See large pre-B cells.

Smallpox (variol) : Once a highly contagious, often fatal disease caused by a poxvirus. Its most noticeable symptom was the appearance of blisters and pustules on the skin. Vaccination has eradicated smallpox throughout the world. ***OR*** Smallpox is an infectious disease, caused by the virus variola, that once killed at least 10% of infected people. It has now been eradicated by vaccination.

SMD : Small molecular drugs.

Smuts (Teliomycetes) : Obligate parasites and pathogens of plants that cause smut on various plant parts such as silene anthers, corn kernels, onion bulbs, and rice grains.

Snapping division : A distinctive type of binary fission resulting in an angular or a palisade arrangement of cells, which is characteristic of the genera *Arthrobacter* and *Corynebacterium*.

snRNA : Small nucleolar *RNA*.

SNP : Single nucleotide polymorphism.

snRNA : Small nuclear RNA.

snRNP : Small (nuclear) ribonucleoprotein.

SNS : Sympathetic nervous system.

SNST : Sciatic nerve stretch test.

SNT : Sinuses, nose and throat.

SO : Supraorbital; sutures out.

SOAP : Subjective, objective, assessment, plan.

SOB : Short of breath.

SOD : Superoxide dismutase.

Sodium bicarbonate : A sodium and a hydrogen ion bonded to a carbonateion ($NaHCO_3$).

Soil : (i) Unconsolidated mineral or material on the immediate surface of the earth that serves as a natural medium for the growth of land plants. (ii) Unconsolidated mineral or organic matter on the surface of the earth that has been subjected to and influenced by genetic and environmental factors of: parent material, climate (including water and temperature effects), macroorganisms and microorganisms, and topography, all acting over a period of time and producing a product soil that differs from the material from which it is derived in many physical, chemical, biological, and morphological properties, and characteristics. ***OR*** The weathered layer of the earth's crust with living organisms, including decomposers, and the products of their decay in termingled.

Soil aggregate : Unit of soil structure generally < 10 mm in diameter and formed by natural forces and substances derived from root exudates and microbial products which cement smaller particles into larger units.

Soil atmosphere : Gases occupying the pore space in soil. Generally characterized as having a greater percentage of carbon dioxide and a lesser percentage of oxygen than the overlying air.

Soil biochemistry : Branch of soil science concerned with enzymes and the reactions, activities, and products of soil microorganisms.

Soil classification : See classification.

Soil crust : A soil surface layer that is either denser, structurally different or more cemented than the material immediately beneath it.

Soil extract : Solution separated from a soil suspension or from a soil by filtration, centrifugation, suction or pressure.

Soil health : See soil quality.

Soil horizon : Layer of soil or soil material approximately parallel to the land surface and differing from adjacent genetically related layers in physical, chemical, and biological properties or characteristics such as color, structure, texture, consistency, kinds and number of organisms present, and degree of acidity or alkalinity.

Soil microbiology : Branch of soil science concerned with soil inhabiting microorganisms and their functions and activities.

Soil organic matter (SOM) : Organic fraction of the soil exclusive of undecayed plant and animal residues. Often synonymous with humus.

Soil population : (i) All the organisms living in the soil, including plants and animals. (ii) Members of the same taxa.

Soil pore : That part of the bulk volume of soil not occupied by soil particles. Soil pores have also been referred to as interstices or voids.

Soil quality : Continued capacity of soil to function as a vital living system to sustain biological productivity, maintain the quality of the environment, and promote plant, animal, and human health.

Soil salinity : Amount of soluble salts in a soil. The conventional measure of soil salinity is the electrical conductivity of a saturation extract.

Soil science : Science dealing with soils as a natural resource on the surface of the earth including soil formation, classification and mapping, and physical, chemical, biological, and fertility properties of soils per se; and these properties in relation to their use and management.

Soil series : Lowest category of U.S. system of soil taxonomy; a conceptualized class of soil bodies (polypedons) that have limits and ranges more restrictive than all higher taxa. The soil series serve as a major vehicle to transfer soil information and research knowledge from one soil area to another.

Soil solution : Aqueous liquid phase of the soil and its solutes.

Soil structure : Combination or arrangement of primary soil particles into secondary particles, units, or peds. The secondary units are characterized and classified on the basis of size, shape, and degree of distinctness into classes, types, and grades, respectively.

Soil texture : Relative proportions of the various soil separates in a soil. The major textural classes are sand, silt, and clay.

Soil water potential (total) : Amount of work that must be done per unit quantity of pure water in order to transport reversibly and isothermically an infinite quantity of water from a pool of pure water, at a specified elevation and at atmospheric pressure, to the soil water (at the point under consideration). Informally, the amount of energy that must be expended to extract water from soil. The total potential (of soil water) consists of the following: gravitational potential, matric potential, and osmotic potential.

Solanaceae : Plant family commonly known as the night shade family which includes many toxic species.

Solanum **:** The genus to which the night shades belong.

Solarization : Method to control pathogens and weeds where moistened soil in hot climates is covered with transparent polyethylene plastic sheets, thereby trapping incoming radiation.

Solitary : Alone.

Solstice : The summer day when the daylight is the longest and the winter day when the night time is the longest.

Soluble : Matter or compounds capable of dissolving into a solution.

Soluble BOD : Soluble BOD is the BOD of water that has been filtered in the standard suspended solids test.

Soluble fibre : Fibre that is almost completely broken down by bacteria in the large intestine.

Soluble protein : See globular protein.

Solute : A smaller amount of a substance which is dissolved in a greater amount of another substance. ***OR*** A liquid mixture of dissolved substances, displaying no phase separation.

Solution : A homogeneous mixture of two or more substances.

Solvent : A substance that can dissolve another substance, for example, water or ethanol.

Somatic cell : Any cell in a plant or animal that is not a gamete (in a germ cell or sex cell).

Somatic diversification theory : When immunologists discovered that antibodies were variable, they entertained various theories, including the somatic diversification theory that postulated that the genes for immunoglobulin were constant, and that they diversified in somatic cells. This turned out to be partly true, as somatic hypermutation is now well established. However, other theories were needed to explain other features of antibody diversity, including somatic gene rearrangement and isotype switching.

Somatic gene therapy : See gene therapy.

Somatic hypermutation : During B-cell responses to antigen, the V-region DNA sequence undergoes somatic hypermutation, resulting in the generation of variant immunoglobulins, some of which bind antigen with a higher affinity. This allows the affinity of the antibody response to increase. These mutations affect only somatic cells, and are not inherited through germline transmission.

Somatic migration : Migration in the host by a parasite through the lungs into the systemic circulation distributed throughout the body and encyst in the tissues.

Somatic recombination : During lymphocyte development, receptor gene segments undergo somatic recombination to generate intact V-region exons that encode the V region of each immunoglobulin and T-cell receptor chain. These events occur only in somatic cells; the changes are not inherited.

Somatic cell : A general body cell with two sets of chromosomes.

Somatic NS : The functional part of the NS which controls voluntary muscles and reflexes.

SOP : Standard operating procedure.

Sordaria (Ascomycetes) : Common fungus on dung. One species, *S. fimicola* is fairly common and is found on other substrates besides dung.

Socarp : The fruiting structure of the Acrasiomycetes.

Sus : A type of fruiting structure composed of a mass of spores or sporangia.

SOS : Stimulation of senses; repeat once if necessary.

SOS repair : A complex, inducible repair process that is used to repair DNA when extensive damage has occurred.

SOS system : A set of DNA repair enzymes and regulatory proteins that regulate their synthesis so that maximum synthesis occurs when the DNA is damaged.

Source : The location or object from which a pathogen is immediately transmitted to the host, either directly or through an intermediate agent.

Southern blot : Hybridization of single stranded nucleic acid (DNA or RNA) to DNA fragments immobilized on a filter.

Southern blotting technique : The procedure used to isolate and identify DNA fragments from a complex mixture. The isolated, denatured fragments are transferred from an agarose electrophoretic gel to a nitrocellulose filter and identified by hybridization with probes. ***OR*** A method for detecting a specific DNA restriction fragment, developed by Edward Southern. DNA from a gel electrophoresis pattern is blotted onto nitrocellulose paper; then the DNA is denatured and fixed on the paper. Subsequently the pattern of specific sequences in the Southern blot can be determined by hybridization to a suitable probe and autoradiography. A Northern blot is similar, except that RNA is blotted instead onto the nitrocellulose paper.

SP : Spine, spinal.

Spacer : See 12/23 rule.

Spadix : The stalk-like portion of an Araceae inflorescence which bears the flowers.

SPAR : Single primer amplification reaction.

Spathe : The hood-like covering over the spadixinthe Araceae.

Spatial variability : Variation in soil properties (i) laterally across the landscape, at a given depth, or with a given horizon, or (ii) vertically downward through the soil.

Spawn : The act of reproduction of fishes. The mixing of the sperm of a male fish and the eggs of a female fish.

Spawning : The laying of eggs by aquatic animals like fish or frogs.

Specification sheet : Detailed information of a product including, tests, color, odor, specific gravity, bacterial strains, other major ingredients, etc.

Specialised cells : Cells that have been allocated a "job" and have made changes to the way they work in order to do that job. In plants, these cells can switch back to being unspecialised cells — or stem cells. In animals, specialised cells cannot revert to stem cells.

Specialized transduction : See restricted transduction.

Species : In microbiology, a collection of closely related strains sufficiently different from all other strains to be recognized as a distinct unit. ***OR*** A group of living organisms that can interbreed to produce viable offspring. ***OR*** In microbiology, a collection of closely related strains sufficiently different from all other strains to be recognized as a distinct unit. ***OR*** A group of actually or potentially interbreeding populations that are reproductively is olated from (cannotinterbreed with) other such groups.

Specific activity : Amount of enzyme activity units per mass of protein. Often expressed as micromoles of product formed per unit time per milligram of protein. Also used in radiochemistry to express the radioactivity per mass of material (radioactive and nonradioactive).

Specific epithet : Designation of a particular organism in the binomial nomenclature system. For example, coli is the specific epithet of *Escherichia coli*.

Specific gravity : Weight of a particle, substance or chemical solution in relation to an equal volume of water.

Specific immune response (acquired specific immunity) : See acquired immunity.

Specific gravity : The density of a substance divided by the density of water at the same temperature.

Specificity : The specificity of an antibody determines its ability to distinguish the immunogen from other antigens.

Spectratyping : One uses spectratyping to define certain types of DNA gene segments that give a repetitive spacing of three nucleotides, or one codon.

Spectrophotometer : An instrument for producing or recording a spectrum and measuring the photometric intensity of each wavelength present.

Spermosphere : The region surrounding a germinating seed where released organic matter stimulates microbial growth. ***OR*** Area of increased microbial activity around a germinating seed.

Sphaerocysts : Globose cells in the flesh of Russulaceae, Hypholoma and Stropharia.

Spheroplast : Osmotically sensitive cell lacking rigidity of and structural containment supplied by cell wall. ***OR*** A relatively spherical cell formed by the weakening or partial removal of the rigid cell wall component (e.g., by penicillin treatment of Gram-negative bacteria). Spheroplasts are usually osmotically sensitive.

Sphygmomanometer : Instrument used to measure bloodpressure.

Spicules : Chitinous structures found in male nematodes, usually paired. They are inserted in the female genital opening during copulation.

Spike : See peplomer.

Spinulose : Covered in small spines.

Spirillum : Any helically shaped (spiral) bacterial cell; *spirilla* (plural) (see also *Bacillus*, *Coccus*).

Spirillum (plural, spirilli) : A rigid, spiral-shaped bacterium. ***OR*** (i) Bacterium with a spiral shape which is relatively rigid. (ii) Bacterium in the genus *Spirillum*.

Spirochete : A flexible, spiral-shaped bacterium with periplasmic flagella.

Spirogyra **:** Genus of algae in Kingdom Protista; Division Chlorophyta which forms colonial strands of cells with spiral chloroplasts.

SPL : Splint.

Spleen (spløen) : A secondary lymphoid organ where old erythrocytes are destroyed and blood-borne antigens are trapped and presented to lymphocytes. ***OR*** The spleen is an organ in the upper left side of the peritoneal cavity containing a red pulp, involved in removing senescent blood cells, and a white pulp of lymphoid cells that respond to antigens delivered to the spleen by the blood.

Splicing : See RNA splicing.

Split interrupted gene : A structural gene with DNA sequences that code for the final RNA product (expressed sequences or exons) separated by regions coding for RNA absent from the mature RNA (intervening sequences or introns).

Spongiform encephalopathies : Degenerative central nervous system diseases in which the brain has a spongy appearance; they appear due to prions.

Spongin : Proteinacous compound of which the spicules in Demospongiae are composed.

Spontaneous generation : The idea that non-living objects could give rise to living organisms, disproved by Redi's and Pasteur's experiments. ***OR*** The hypothesis that living organisms can arise from nonliving matter.

Sporadic cases : Infections and disease occuring randomly throughout a population (see also point source outbreak)

Sporadic disease : A disease that occurs occasionally and at random intervals in a population.

Sporangiolum : A small sporangium producing a small number of sporangiospores.

Sporangiophore : A specialized hypha that bears a sporangium.

Sporangiospore : Spore formed within a sporangium by fungi in the phylum Zygomycota. ***OR*** An asexual spore produced within a sporangium. ***OR*** A spore born within a sporangium.

Sporangium : A sac-like structure producing asexual spores endogenously by cytoplasmic cleavage. ***OR*** Fungal structure which converts its cytoplasm into a variable number of sporangiospores; formed by fungi in the phylum Zygomycota. ***OR*** Area(s) on the sporophyte generation of a plant in which spores are produced.

Spore : The reproductive part of some plants and fungi (instead of a seed), which an develop into a new organism. ***OR*** Reproductive structure often of a single rounded cell with a thickened wall that disseminates the fungus. ***OR*** A reproductive propagule formed by either meiosis or mitosis. However, if by asexual means, cleavage of cytoplasm is usually involved.

Spore-producing structures : Fungal parts involved in spore production such as hyphae, conidiophores, phialides, and fruiting bodies.

Sporobolomyces (Yeast) : Can be commonly detected in air samples. Frequently encountered indoors in water tanks, humidifiers, drain pans, etc.

Sporobolomyces sp. **:** Reported to be allergenic.

Sporocyst : In the trematode life cycle, the miracidium will invade a snail, lose the cilia and develop into an elongated sac, called the sporocyst, from which the redia will develop (in some species the sporocyst gives rise to a daughter sporocyst) Sporozoite. The initial invasive stage of most of the Apicompexan protozoa.

Sporodochium (pl. sporodochia) : A cushion-shaped mass of hyphae bearing conidiophores.

Sporogenesis : See sporulation.

Sporophore : Fungal fruiting body.

Sporophyte : In plants and algae, the 2n generation which produces spores.

Sporoschisma (Hyphomycetes) : Found on rotten wood and dead stems.

Sporothrix (Hyphomycetes) : Teleomorph: Ophiostoma (Ascomycetes) Isolated from soil, live or dead plants, and peat moss. *S. schenckii* is an agent of human sporotrichosis, cutaneous infection, and ocular mycosis, usually in immuno-compromised people.

Sporothrix sp. **:** Can cause sporotrichosis. Usually only in populations which are immune compromised.

Sporotrichosis : A subcutaneous fungal infection caused by the dimorphic fungus *Sporothrix schenckii*.

Sporotrichum (Hyphomycetes) : Teleomorph: Phanerochaete (Basidiomycetes) Can get airborne and be inhaled where it can form giant cells in the lungs.

Sporotrichum sp. **:** Reported to be allergenic. See also *Sporothrix* sp. There is some taxonomic confusion between these two genera. This genera does not cause sporotrichosis.

Sporulation : The process of spore formation. ***OR*** Formation from vegetative cells of metabolically inactive cells that can resist extreme environmental conditions.

SPR : Surface plasmon resonance.

SPR : Sprain.

Spread plate : Method for performing a plate count of microorganisms. A known amount of a serial dilution is spread over the surface of an agar plate. After growth the number of colony forming units is counted. ***OR*** A Petri dish of solid culture medium with isolated microbial colonies growing on its surface, which has been prepared by spreading a dilute microbial suspension evenly over the agar surface. ***OR*** Method for performing a plate count of microorganisms. A known amount of a serial dilution is spread over the surface of an agar plate. After growth the number of colony forming units is counted.

Sputum : The mucous secretion from the lungs, bronchi, and trachea that is ejected (expectorated) through the mouth.

SQ : Squamous; subcutaneous.

Squamose : Covered with scales.

Squamous cells : Flat, scale-like cells.

Squamulose : Covered with very small scales.

SR : Sinus rhythm, stretch reflex; systems review.

Src-family tyrosine kinases : The Src-family tyrosine kinases are receptor-associated protein tyrosine kinases. They have several domains, called Src-homology 1, 2, and 3. The SH1 domain contains the active site of the kinase, the SH2 domain can bind to phosphotyrosine residues, and the SH3 domain is involved in interactions with proline-rich regions in other proteins.

SRP : Signal recognition particle.

SS : Suspended solids - Solids in suspension in water which can be filtered out on a lab filter.

SS : Sacrosciatic; signs and symptoms; sitting and standing.

SSC : Size, shape and consistency.

ssDNA : Single-stranded *DNA*.

SSN : Severely subnormal.

SSP : Semi synthetic penicillin.

SSR : Simple sequence repeat.

ST : Shock therapy; slight trace; stomach.

Stabilize : To convert to a form that resists change. Organic material is stabilized by bacteria which convert the material to gases and other relatively inert substances. Stabilized organic material generally will not give off obnoxious odors.

Stachybotrys (Hyphomycetes) : A common fungal saprophyte found on many substrates like grains, decaying plant materials, textiles, and tobacco. Grows indoors on water-damaged cellulose rich materials, such as sheet rock, paper, ceiling tiles, insulation backing, gypsum board, and wallpaper. The presence of this fungus can be significant due to its ability to produce mycotoxins under certain environmental conditions. Exposure to the toxins can occur through inhalation, ingestion, or skin exposure. It is possible that *Stachybotrys* may play a role in the development of sick building syndrome, but probably only in conjunction with other factors. Until more information is available on the health risks of environmental exposure to *Stachybotrys*, caution should be taken when dealing with this fungus.

Stacking energy : The energy of interaction that favors the face-to-face packing of purine and pyrimidine base pairs.

Stalk (stawk) : A nonliving bacterial appendage produced by the cell and extending from it.

Stamen : A "male" plant part consisting of a filament and an anther.

Standard free energy change : The free energy change of a reaction at 1 atmosphere pressure when all reactants and products are present in their standard states; usually the temperature is 25°C.

Standard reduction potential : A measure of the tendency of a reductant to lose electrons in an oxidation-reduction (redox) reaction. The more negative the reduction potential of a compound, the better electron donor it is.

Stanley Miller : Person who,in 1953, tested Oparin's hypothesis on the abiotic synthesis of organic monomers on early earth.

Stapes : One of the bones in the middle ear.

Staphylococcal enterotoxins (SEs) : Staphylococcal enterotoxins (SEs) cause food poisoning and also stimulate many T cells by binding to MHC class II molecules and the Vp domain of certain T-cell receptors; the staphylococcal enterotoxins are thus super antigens.

Staphylococcal food poisoning : A type of food poisoning caused by ingestion of improperly stored or cooked food in which *Staphylococcus aureus* has grown. The bacteria produce exotoxins that accumulate in the food.

Staphylococcal scalded skin syndrome (SSSS) : A disease caused by *Staphylococci* that produce an exfoliative toxin. The skin becomes red (erythema) and sheets of epidermis may separate from the underlying tissue.

Starch : A complex carbohydrate commonly found in foods, such as potatoes, wheat, rice and corn. ***OR*** A complex polymer of glucose, used by plants and green algae to store surplus sugar for later use.

Starter culture : An inoculum, consisting of a mixture of carefully selected microorganisms, used to start a commercial fermentation.

Stasis : Stagnation or inactivity of the life processes within organisms.

STAT : Immediately.

Stationary phase : Period during the growth cycle of a population in which growth rate equals the death rate. ***OR*** Of microbial growth in a batch culture when population growth ceases and the growth curve levels off.

Statistically significant difference : A mathematical term that indicates that the difference between two (or more) results is greater than the difference you might expect due to chance.

Statistics : The mathematics of the collection, organization, and interpretation of numerical data.

STATs : See Janus kinases.

STD : Sexually transmitted disease.

Steady state : In enzyme-kinetic analysis, the time interval when the rate of reaction is approximately constant with time. The term is also used to describe the state of a living cell where the concentrations of many molecules are approximately constant because of a balancing between their rate of synthesis and breakdown.

Stellate : Star-shaped.

Stem : Stipe.

Stem cell : A cell that has the potential to become any of the specialised cell types that make up an organism. ***OR*** In developmental biology, a cell that proliferates indefinately and that, by differentiating into specialized cells, can serve as a continuous source of new tissue and blood cells (e.g., hematopoietic stem cells).

Stem necrosis : Black areas on stems, usually in an elongated oval or circle or band.

Stem-cell factor, SCF : Stem-cell factor (SCF) is a transmembrane protein on bone marrow stromal cells that binds to c-Kit, a signaling receptor carried on developing B cells and other developing white blood cells.

Stem-nodulating rhizobia : Rhizobia (members of the genera *Rhizobium, Bradyrhizobium*, and *Azorhizobium*) that produce nitrogen-fixing structures above the soil surface on plant stems. These most often are observed in tropical plants and produced by *Azorhizobium*.

Stemphylium (Hyphomycetes) : A common saprobe typically is found on dead plants and wood. It has been also isolated from air, paper, and cellulosic materials.

Stenohaline : Having a narrow range of salt concentration tolerance.

Stenothermal : Having a narrow range of temperature tolerance.

Stereoisomers : Isomers that are nonsuper imposable mirror images of each other.

Sterigma : The stem at the top of the basidium which bears each spore. ***OR*** A small pointed structure upon which a basidiospore forms.

Sterile : Free of any microorganisms, viruses or other pathogens. ***OR*** The most stringent category used during Leg 201. Sterile indicates the absence of contaminating prokaryotic cells that are capable of metabolism and growth and their nucleic acids.

Sterilization : Rendering an object or substance free of viable microbes. ***OR*** The removal or destruction of all living microorganisms, including pathogenic and other bacteria, vegetative forms and spores.

Sternum : The breast bone.

Steroids : Compounds that are derivatives of a tetracyclic structure composed of a cyclopentane ring fused to a substituted phenanthrene nucleus.

Stethoscope : An instrument with a diaphragm or bell and ear pieces for listening to heart and breathing sounds.

Sticky cards : Sticky cards are a monitoring tool for insect pest levels in plant growing areas. Typically blue or yellow, the cards have a very sticky surface that traps thrips and other insects for regular tracking of pest levels.

Stigma : A light-sensitive eyespot, which is found in some algae and photosynthetic protozoa; it is believed to be involved in phototaxis, at least in some cases.

Stigma : Can mean a reddish-colored, light-sensitive organelle in Euglena or the sticky tip of the pistilina flower.

Stipe : Stem usually of a toadstool.

Stipitate : With a stem.

Stipules : Enlarged, leaflet- like areas at the base of a leaf petiole.

STL : Swelling, tenderness, limitation.

STLOM : Swelling, tenderness, limitation of motion.

Stolon : A running hypha from which rhizoids and sporangiospores arise.

Stoma : Mouth opening of predacious and bacteriovorous nematodes.

Stomate : A pore for air exchange in a leaf.

Stoneworts : A group of approximately 250 species of algae that have a complex growth pattern, with nodal regions from which whorls of branches arise; they are abundant in fresh to brackish waters.

Stage polysaccharide : Energy reserve deposited in the cell when there is an excess of carbon available. These are usually deposited as large granules in the cell. The most common example of a storage polysaccharide in plants is starch. Its counterpart in animal cells is glycogen.

Storm sewer : A separate pipe, conduit or open channel (sewer) that carries runoff from storms, surface drainage and street wash, but does not include domestic and industrial wastes. Storm sewers are often the recipients of hazardous or toxic substances due to the illegal dumping or hazardous wastes or spills created by accidents involving vehicles and trains transporting these substances.

STP : Standard temperature (25 ºC) and pressure (300 mm Mercury).

STP : Standard temperature and pulse.

STP test : Laboratory test for nitrifiers. For CF1100, 7110 & 1400 to be within specifications, this test must show a positive color change result (from fuschia to yellow) within 7 days. When testing the 50X concentrate of these three products, a positive color change is expected within 24 hours.

STR : Sequence tandem repeat.

STR : Straight; strain; strength.

Strain : Population of cells all descended from a single pure isolate. ***OR*** A population of organisms that descends from a single organism or pure culture isolate. ***OR*** A group of organisms of the same species that are more like one another than others in the species.

Stratification : Layering of something such as the vegetation in a community, rocks, or the cells in our bodies.

Stratum : A layer of something,such as rocks or cells—"stratified" cells or rocks have several layers.

Streak plate : A Petri dish of solid culture medium with isolated microbial colonies growing on its surface, which has been prepared by spreading a microbial mixture over the agar surface, using an inoculating loop.

Streptococcal pneumonia **:** An endogenous infection of the lungs caused by *Streptococcus pneumoniae* that occurs in predisposed individuals.

Streptococcal strep throat : One of the most common bacterial infections of humans. It is commonly referred to as "strep throat." The disease is spread by droplets of saliva or nasal secretions and is caused by *Streptococcus* spp. (particularly group A *Streptococci*).

Streptokinase (fibrinolysin) : Extracellular enzyme of *Streptococci* that catalyzes the conversion of plasminogen to plasmin; used clinically to dissolve blood clots (thrombi).

Streptolysin-O (SLO) : A specific hemolysin produced by *Streptococcus pyogenes* that is inactivated by oxygen (hence the "O" in its name). SLO causes beta-hemolysis of blood cells on agar plates incubated anaerobically.

Streptolysin-S (SLS) : A product produced by *Streptococcus pyogenes* that is bound to the bacterial cell but may sometimes be released. SLS causes beta hemolysis on aerobically incubated blood-agar plates and can act as a leukocidin by killing leukocytes that phagocytose the bacterial cell to which it is bound.

Streptomycin : A bactericidal aminoglycoside antibiotic produced by *Streptomyces griseus*.

Stress : An organism's total response to environmental demands or pressures and includes any event hat a person considers a threat. The symptoms of stress can be either physical and psychological. Irritable Bowel Syndrome (IBS) is regarded as a stress-related physical illness.

Striate : Having lines or minute furrows. ***OR*** With fine radiating lines or furrows around the cap margin.

Strict anaerobes : See obligate anaerobes.

Strobila : The string of proglottides which make up the "body" of a tapeworm.

Strobilocercus : A type of larva of cyclophyllidian tapeworms (i.e. The larva of *Taenia taeniaeformis*).

Stroke : When part of the brain is damaged due to its blood supply being reduced or stopped.

Stroma : The chloroplast matrix that is the location of the photosynthetic carbon dioxide fixation reactions. ***OR*** A hard conglomerate of hyphae, found in some Ascomycetes.

Stroma : The fluid within chloroplasts.

Stromal cells : The development of B lymphocytes and T lymphocytes occurs in association with stromal cells, which provide various soluble and cell-bound signals to the developing lymphocyte.

Stromatolite : Dome-like microbial mat communities consisting of filamentous photosynthetic bacteria and occluded sediments (often calcareous or siliceous). They usually have a laminar structure. Many are fossilized, but some modern forms occur.

Structural domain : An element of protein tertiary structure that forms an independent folding unit.

Structural gene : A gene that codes for the synthesis of a polypeptide or polynucleotide with a nonregulatory function. ***OR*** A gene encoding the amino acid sequence of a polypeptide chain.

Structural polysaccharide : Polysaccharide that serves primarily as a structural element in cell walls and coats and intercellular spaces, and connective tissue where they give shape, elasticity, or rigidity to plant or animal tissues and protection and support to unicellular organisms. Cellulose is the predominant structural polysaccharide in plant cell walls and chitin is abundant in fungal cell walls and insect exoskeletons.

Structural protein : A protein that serves a structural function.

Structure : See soil structure.

STS : Sequence-tagged site.

Stud RAM : A male sheep that is selected in a breeding programme to father lots of lambs because of its desirable physical or genetic characteristics.

Stunting : Smaller than normal plant size. This symptom may go unnoticed unless healthy plants are available for comparison.

Style : A flower part that is the stalk-like,middle portion of the "female" part.

Stylet : Small, spearlike mouth structure used by some nematodes to pierce cells to feed on protoplasm; usually associated with herbivorous or fungivorous nematodes.

Subacute : Course of a disease that falls somewhere between acute and chronic

Subacute sclerosing panencephalitis : Diffuse inflammation of the brain resulting from virus and prion infections.

Subatomic particles : Small particles like protons,electrons,and neutrons, which make up atoms.

Subcutaneous : Antigens can be injected into the subcutaneous layer to induce an adaptive immune response. See also intracutaneous.

Subdecurrent : Where gills run down the stem for only a short distance.

Subgingival plaque : The plaque that forms at the dentogingival margin and extends down into the gingival tissue.

Subglobose : Almost spherical, or not quite round or spherical.

Subhymenium : Differentiated tissue that gives rise to the hymenium.

Subovoid : Not quite round.

SUBQ : Subcutaneous.

Subsidence : Decline in soil volume due to oxidation of organic matter.

Subspecies (ssp.) : Taxonomic level below species

Substrate : The material on which a fungus grows e.g. On wood. ***OR*** (i) Substance, base, or nutrient on which an organism grows. (ii) Compounds or substances that are acted upon by enzymes or catalysts and changed to other compounds in the chemical reaction. ***OR*** A molecule undergoing reaction with an enzyme. Any one of a number of essential nutrients for microbial growth. Specific substrates vary with specific physiology of microorganisms (bacteria or eukaryotic cells).

Substrate,Substratum : Surface in / on which an organism lives.

Substrate level phosphylation : The synthesis of ATP from ADP by phosphorylation coupled with the exergonic breakdown of a high-energy organic substrate molecule. ***OR*** Synthesis of highenergy phosphate bonds through reaction of inorganic phosphate with an activated (usually) organic substrate.

Substratum : A surface within a liquid matrix that can serve as a physical support upon which biofilms can accumulate.

Subsurface biosphere : The region below the plant root zone where microbial populations can grow and function.

Subtmentose : More or less woolly.

Subunit : Individual polypeptide chains in a protein.

Succession : Gradual process brought about by the change in the number of individuals of each species of a community and by the establishment of new species that gradually replace the original inhabitants.

Sucrose : Common table,beet,or cane sugar :a disaccharide consisting of glucose + fructose.

SUD : Sudden unexpected or unexplained death.

Suffix "itis" : Generally indicates inflammation.

Sugar : Any of several small carbohydrates, such as glucose, which are "sweet" to the taste.

Sulcate : Grooved.

Sulfate reduction : The process of sulfate use as an oxidizing agent, which results in the accumulation of reduced forms of sulfur such as sulfide, or incorporation of sulfur into organic molecules, usually as sulfhydryl groups.

Sulfonamide : A chemotherapeutic agent that has the SO_2-NH_2 group and is a derivative of sulfanilamide.

Sulfur cycle : Sequence of transformations undergone by sulfur where it is taken up by living organisms, transformed upon death and decomposition of the organism, and converted ultimately to its original state of oxidation.

SUP : Superior; supination; supine.

Superantigen : Superantigens are bacterial proteins that stimulate the immune system much more extensively than do normal antigens. They stimulate

T cells to proliferate nonspecifically through simultaneous interaction with class II MHC proteins on antigen-presenting cells and variable regions on the b chain of the T-cell receptor complex. Examples include streptococcal scarlet fever toxins, staphylococcal toxic shock syndrome toxin-1, and streptococcal M protein. ***OR*** Superantigens are molecules that stimulate a subset of T cells by binding to MHC class II molecules and Vb domains of T-cell receptors, stimulating the activation of T cells expressing particular Vb gene segments. The staphylococcal enterotoxins are one of the sources of superantigens.

Superbug : A bacteria that is resistant to treatment with antibiotics.

Supercoiled DNA : Supertwisted, covalently-closed duplex DNA.

Superinfection : A new bacterial or fungal infection of a patient that is resistant to the drug(s) being used for treatment.

Superi : Where the ring is attached above the middle of the stem.

Supernatant : Liquid removed from a tank once the solids have settled. ***OR*** Supernatant commonly refers to the liquid between the sludge on the bottom and the scum on the surface of an anaerobic digester. This liquid is usually returned to the influent wet well or to the primary clarifier. ***OR*** The liquid layer that remains after a solution is spun in a centrifuge. ***OR*** The liquid above some solid precipitate which has settled from it.

Superoxide dismutase (SOD) : An enzyme that protects many microorganisms by catalyzing the destruction of the toxic superoxide radical.

Supersaturate : When a solution contains a substance at a very high concentration, higher than is normally thought to be possible.

Supply chain : All of the steps needed to take a basic material and turn it into a finished product ready for sale.

Suppress gene : A gene that can reverse the phenotype of a mutation in another gene.

Suppress mutation : A mutation that overcomes the effect of another mutation and produces the normal phenotype. ***OR*** A mutation that restores a function lost by an initial mutation and that is located at a site different from the initial mutation.

Suppress T cells : See regulatory T cells.

Suppurative : Producing pus.

Surface area : Area of the solid particles in a given quantity of soil or porous medium.

Surface immunoglobulin, SiG : The membrane-bound immunoglobulin that acts as the antigen receptor on B cells is often known as surface immunoglobulin (SiG).

Surface soil : Uppermost part of the soil, ordinarily moved in tillage, or its equivalent in uncultivated soils ranging in depth from 7 to 20 cm. Frequently designated as the surface layer, the Ap layer, or the Ap horizon.

Surfactant : A substance that lowers the surface tension of a liquid. ***OR*** Surface-active agent. The active agent in detergents that possesses a high cleaning ability. Used in a spray solution to improve its sticking and wetting properties when applied to plants, algae, or petroleum.

Surrogate light chain : The surrogate light chain is made up of two molecules called vpreb and ë5. Together, this chain can pair with an in-frame heavy chain, move to the cell surface, and signal for pre-B-cell growth.

Susceptibility : The likelihood, or chance, that someone will be affected by something. For example, someone who is very susceptible to a particular disease may be more likely to get that disease than others, who are less susceptible.

Sustainable : A resource or system that meets present needs without compromising those of future generations.

SV : Sludge volume - a settling test using a two liter settleometer to measure sludge quality expressed in percent and related to time, ie. 80% in five minutes or 30% in 30 minutes. Also used to determine the rate of settling.

SV : Sedimentation velocity.

SV30 : The value obtained in a 30 minute settleometer test.

Svedberg unit : The unit used in expressing the sedimentation coefficient; the greater a particle's Svedberg value, the faster it travels in a centrifuge.

Svedberg unit (S) : The unit used to express the sedimentation constant (S = 10-l3 sec). The sedimentation constant S is proportional to the rate of sedimentation of a molecule in a given centrifugal field and is related to the size and shape of the molecule.

SVI : Sludge volume index - a settling test used to measure sludge quality.

SVR : The volume of sludge blanket divided by the daily volume of sludge pumped from the thickener.

SW : Stab wound; swollen, swelling.

Swab : A wad of absorbent material usually wound around one end of a small stick and used for applying medication or for removing material from an area; also, a dacron-tipped polystyrene applicator.

Swarm cell : A flagellated cell; the term is usually applied to the motile cells of the Myxomycota.

Switch region : When isotype switching occurs, the active heavy-chain V-region exon undergoes somatic recombination with a 3' constant-region gene at a switch region of DNA. These DNA joints do not need to occur at precise sites, because they occur in intronic DNA. Thus, all switch recombinations are productive.

SX : Symptoms.

Syk : See tyrosine kinase.

SYM : Symmetric; symptoms.

Symbiosis : Living together in intimate association of two dissimilar organisms. The interactions between the organisms can be commensal or mutualistic. ***OR*** a relationship in which two dissimilar organisms (symbiotes, symbionts) live in close association with one another. ***OR*** The close relationship of two organisms in proximity, with one benefiting and the other either benefiting (mutualism), not being significantly affected (commensalism), or being harmed (parasitism) by the relationship. ***OR*** Two (or more) species living together and interacting.

Symbiosome : The final nitrogen-fixing form of *Rhizobium* that is active within root nodule cells.

Sympathetic NS : The functional part of the NS which increases energy expenditure and prepares for action.

Sympathetic ophthalmia : When one eye is damaged, there is often an autoimmune response that damages the other eye, a syndrome known as sympathetic ophthalmia.

Sympatric species : Species occupying the same area at the same time.

Sympetalous : A plant with more advanced flowers with at least some united flower parts.

Sympodial : A mode of conidiogenous cell growth which results in the development of conidia on a geniculate or zig-zag rachis.

Symptom : A change during a disease that a person subjectively experiences (e.g., pain, bodily discomfort, fatigue, or loss of appetite). Sometimes the term symptom is used more broadly to include any observed signs. ***OR*** Subjective evidence of disease as reported by the patient (host). Note, therefore, that animals do not have symptoms, only signs, since they cannot talk to the veterinarian.

Synapse : The chemical connection for communi-cation between two nerve cells or between a nerve cell and a target cell such as a muscle cell. ***OR*** A junction between two nerve cells or a nerve and a muscle cell.

Synapsis : The pairing of homologous chromosomes, seen during the first meiotic prophase. ***OR*** The process of pairing of the homologous chromosomes during meiosis.

Syncephalastrum (Zygomycete) : Often isolated from soil and dung in tropical and subtropical regions. Can also be a persistent laboratory contaminant.

Syndrome : See disease syndrome.

Synecology : The study of organisms grouped or associated in the same habitat or environment.

Synergism : Association between organisms that is mutually beneficial. Both populations, however, are capable of surviving in their natural environment on their own. ***OR*** Association between organisms that is mutually beneficial. Both populations are capable of surviving in their natural environment on their own although, when formed, the association offers mutual advantages. ***OR*** Combine effect greater than sum of individual effects.

Synergistic : When two or more organisms coexist in a relationship that is strongly co-operative, so that their combined effect exceeds a simple sum of their individual effects.

Syngamy : Fertilization,union of egg and sperm,to restore the 2n condition.

Syngeneic graft : A syngeneic graft is a graft between two genetically identical individuals. It is accepted as self.

Synnema (pl. synnemata) : A group of erect conidiophores that are cemented together producing conidia at the apex and/or along the sides.

Synthesis : Breaking down organic compounds and converting the degradation products into new cell growth. An energy using process.

Synthetic : Made in a laboratory or factory by a chemical process, usually to imitate a natural process.

Synthetic medium : See defined medium.

Syntrophism : The association in which the growth of one organism either depends on, or is improved by, the provision of one or more growth factors or nutrients by a neighboring organism. Sometimes both organisms benefit. This type of mutualism is also known as cross-feeding or the satellite phenomenon.

Syntrophy : Interaction of two or more populations that supply each other's nutritional needs.

Syphilis : See venereal syphilis.

SYS : System; systolic.

SystemaNaturae **:** Linnaeus'classification scheme published in 1758.

Systematic epidemiology : The field of epidemiology that focuses on the ecological and social factors that influence the development of emerging and reemerging infectious diseases.

Systematics : The scientific study of organisms with the ultimate objective being to characterize and arrange them in an orderly manner; often considered synonymous with taxonomy.

Systemic : Not localized in a particular place of the body; an infection disseminated widely through the body is said to be systemic.

Systemic anaphylaxis : Systemic anaphylaxis is the most dangerous form of immediate hypersensitivity reaction. It involves antigen in the bloodstream triggering mast cells all over the body. The activation of these mast cells causes widespread vasodilation, tissue fluid accumulation, epigiottal swelling, and often death.

Systemic lupus erythematosus : An autoimmune, inflammatory disease that may affect every tissue of the body. ***OR*** Systemic lupus erythematosus (SLE) is an autoimmune disease in which autoantibodies against DNA, RNA, and proteins associated with nucleic acids form immune complexes that damage small blood vessels, especially of the kidney.

Systole : Blood pressure when the ventricles are contracting and putting pressure on the Blood.

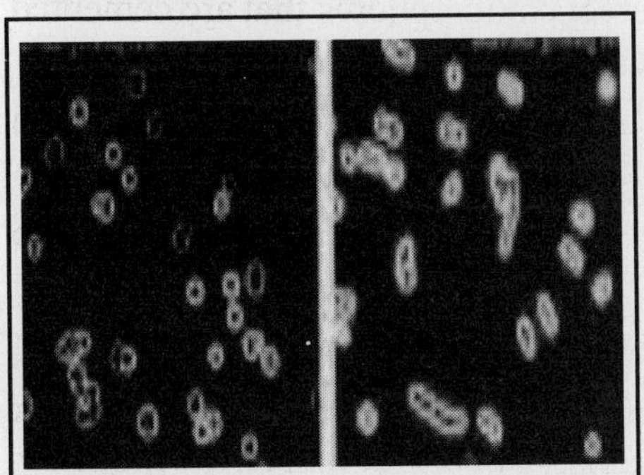

Saccharomyces

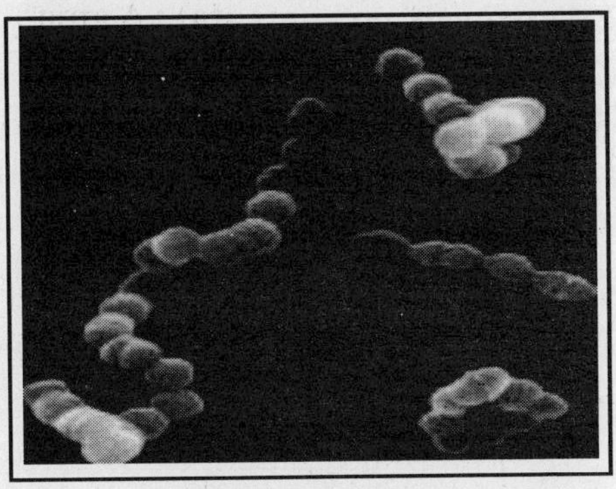

Streptococcus

T

T : Thymine

T : TH1 cells are a subset of CD4 T cells that are characterized by the cytokines they produce. They are mainly involved in activating macrophages, and are sometimes called inflammatory CD4 T cells. ***OR*** TH2 cells are a subset of CD4 T cells that are characterized by the cytokines they produce. They are mainly involved in stimulating B cells to produce antibody, and are often called helper CD4 T cells. ***OR*** The term TH3 cell has been used to describe unique cells that produce mainly transforming growth factor-â in response to antigen; they develop predominantly in the mucosal immune response to antigens that are presented orally. ***OR*** See regulatory T cells.

T cell T lymphocyte : A type of lymphocyte derived from bone marrow stem cells that matures into an immunologically competent cell under the influence of the thymus. T cells are involved in a variety of cell-mediated immune reactions.

T cells : A type of white blood cell that has an important role in activating or directing the immune system to fight infections.

T cells, T lymphocytes : T cells, or T lymphocytes, are a subset of lymphocytes defined by their development in the thymus and by heterodimeric receptors associated with the proteins of the CD3 complex. Most T cells have α:β heterodimeric receptors but ã:ä T cells have a γ:δ heterodimeric receptor.

T lymphocyte : See T cell.

T&A : Tonsils and adenoids.

T&R : Tenderness and rebound.

t.i.d. : Three times daily (Latin, ter in die) (see also b.i.d; q.d.; q.i.d.)

TA : Traffic accident.

TAB : Therapeutic abortion.

TACH : Tachycardia.

Tachyzoite : Extra-intestinal stage of Toxoplasma that are found in in vacuoles muscle, liver, brain, and lung found in intermediate and definitive hosts. Sporulated oocysts are ingested and liberated sporozoites rapidly penetrate the intestinal wall and through the blood. It is the invasive and proliferative stage of Toxoplasma.

TACI, transmembrane activator and CAML-interactor : TACI (transmembrane activator and CAML-interactor) is a member of the TNF-receptor family and is one of two major receptors for BLYS. It is found on B cells, dendritic cells, and T cells and is probably the important receptor for receiving signals from BLYS.

Tacrolimus : Tacrolimus, or FK506, is an immuno-suppressive polypeptide drug that inactivates T cells by inhibiting signal transduction from the T-cell receptor. Tacrolimus and cyclosporin A are the most commonly used immunosuppressive drugs in organ transplantation.

Taeniolella (Hyphomycetes) : Common fungus on dead branches, wood, and senescent leaves. Was isolated from human cutaneous and subcutaneous lesions.

TAH : Total abdominal hysterectomy.

TAL : Tendon achilles lengthening.

Tandem duplication : A duplication in which the repeated regions are immediately adjacent to one another.

TAP1, TAP2 : TAP1 and TAP2 (transporters associated with antigen processing) are ATP-binding cassette proteins involved in transporting short peptides from the cytosol into the lumen of the endoplasmic reticulum, where they associate with MHC class I molecules.

Tapasin, TAP-associated protein : Tapasin, or the TAP-associated protein, is a key molecule in the assembly of MHC class I molecules; a cell deficient in this protein has only unstable MHC class I molecules on the cell surface.

Target cells : The functions of effector T cells are always assayed by the changes that they produce in antigen-bearing target cells. These cells can be B cells, which are activated to produce antibody; macrophages, which are activated to kill bacteria or tumor cells; or labeled cells that are killed by cytotoxic T cells.

Tarsals, Tarsi : The heel bones.

Tat : The Tat protein is a product of the tat gene of HIV. It is produced when latently infected cells are activated, and it binds to a transcriptional enhancer in the long terminal repeat of the provirus, increasing the transcription of the proviral genome.

TAT : Tetanus antitoxin.

Taxon : A group into which related organisms are classified.

Taxonomic hierarchy : The classification system developed by Linnaeus including Kingdom, Phylum, Class, Order, Family, Genus, and Species levels.

Taxonomy : Study of scientific classification and nomenclature. ***OR*** The system of description, naming and classification of organisms into orderly categories based upon organismal interrelationships. ***OR*** Study of scientific classification and nomenclature. ***OR*** The science of biological classification; it consists of three parts: classification, nomenclature, and identification.

TB : Total body; tuberculosis.

TB skin test : Tuberculin hypersensitivity test for a previous or current infection with Mycobacterium tuberculosis.

TBA : To be arranged.

TBF : Total body fat.

TBLC : Term birth, living child.

TBSV : Tomato bushy stunt virus.

TBV : Total blood volume.

TBW : Total body weight.

TC : Throat culture; total cholesterol.

TCA : Tricarboxylic acid cycle.

TCA cycle : See tricarboxylic acid cycle.

T-cell antigen receptor (TCR) : The receptor on the T cell surface consisting of two antigen-binding peptide chains; it is associated with a large number of other glycoproteins. Binding of antigen to the TCR, usually in association with MHC, activates the T cell. ***OR*** See T-cell receptor.

T-cell clone : A T-cell clone is derived from a single progenitor T cell. See also cloned T-cell line.

T-cell hybrids : T-cell hybrids are formed by fusing an antigen-specific, activated T cell with a T-cell lymphoma. The hybrid cells bear the receptor of the specific T-cell parent and grow in culture like the lymphoma.

T-cell lines : T-cell lines are cultures of T cells grown by repeated cycles of stimulation, usually with antigen and antigen-presenting cells. When single T cells from these lines are propagated, they can give rise to T-cell clones or cloned T-cell lines.

T-cell receptor, TCR : The T-cell receptor (TCR) consists of a disulfide-linked heterodimer of the highly variable a and b chains expressed at the cell membrane as a complex with the invariant CD3 chains. T cells carrying this type of receptor are often called a : b T cells. An alternative receptor made up of variable g and d chains is expressed with CD3 on a subset of T cells. Both of these receptors are expressed with a disulfide-linked homodimer of æ chains.

T-cell zones : The T-cell zones in lymphoid tissues are enriched in T cells and are distinct from the B-cell zones and the stromal elements.

TCI : Transient cerebral ischemia.

TCN : Tetracycline.

TCR : T-cell receptor.

TD : Total disability.

T-dependent antigen : An antigen that effectively stimulates B-cell response only with the aid of T-helper cells that produce interleukin-2 and B-cell growth factor.

T-DNA : Transferred *DNA*.

TDS : Total dissolved solids is commonly estimated from the electrical conductivity of the water. Pure water is a poor conductor of electricity. Impurities dissolved in the water cause an increase in the ability of the water to conduct electricity. Conductivity, usually expressed in units of microsimens, formerly micromhos or in mg/l, thus becomes an indirect measure of the level of impurities in the water.

TE : Therapeutic exercises.

Tec kinase : Activation of the lymphocyte antigen receptors is linked to activation of PLC-g through members of the Tec kinase family of src-like tyrosine kinases. Other Tec kinases are Btk in B cells, which is mutated in the human immunodeficiency disease X-linked agammaglobulinemia (XLA), and Itk in T lymphocytes.

Teichoic acids : All wall, membrane, or capsular polymers containing glycerophosphate or ribitol phosphate residues.

Teichoic acids : All wall, membrane, or capsular polymers containing glycerophosphate or ribitol phosphate residues. ***OR*** Polymers of glycerol or ribitol joined by phosphates; they are found in the cell walls of Gram-positive bacteria.

Teleomph : The perfect stage or sexual state of the fungus involved in producing meiotic or sexual spores. ***OR*** The sexual state of a fungus. ***OR*** Sexual stage in reproduction in which cells are formed by the process of meiosis and genetic recombination.

Telophase : The last stage of mitosis or meiosis where the chromosomes begin to uncoil, the new daughter nuclei form,and cytokinesis usually occurs.

Temperate phages : Bacteriophages that can infect bacteria and establish a lysogenic relationship rather than immediately lysing their hosts.

Temperate virus : Virus which upon infection of a host does not necessarily cause lysis but whose genome may replicate in synchrony with that of the host.

Template : A polynucleotide chain that serves as a surface for the absorption of monomers of a growing polymer and thereby dictates the sequence of the monomers in the growing chain.

Template strand : A strand of DNA or RNA that specifies the base sequence of a newly synthesized complementary strand of DNA or RNA.

TEN : Tension.

TEND : Tender.

Tenesmus : Straining to defecate.

TENS : Transelectrical nerve stimulator.

Tep : Tissue engineered product.

Teratogen : A substance which causes birth defects

Terminal bud death : Death of the growing point of a shoot, often including surrounding leaves and stems.

Terminal complement components : The complement system can be activated directly or by antibody, but both pathways converge with the activation of the terminal complement components, which may assemble to form the membrane attack complex.

Terminal deoxynucleotidyl transferase, TDT : The enzyme terminal deoxynucleotidyl transferase (TDT) inserts nontemplated or N-nucleotides into the junctions between gene segments in T-cell receptor and immunoglobulin V-region genes. The N-nucleotides contribute greatly to junctional diversity in V regions.

Terminal electron accept : External oxidant (often oxygen) that accepts the electrons as they exit from the electron transport chain.

Termination factors : Proteins that are exclusively involved in the termination reactions of protein synthesis on the ribosome.

Terminator : A sequence that marks the end of a gene and stops transcription.

Terpenes : A diverse group of lipids made from isoprene precursors.

Tertiary response, tertiary immunization : When the same antigen is injected a third time, the response elicited is called a tertiary response and the injection a tertiary immunization.

Tertiary structure : In a protein or nucleic acid, the final folded form of the polymer chain.

Tertiary treatment : The removal from sewage of inorganic nutrients, heavy metals, viruses, etc., by chemical and biological means after microorganisms have degraded dissolved organic material during secondary sewage treatment.

Test : Hard external covering or shell. ***OR*** A loose-fitting shell of an amoeba.

Testa : Hard external covering or shell.

Testosterone : The primary male sex hormone (it is also found in low levels in females).

Tetanolysin : A hemolysin that aids in tissue destruction and is produced by *Clostridium tetani*.

Tetanospasmin : The neurotoxic component of the tetanus toxin, which causes the muscle spasms of tetanus. Tetanospasmin production is under the control of a plasmid gene.

Tetanus : An often fatal disease caused by the anaerobic, spore-forming bacillus *Clostridium tetani*, and characterized by muscle spasms and convulsions.

Tetracyclines : A family of antibiotics with a common four-ring structure, which are isolated from the genus *Streptomyces* or produced semisynthetically; all are related to chlortetracycline or oxytetracycline.

Tetrad : A pair of homologous chromosomes (so four sister chromatids) during meiosis,also known as a bivalent.

Tetrahedron : A geometric figure with four sides /faces.

Tetramer : Structure resulting from the association of four subunits.

Tetrapartite associations : A mutualistic association of the same plant with three different types of micro-organisms.

Tetraploa (Hyphomycetes) : Teleomorph: Massarina (Ascomycetes) Found on stems and leaf bases of many plants. *Tetraploa aristata* has been reported to cause keratitis.

Tetrathyridium : Worm like larva with an invaginated scolex found only in Mesocestoididae.

Texture : See soil texture.

TF : Transcription factor.

TGGE : Thermal gel gradient electrophoresis.

TGS : Transcriptional gene silencing.

TH : Thoracic.

TH0 cell : See regulator T cell.

TH1 cell : See regulator T cell.

TH2 cell : See regulator T cell.

Thallic : A mode of conidial ontogeny where a conidium is formed from a pre-existing hyphal segment or cell.

Thallus : A type of body that is devoid of root, stem, or leaf; characteristic of some algae, many fungi, and lichens. ***OR*** Vegetative body that is not differentiated into tissue systems or organs.

THC : Tetrahydrocortisone.

T-helper (TH) cell : A cell that is needed for T-cell-dependent antigens to be effectively presented to B cells. It also promotes cell-mediated immune responses.

Theory : A set of principles and concepts that have survived rigorous testing and that provide a systematic account of some aspect of nature.

Therapeutics : The branch of medicine focusing on the treatment of disease and how medicines work is called therapeutics.

Thermal death time (TDT) : The shortest period of time needed to kill all the organisms in a microbial population at a specified temperature and under defined conditions.

Thermo receptors : Temperature – sensitive sensory receptors.

Thermoacidophiles : A group of bacteria that grow best at acid pHs and high temperatures; they are members of the Archaea.

Thermocline : Organism whose optimum temperature for growth is between 45 and 85ºC. ***OR*** In a lake is the point where there is a rapid temperature drop with depth.

Thermometer : A calibrated, liquid-filled tube used to monitor the temperature of a solution or substance.

Thermomyces (Hyphomycetes) : A thermophilic fungus that grows rapidly at 40° C.

Thermophile : Organism whose optimum temperature for growth is between 45 and 85ºC. ***OR*** A microorganism that grows at elevated temperatures. ***OR*** A microorganism that can grow at temperatures of 55°C or higher; the minimum is usually around 45°C.

Thermophilic : Requiring very high temperatures for normal growth and development (thermo=heat; philic=fond).

Thermophilic bacteria : Hot temperature bacteria. A group of bacteria that grow and thrive in temperatures above 113 F (45 ºC), such as *Bacillus licheniformis*. The optimum temperature range for these bacteria in anaerobic decomposition is 120 F (49 ºC) to 135 F (57 ºC).

THF : Tetrahydrofolate.

Thiamine : Vitamin B_1 ; a deficiency causes beriberi

Thigmotaxis : Moving in response to touch

Thigmotropism : Stereotropism growing or leaning in response to touch

Thioester : An ester of a carboxylic acid with a thiol or mercaptan.

Thixotropic : A substance that is gel-like when still, but liquid when agitated.

Thomas Malthus : Person who, in 1798, published a paper on sociological conditions in the London slums in which he developed the idea of "survival of the fittest".

TH : Thoracic.

Thacic vertebrae : Those vertebrae which are attached to the ribs.

Thacic duct : The lymph from most of the body, except the head, neck, and right arm, is gathered in a large lymphatic vessel, the thoracic duct, which runs parallel to the aorta through the thorax and drains into the left subclavian vein. The thoracic duct thus returns the lymphatic fluid and lymphocytes back into the peripheral blood circulation.

Thr : Threonine.

THR : Total hip replacement.

Thrips : Thrips are tiny (1/16th of an inch) winged insects that feed on plants by sucking sap from cells of leaves and flowers.

Thrombolytic : Capable of lysing (dissolving) a thrombus.

Thrombosis : Formation, development or presence of a thrombus.

Thrombus (singular) : "clot" in a blood vessel; aggregation of blood factors, including platelets, fibrin, clotting factors, and other cellular elements attached to the inside wall of a blood vessel that often causes occlusion (blockage); Thrombi (plural).

Thrush (thrush) : Infection of the oral mucous membrane by the fungus *Candida albicans*; also known as oral candidiasis.

Thylakoid : A flattened sac in the chloroplast stroma that contains photosynthetic pigments and the photosynthetic electron transport chain; light energy is trapped and used to form ATP and NAD(P)H in the thylakoid membrane. ***OR*** One of the sacs with in a chloroplast; chlorophyll is built in to the Thylakoid membranes.

Thymectomy : Surgical removal of the thymus is called thymectomy.

Thymic anlage : The thymic anlage is the tissue from which the thymic stroma develops during embryogenesis.

Thymic cortex, thymic cortical epithelial cells : The thymic cortex is the outer region of each thymic lobule in which thymic progenitor cells proliferate, rearrange their T-cell receptor genes, and undergo thymic selection, especially positive selection on thymic cortical epithelial cells.

Thymic stroma : The thymic stroma consists of epithelial cells and connective tissue that form the essential microenvironment for T-cell development.

Thymidine : A molecule formed when the nucleotide thymine (T) is attached to a sugar group. Thymidine is one of the building blocks of DNA. ***OR*** One of the four nucleosides found in DNA.

Thymine : A pyrimidine base found in DNA. ***OR*** The pyrimidine 5-methyluracil that is found in nucleosides, nucleotides, and DNA.

Thymocytes : Thymocytes are lymphoid cells found in the thymus. They consist mainly of developing T cells, although a few thymocytes have achieved functional maturity.

Thymus : The thymus, the site of T-cell development, is a lymphoepithelial organ in the upper part of the middle of the chest, just behind the breastbone. ***OR*** A primary lymphoid organ in the chest that is necessary in early life for the development of immunological functions. T-cell maturation takes place here.

Thymus-dependent antigens, TD antigens, thymus-independent antigens, TI antigens, TI-1 antigens, TI-2 antigens : Some antigens elicit responses only in individuals that have T cells; they are calledthymus-dependent antigens or TD antigens. Other antigens can elicit antibody production in the absence of T cells and are called thymus-independent antigens or TI antigens. There are two types of TI antigen: the TI-1 antigens, which have intrinsic B-cell activating activity, and the TI-2 antigens, which seem to activate B cells by having multiple identical epitopes that cross-link the B-cell receptor.

Thymus-dependent lymphocyte : T cell and T lymphocyte are short designations for thymus-dependent lymphocyte, the lymphocyte population that fails to develop in the absence of a functioning thymus.

Ti plasmid : A plasmid obtained from *Agrobacterium tumefaciens* that is used to insert genes into plant cells. ***OR*** Conjugative tumor inducing plasmid present in the bacterium *Agrobacterium tunefaciens* which can transfer genes into plants.

TIA : Transient ischemia attacks.

Tibia : The thick, inside (big-toe side) shin bone.

TID : Three times a day.

TIM : Triose phosphate isomerase.

Time-lapse video microscopy : One uses time-lapse video microscopy to examine processes in biology, that are anywhere from cell migration (fast) to a flower blossoming (slow).

TIN : Three times a night.

T-independent antigen : An antigen that triggers a B cell into immunoglobulin production without T-cell cooperation.

Tinea : A name applied to many different kinds of superficial fungal infections of the skin, nails, and hair, the specific type (depending on characteristic appearance, etiologic agent, and site) usually designated by a modifying term.

Tinea capitis : A fungal infection of scalp hair caused by species of *Trichophyton* or *Microsporum*.

Tinea cpis : A fungal infection of the smooth parts of the skin caused by either *Trichophyton rubrum*, *T. mentagrophytes*, or *Microsporum canis*.

Tinea cruris : A fungal infection of the groin caused by either *Epidermophyton floccosum*, *Trichophyton mentagrophytes*, or *T. rubrum*; also known as jock itch.

Tinea manuum : A fungal infection of the hand caused by *Trichophyton rubrum*, *T. mentagrophytes*, or *E. floccosum*.

Tinea pedis : A fungal infection of the foot caused by *Trichophyton rubrum*, *T. mentagrophytes*, or *E. floccosum*; also known as athlete's foot.

Tinea unguium : A fungal infection of the nail bed caused by either *Trichophyton rubrum* or *T. mentagrophytes*.

Tinea versicol : A fungal infection caused by the yeast, *Malassezia furfur*, that forms brownish-red scales on the skin of the trunk, neck, face, and arms.

Tingible body macrophages : During the process of germinal center formation, cells called tingible body macrophages appear. These are phagocytic cells engulfing apoptotic B cells, which are produced in large numbers during the height of the germinal center response.

Tissue : A group of the same kind of cells with a common structure and function.

Tissue culture : The production of new plants from small amounts of plant tissue under carefully controlled laboratory conditions are said to have been produced by tissue culture.

Tissue dendritic cells : All dendritic cells arise from hematopoietic progenitors that migrate to various locations all over the body. Here, they are referred to as tissue dendritic cells.

Tissue grafts : Transplantation of organs or tissue grafts such as skin grafts is used medically to repair organ or tissue deficits.

Tissue-specific autoimmune disease : Some autoimmune diseases attack particular tissues, such as the b cell in the islets of Langerhans in autoimmune diabetes mellitus; such diseases are called tissue-specific autoimmune disease.

Titer : The titer of an antiserum is a measure of its concentration of specific antibodies based on serial dilution to an end point, such as a certain level of color change in an ELISA assay. ***OR*** Reciprocal of the highest dilution of an antiserum that gives a positive reaction in the test being used.

Titoki : New Zealand native tree, also know as the New Zealand ash.

TIW : Three times a week.

TJ : Triceps jerk.

TKO : To keep open.

TL : Tubal ligation.

TLC : Thin layer chromatography. ***OR*** Total lung capacity.

TLR : Toll-like receptor.

TLR1-10 : See Toll-like receptors.

TM : Temperature by mouth; tympanic membrane.

TMA : Tissue microarray.

TMAS : Taylor manisfest anxiety scale.

TMJ : Temporo mandibular joint.

TML : Tender midline.

TMT : Tarso-metatarsal joint.

TMV : Tobacco mosaic virus.

TNC : Too numerous to count.

TND : Test not done.

TNF : See lymphotoxin; tumor necrosis factor-a. Tumor necrosis factor.

TNF receptor, TNFR : There are several members of the TNF receptor (TNFR) family. Some lead to apoptosis of the cell on which they are expressed (TNFR-I, II, Fas), while others lead to activation (CD40, 4-1BB). All of them signal as trimeric proteins.

TNS : Tension.

TOC : Total organic carbon - a measure of the amount of organic carbon in water.

TOC : Total organic carbon.

Tocopherol : Another name for vitamin E.

TOD : Another method of measuring organic matter in wastewater involves the oxidation of the sample to stable end products in a platinum-catalyzed combustion chamber at 900 degrees C. Total oxygen demand is determined by measuring the oxygen content of the inert carrier gas, nitrogen. TOD measurements are becoming more popular because of their quickness in determining what is entering the plant and how the plant is responding. Analysis time is approximately 5 minutes. The results obtained generally will be equivalent to those obtained in the COD test.

Tolerance, tolerant : Tolerance is the failure to respond to an antigen; the immune system is said to be tolerant to self antigens. Tolerance to self antigens is an essential feature of the immune system; when tolerance is lost, the immune system can destroy self tissues, as happens in autoimmune disease.

Toll pathway : The Toll pathway is an ancient signaling pathway that activates the transcription factor nfêb by degrading its inhibitor iêb.

Toll-like receptors, TLR, Toll-like receptor 4, TLR-4 : All members of the Toll family of receptors so far discovered have been homologous to the original Toll in Drosophila. They have been named Toll-like receptors (TLR) followed by a number, as in Toll-like receptor 4 or TLR-4.

Tomato spotted wilt virus (TSWV) : Tomato spotted wilt virus was first discovered causing wilt and spotting of tomatoes. The virus is thripstransmitted and can infect a range of plants including tomatoes, peppers and peanuts usually causing losses in field plantings. The "tospovirus" group name is based on tomato spotted wilt virus.

Tonsillitis : Inflammation of the tonsils, especially the palatine tonsils often due to *S. pyogenes* infection.

Tonsils : The palatine tonsils that lie on either side of the pharynx are large aggregates of lymphoid cells organized as part of the mucosal- or gut-associated immune system.

Topoisomerase : An enzyme that changes the extent of supercoiling of a DNA duplex.

Topological map : A map with concentric lines representing changes in elevation

Topsoil : (i) Layer of soil moved in cultivation. (ii) The A horizon. (iii) Presumably fertile soil material used to topdress roadbanks, gardens, and lawns.

TOPV : Trivalent oral polio vaccine.

Tmentose : Densely woolly; velvety or with thick soft hairs.

Torula (Hyphomycetes) : Cosmopolitan fungi commonly found on wood, leaves, plant roots, and plant litter. Has also been isolated from air and soil. Some species cause stains in hardwoods.

Toruloid : Having swellings at intervals.

TOS : Thoracic outlet syndrome.

Tospoviruses : A group of plant viruses that are vectored by thrips. Two well known members are tomato spotted wilt virus (TSWV) and impatiens necrotic spot virus (INSV).

TOT : Total.

Toxemia : The condition caused by toxins in the blood of the host.

Toxic : Poisonous substances are toxic substances. ***OR*** A substance which is poisonous to a living organism.

Toxic shock syndrome : A staphylococcal disease that most commonly affects females who use certain types of tampons during menstruation. It is associated with the production of toxic shock syndrome toxin by certain strains of *Staphylococcus aureus.* ***OR*** Toxic shock syndrome is a systemic toxic reaction caused by the massive production of cytokines by CD4 T cells activated by the bacterial superantigen toxic shock syndrome toxin-1 (TSST-1), which is secreted by *Staphylococcus aureus*.

Toxic shock-like syndrome (TSLS) : A disease caused by an invasive group. A streptococcus infection that is characterized by a rapid drop in blood pressure, failure of many organs, and a very high fever. It probably results from the release of one or more streptococcal pyrogenic exotoxins.

Toxicity : The relative degree of being poisonous or toxic. A condition which may exist in wastes and will inhibit or destroy the growth or function of certain organisms.

Toxicology : The study of the toxic effect of substances.

Toxigenicity : The capacity of an organism to produce a toxin.

Toxin : Microbial substance able to induce host damage. ***OR*** A poison, usually produced by a living organism. ***OR*** A microbial product or component that can injure another cell or organism at low concentrations. Often the term refers to a poisonous protein, but toxins may be lipids and other substances.

Toxin neutralization : The inactivation of toxins by specific antibodies, called antitoxins, that react with them.

Toxoid : A bacterial exotoxin that has been modified so that it is no longer toxic but will still stimulate antitoxin formation when injected into a person or animal. ***OR*** A protein toxin that has been altered with formaldehyde to lose physiological toxicity while retaining antigenicity; useful in immunization and common method for early vaccine formulations.

Toxoids : Inactivated toxins called toxoids are no longer toxic but retain their immunogenicity so that they can be used for immunization.

Toxoplasmosis : A disease of animals and humans caused by the parasitic protozoan, *Toxoplasma gondii*.

TPA : Tissue plasminogen activator.

TPD : Temporary partial disability.

TPMT : Thiopurine methyltransferase.

TPP : Thiamine pyrophosphate.

TPR : Temperature, pulse, respiration.

TQ : Tourniquet.

TQM : Total quality management.

TR : Temperature, rectal; Trace; trauma.

Trace gas : Gas other than nitrogen and oxygen in the atmosphere, particularly those gases that are active in the chemistry or radiation balance of the atmosphere.

Trace ability : The ability to trace something back to its source.

Trachea : The wind pipe.

Tracheal migra : Tion migration in the host by a parasite into alveoli up the respiratory tree and then swallowed and then mature in the intestine.

Trachoma : A chronic infectious disease of the conjunctiva and cornea, producing pain, inflammation and sometimes blindness. It is caused by *Chlamydia trachomatis* serotypes A-C.

TRAFs : The family of proteins known as TNF receptor-associated factors, consists of at least six members that bind to various TNF family receptors or TNFRS. They share a domain known as a TRAF domain, and have a crucial role as signal transducers between upstream members of the TNFR family and downstream transcription factors.

Trait : Characteristic, usually a physical characteristic of a living organism, such as the height of a plant or the hair colour of a mammal.

Trama : Name given to the flesh or context of the cap or gills.

Transamination : The removal of amino acid's amino group by transferring it to an a-keto acid acceptor. ***OR*** Enzymatic transfer of an amino group from an ?-amino acid to an ?-keto acid.

Trans bond : A C=C double bond where the attached pieces of the carbon chain are on opposite sides of the double bond

Transcriptase : An enzyme that catalyzes transcription; in viruses with RNA genomes, this enzyme is an RNA-dependent RNA polymerase that is used to make RNA copies of the RNA genomes.

Transcription : The process in which single-stranded RNA with a base sequence complementary to the template strand of DNA or RNA is synthesized. ***OR*** Synthesis of an RNA molecule complementary to one of the two strands of a DNA doublestranded molecule. ***OR*** The process in which DNA is copied to form a complementary RNA sequence. The first step in protein synthesis. ***OR*** RNA synthesis that occurs on a DNA template. ***OR*** Transfer of the genetic code sequence from the DNA "alphabet" to the RNA.

Transcytosis : The active transport of molecules across epithelial cells is called transcytosis. Transcytosis of IgA molecules involves transport across intestinal epithelial cells in vesicles that originate on the basolateral surface and fuse with the apical surface in contact with the intestinal lumen.

Transduction : Transfer of host genetic information via a virus or bacteriophage particle. ***OR*** Transfer of host genetic information via a virus or bacteriophage particle. ***OR*** The transfer of genes between bacteria by bacteriophages. ***OR*** the transfer and recombination of genetic material from one bacterial cell (during the viral lytic cycle) into the genome of another bacterial cell (lysogeny) by a bacteriophage.

Transfection : The insertion of small pieces of DNA into cells is called transfection. If the DNA is expressed without integrating into host call DNA, this is called a transient transfection; if the DNA integrates into host cell DNA, then it replicates whenever host cell DNA is replicated, producing a stable transfection. ***OR*** An artificial process of infecting cells with naked viral DNA.

Transfer host : A host that is not necessary for the completion of a parasite's life cycle, but is used as a vehicle for reaching a final host.

Transfer RNA (tRNA) : A small RNA that binds an amino acid and delivers it to the ribosome for incorporation into a polypeptide chain during protein synthesis. ***OR*** Type of RNA that carries amino acids to the ribosome during translation. ***OR*** Any of a family of low-molecular weight RNAs that transfer amino acids from the cytoplasm to the template for protein synthesis on the ribosome.

Transferase : An enzyme that catalyzes the transfer of a molecular group from one molecule to another.

Transformation : Transfer of genetic information into living cells as free DNA. ***OR*** Transfer of genetic information into living cells as free DNA. ***OR*** The uptake and incorporation of foreign (donor) "naked" DNA from the environment into the genome of a recipient cell.

Transformation : The process where by one strain of bacteria absorbs genetic material from another strain and turns into that strain

Transgenesis, transgenic mice : Foreign genes can be placed in the mouse genome by transgenesis. This generates transgenic mice that are used to study the function of the inserted gene, or transgene, and the regulation of its expression.

Transgenic : Describes genetically modified plants or animals containing foreign genes inserted by means of recombinant DNA techniques.

Transgenic : Describing an organism that contains transfected DNA in the germ line. ***OR*** Organisms that contain genes from another organism.

Transgenic animal plant : An animal or plant that has gained new genetic information from the insertion of foreign DNA. It may be produced by such techniques as injecting DNA into animal eggs, electroporation of mammalian cells and plant cell protoplasts, or shooting DNA into plant cells with a gene gun.

Transient carrier : See casual carrier.

Transient flora : Microbial flora only temporarily associated with a particular niche (see also normal flora).

Transition mutations : Mutations that involve the substitution of a different purine base for the purine present at the site of the mutation or the substitution of a different pyrimidine for the normal pyrimidine.

Transition state : The activated state in which a molecule is best suited to undergoing a chemical reaction.

Translation : The process in which a messenger RNA (mRNA) sequence is used as a template to form a sequence of amino acids (a protein). Occurs after transcription. ***OR*** Synthesis of proteins using the genetic information in mrna as a template. ***OR*** Protein synthesis; the process by which the genetic message carried by mrna directs the synthesis of polypeptides with the aid of ribosomes and other cell constituents. ***OR*** Transfer of the genetic code sequence from the RNA "language" to the protein /aminoacid.

Translocations : Some cancers have chromosomal translocations, in which a piece of one chromosome is abnormally linked to another chromosome.

Transmission : The passage or transfer of a disease from one host to another. ***OR*** The movement and infection of disease agents, for example thrips transmit tospoviruses and cause new infections throughout a greenhouse crop.

Transmission electron microscope : A microscope in which an image is formed by passing an electron beam through a specimen and focusing the scattered electrons with magnetic lenses.

Transovarian passage : The passage of a microorganism such as a rickettsia from generation to generation of hosts through tick eggs (No humans or other mammals are needed as reservoirs for continued propagation of the rickettsias).

Transpeptidation : 1. The reaction that forms the peptide cross-links during peptidoglycan synthesis. 2. The reaction that forms a peptide bond during the elongation cycle of protein synthesis.

Transpiration : The process by which water vapor is released to the atmosphere by living plants, a process similar to people sweating. ***OR*** When a cell, tissue or organ is removed from one organism and placed in another.

Transplantation, transplanted organs : The grafting of organs or tissues from one individual to another is called transplantation. The transplanted organs or grafts can be rejected by the immune system unless the host is tolerant to the graft antigens or immunosuppressive drugs are used to prevent rejection.

Transport host : See paratenic host.

Transport protein : A protein whose primary function is to transport a substance from one part of the cell to another, from one cell to another, or from one tissue to another.

Transpters associated with antigen processing : See TAP1 and TAP2.

Transposable element : Genetic element that can move (transpose) from one site on a chromosome to another.

Transposition : The movement of a piece of DNA around the chromosome. ***OR*** Movement of a piece of DNA around the chromosome, usually through the function of a transposable element.

Transposon : Transposable element of which, in addition to genes involved in transposition, carries other genes; often confers selectable phenotypes such as antibiotic resistance. ***OR*** Transposable genetic element; DNA sequence capable of replicating and inserting new copy in different site in genome; carries genes for its own transpositon as well as for other functions (e.g., antibiotic resistance).

Transposon A DNA segment : That carries the genes required for transposition and moves about the chromosome; if it contains genes other than those required for transposition, it may be called a composite transposon. Often the name is reserved only for transposable elements that also contain genes unrelated to transposition.

Transposon mutagenesis : Insertion of a transposon into a gene; this inactivates the host gene leading to a mutant phenotype and also confers the phenotype associated with the transposon gene.

Transversion mutations : Mutations that result from the substitution of a purine base for the normal pyrimidine or a pyrimidine for the normal purine.

Trapezius : The muscle which extends from the back of the head,down the back of the neck and edge of the shoulders, and part way down the upper back.

Traveler's diarrhea : A type of diarrhea resulting from ingestion of certain viruses, bacteria, or protozoa normally absent from the traveler's environment. One of the major pathogens is enterotoxigenic *Escherichia coli*.

Tree of life : A diagrammatic representation of the phylogenetic relationships among organisms.

TRFLP : Terminal restriction-fragment length polymorphism.

Tricarboxylic acid (TCA) cycle : The cyclical process whereby acetate is completely oxidized to CO_2 and water, and electrons are transferred to NAD+ and flavine. The TCA cycle is localized to the mitochondria in eukaryotic cells and to the plasma membrane in prokaryotic cells. Also called the Krebs or citric acid cycle.

Tricarboxylic acid cycle (TCA cycle, citric acid cycle, Krebs cycle) : Series of metabolic reactions by which pyruvate is oxidized completely to carbon dioxide, also forming NADH, which allows ATP production. ***OR*** The cycle that oxidizes acetyl coenzyme A to CO_2 and generates NADH and $FADH_2$ for oxidation in the electron transport chain; the cycle also supplies carbon skeletons for biosynthesis.

Triceps brachii : The muscle on the back side of the upper arm.

Trichoderma (Hyphomycetes) : Very common fungus especially in soil and decaying wood, dead leaves, fallen timber, compost heaps, and activated sludge. Can be found indoors in water-damaged buildings. Has occasionally been associated with disease in immunocompromised individuals.

Trichome : Row of cells which have remained attached to one another following successive cell divisions. Trichomes are formed by many cyanobacteria and by species of *Beggiatoa.* ***OR*** A row or filament of bacterial cells that are in close contact with one another over a large area.

Trichomoniasis : A sexually transmitted disease caused by the parasitic protozoan *Trichomonas vaginalis*.

Trichophyton (Hyphomycetes) : Some species are dermatophytes (fungus growing on the skin) of humans or animals.

Trickling filter : A bed of rocks covered with a microbial film that aerobically degrades organic waste during secondary sewage treatment. ***OR*** An attached culture wastewater treatment system. A large tank generally filled with rock or rings (see Bio-Tower). Wastewater is sprayed over the top of the media, providing the opportunity for the formation of slimes or biomass to remove wastes from the wastewater, through revolving arms which have spray nozzles. Water is pumped from the bottom of a trickle filter to a secondary clarifier.

Triglyceride : A combination of glycerol and fatty acid that circulates in the blood.

Triglyceride : The chemical name for fats and oils ; made from glycerol plus three fatty acids.

Trigonometry : The mathematical study of the relationships among the sides of triangles.

Trihalomethanes (THMs) : Halogenated one-carbon compounds formed during water disinfection; many of these compounds are potential carcinogens.

Tripartite associations : A mutualistic association of the same plant with two types of microorganisms.

Trisomy : Having three, rather than two, of a particular pair of chromosomes.

Tritirachium (Hyphomycetes) : A saprobe fungus commonly isolated from decaying plant materials. Easily gets airborne. Can cause corneal ulcers.

tRNA : Transfer RNA.

TRO : To return to office.

Trommel washer : Large piece of manufacturing equipment that rotates and can be used for washing and sieving materials.

Trophic level : Describes the residence of nutrients in various organisms along a food chain ranging from the primary nutrient assimilating autotrophs to the predatory carnivorous animals. ***OR*** A group of organisms with the same feeding habit in the food chain, ranging from the primary nutrient assimilating autotrophs to predatory carnivorous animals.

Trophic levels : The various levels in a food chain or web.

Trophozoite : The stage of the protozoa in the host which feeds and grows until division commences.

Trophozoite : The active, motile feeding stage of a protozoan organism; in the malarial parasite, the stage of schizogony between the ring stage and the schizont.

Tropism : The movement of living organisms toward or away from a focus of heat, light, or other stimulus. ***OR*** The ability of a virus to infect specific cell or tissue types.

Trp : Tryptophan.

TRT : Treatment.

True pathogen : Any microorganism capable of causing disease; an infecting agent (see also opportunistic pathogen).

Truncate : Cut off sharply. ***OR*** Ending abruptly as if cut off.

Trypanosomagambiense **:** Parasitic member of Kingdom Protista; Division Zoomastigophora which causes African sleeping sickness and is spread by the tsetse fly.

Trypanosome : A protozoan of the genus *Trypanosoma*. Trypanosomes are parasitic flagellate protozoa that often live in the blood of humans and other vertebrates and are transmitted by insect bites.

Trypanosomiasis : An infection with trypanosomes that live in the blood and lymph of the infected host.

Trypsin : A proteolytic enzyme that cleaves (cuts) peptide chains next to the basic amino acids arginine and lysine.

Tryptic peptide mapping : The technique of generating a chromato-graphic profile characteristic of the fragments resulting from trypsin enzyme cleavage of the protein.

TS : Thoracic spine.

TSD : Tay-Sachs disease.

TSP : Thoracic spine.

TSS : Total suspended solids.

TT : Tactile tension; total time.

TTD : Temporary total disability.

TTP : Thymidine triphosphate.

Tubercle : A small, rounded nodular lesion produced by *Mycobacterium tuberculosis*.

Tuberculate : Having small wart-like structures.

Tuberculin test : The tuberculin test is a clinical test in which a purified protein derivative (PPD) of *Mycobacterium tuberculosis*, the causative agent of tuberculosis, is injected subcutaneously. PPD elicits a delayedtype hypersensitivity reaction in individuals who have had tuberculosis or have been immunized against it.

Tuberculoid (neural) leprosy : A mild, nonprogressive form of leprosy that is associated with delayed-type hyper-sensitivity to antigens on the surface of *Mycobacterium leprae*. It is characterized by early nerve damage and regions of the skin that have lost sensation and are surrounded by a border of nodules.

Tuberculoid leprosy : See leprosy.

Tuberculosis : An infection in the lungs.

Tuberculosis (TB) : A disease spread through the air by bacteria, affecting mammals (including humans). It usually attacks the lungs, and coughing spreads the bacteria; however, it can also attack other organs such as the kidneys and brain. If not treated properly, TB can cause death. ***OR*** An infectious disease of humans and other animals resulting from an infection by a species of *Mycobacterium* and characterized by the formation of tubercles and tissue necrosis, primarily as a result of host hypersensitivity and inflammation. Infection is usually by inhalation, and the disease commonly affects the lungs (pulmonary tuberculosis), although it may occur in any part of the body.

Tuberculous cavity : An air-filled cavity that results from a tubercle lesion caused by *M. tuberculosis*.

Tubes : Hollow cylindrical structures that bear the hymenium. They replace the gills in polypores

Tularemia : A plaguelike disease of animals caused by the bacterium *Francisella tularensis* subsp. *tularensis* (Jellison type A), which may be transmitted to humans.

Tumble : Random turning or tumbling movements made by bacteria when they stop moving in a straight line.

Tumor : A growth of tissue resulting from abnormal new cell growth and reproduction (neoplasia).

Tumor immunology : Tumor immunology is the study of host defenses against tumors, usually studied by tumor transplantation.

Tumor necrosis factor-a, TNF-a, tum necrosis fact recept, TNFR : Tumor necrosis factor-a (TNF-a) is a cytokine produced by macrophages and T cells that has multiple functions in the immune response. It is the defining member of the TNF family of cytokines. These cytokines function as cell-associated or secreted proteins that interact with receptors of the tumor necrosis factor receptor (TNFR) family, which in turn communicates with the interior of the cell via components known as trafs (tumor necrosis factor receptor associated factors).

Tumor necrosis factor-b, TNF-b : See lymphotoxin.

Tumigenesis : The mechanism of tumor formation.

Tumor-specific transplantation antigens, TSTA, tumor rejection antigens : Tumors transplanted into syngeneic recipients can grow progressively or can be rejected through T-cell recognition of tumor-specific transplantation antigens (TSTA) or tumor rejection antigens. TSTA are peptides of mutant or overexpressed cellular proteins bound to MHC class I molecules on the tumor cell surface.

Tumour : A mass of cells resulting from abnormal cell growth and division. ***OR*** An abnormal growth of tissue. It may be localised (benign) or invade nearby tissues (malignant) or distant tissues (metastatic).

Tunel assay : The tunel assay (tdt-dependent dutp-biotin nick end labeling) identifies apoptotic cells in situ by the characteristic fragmentation of their DNA. Biotin-tagged DUTP added to the free 3' ends of the DNA fragments by the enzyme TDT can be detected by immunohistochemical staining with enzyme-linked streptavidin.

TUR : Transurethral resection.

TURB : Transurethral resection of bladder; turbinate.

Turbid : A description for a liquid that is muddy, thick, or not clear.

Turbidity : The amount of suspended matter in wastewater, obtained by measuring its light scattering ability.

Turbidostat : A continuous culture system equipped with a photocell that adjusts the flow of medium through the culture vessel so as to maintain a constant cell density or turbidity.

Turnover number : The maximum number of molecules of substrate that can be converted to product per active site per unit time.

TURP : Transurethral resection of prostate gland.

TV : Tinea versicolor; total volume.

TVC : Timed vital capacity.

TVH : Total vaginal hysterectomy.

TVR : Thoracic ventral root.

TWHW : Toe walking, heel walking.

Twiddle : See tumble.

Two-component phosphelay system : A signal transduction regulatory system that uses the transfer of phosphoryl groups to control gene transcription and protein activity. It has two major components: a sensor kinase and a response regulator.

Two-dimensional gel electrophesis : In two-dimensional gel electrophoresis, proteins are separated by isoelectric focusing in one dimension, followed by SDS-PAGE on a slab gel at right-angles to the first dimension. This can separate and identify large numbers of distinct proteins.

TX : Traction; transplant; treatment.

Tympanum : The ear drum.

TYP : Typical.

Type 1 diabetes : Type 1 diabetes (formerly known as "childhood" or "juvenile" diabetes) is a chronic (lifelong) disease that occurs when the pancreas produces too little insulin to regulate blood sugar levels appropriately. Insulin is a hormone that is needed to convert sugar (glucose), starches and other food into energy needed for daily life.

Type I hypersensitivity : A form of immediate hypersensitivity arising from the binding of antigen to IgE attached to mast cells, which then release anaphylaxis mediators such as histamine. Examples: hay fever, asthma, and food allergies.

Type I hypersensitivity reactions, type II hypersensitivity reactions, type III hypersensitivity reactions, type IV hypersensitivity reactions: Hypersensitivity reactions are classified by mechanism: type I hypersensitivity reactions involve IgE antibody triggering of mast cells; type II hypersensitivity reactions involve IgG antibodies against cell surface or matrix antigens; type III hypersensitivity reactions involve antigen:antibody complexes; and type IV hypersensitivity reactions are T cell-mediated.

Type II hypersensitivity : A form of immediate hypersensitivity involving the binding of antibodies to antigens on cell surfaces followed by destruction of the target cells (e.g., through complement attack, phagocytosis, or agglutination).

Type III hypersensitivity : A form of immediate hypersensitivity resulting from the exposure to excessive amounts of antigens in which antibodies bind to the antigens and produce antibody-antigen complexes. These activate complement and trigger an acute inflammatory response with subsequent tissue damage. Examples: poststreptococcal glomerulonephritis, serum sickness, and farmer's lung disease.

Type IV hypersensitivity : A delayed hypersensitivity response (it appears 24 to 48 hours after antigen exposure). It results from the binding of antigen to activated T lymphocytes, which then release cytokines and trigger inflammation and macrophage attacks that damage tissue. Type IV hypersensitivity is seen in contact dermititis from poison ivy, leprosy, and tertiary syphilis.

Type-V hypersensitivity : A stimulatory hypersensitivity response. It is mediated by autoantibodies. Autoantibodies are directed to harmone receptors and this autoantibodies acts as a hormone and stimulates the gland.

Typhoid fever : A bacterial infection transmitted by contaminated food, water, milk, or shellfish. The causative organism is *Salmonella typhi*, which is present in human feces.

Tyr : Tyrosine.

Tyrosine kinase : A tyrosine kinase is an enzyme that specifically phosphorylates tyrosine residues in proteins. They are critical in T- and B-cell activation. The kinases that are critical for B-cell activation are Blk, Fyn, Lyn, and Syk. The tyrosine kinases that are critical for T-cell activation are called Lck, Fyn, and ZAP-70.

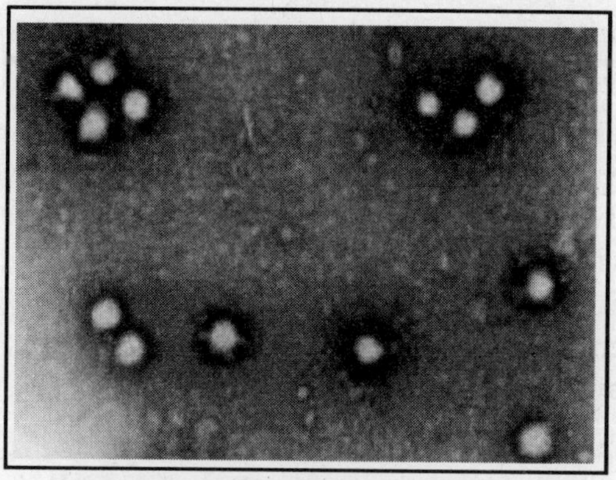

Tomato spotted wilt virus

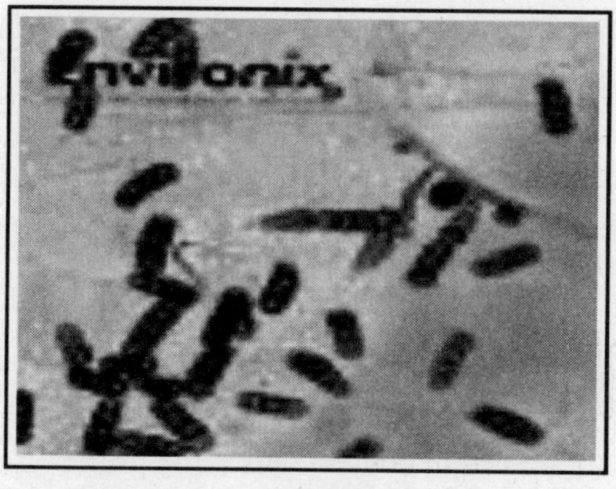

Torula

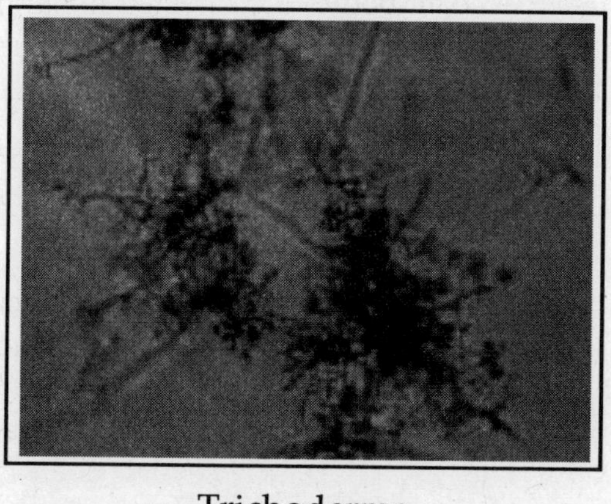

Trichoderma

U

U : Units.

U&L : Upper and lower.

U&MNP : Ulnar and median nerve palsy.

U/O : Under observation for.

U/WB : Unit of whole blood.

UA : Umbilical artery; uric acid; urinalysis.

UC : Ulcerative colitis; unchanged; unconscious.

UCD : Usual childhood diseases.

UCG : Urinary chorionic gonadotrophin.

UCR : Unconditioned reflex, response.

UCS : Unconditioned stimulus; unconscious.

UD : Ulnar deviation; under developed; urethral discharge, drainage.

UDP : Uridine diphosphate.

UDPG : UDP-glucose.

UE : Undetermined etiology; upper extremity.

UGA : Under general anesthesia.

UHF : Ultra high frequency.

UHTS : Ultra-high-throughput system.

UIQ : Upper inner quadrant.

UK : Unknown.

UL : Upper lobe.

Ulcerative colitis : A disease where the large intestine or colon has ulcers.

ULN : Upper limits of normal; ulnar.

Ulna : The bone on the little finger side of the lower arm which also forms the elbow.

Ulocladium (Hyphomycetes) : A fungus found as a saprobe in soil, plant materials, rotten wood, paper, textiles and cellulose materials. Frequently collected in air and dust samples. Can grow indoors on water-damaged building materials. Has not been associated with disease in humans but can be very allergenic.

ULOQ : Upper left outer quadrant.

ULQ : Upper left quadrant.

Ultracentrifuge : A high-speed centrifuge that can attain speeds up to 60,000 rpm and centrifugal fields of 500,000 times gravity. Useful for characterizing and/or separating macromolecules.

Ultramicrobacteria : Bacteria that can exist normally in a miniaturized form or which are capable of miniaturization under low-nutrient conditions. ***OR*** Small (0.3 micrometer), metabolically dormant bacteria produced by starvation. These bacteria produce little or no exopolysaccharide. In the presence of nutrients, the bacteria become active again, regaining normal size, metabolic activity, and EPS production.

Ultrastructure : The detailed structure of a biological specimen, such as a cell, tissue, or organ, which can only be observed using an electron microscope.

Ultraviolet (UV) radiation : Radiation of fairly short wavelength, about 10 to 400 nm, and high energy.

Ulva : A multi cellular member of Kingdom Protista; Division Chlorophyta which is Commonly known as sea lettuce.

Umbelliferae : Also known as Apiaceae, the plant family commonly known as the parsley. Family whose flowers are arranged in an umbel in which numerous flowers arise from the same point like umbrella ribs.

Umbilical cord : The cord that connects a fetus to the placenta during pregnancy – it carries blood and nutrients to the growing fetus.

Umbilicate : Used to describe cap having a central depression.

Umbo : A raised conical mound on the centre of the cap.

Umbonate : Cap with a raised knob in the centre.

UMF number : A rating of the antibacterial action of Manuka honey. The number is equivalent to the percentage of phenol it takes to have the same effect. Manuka honey with UMF 18+ for example can be used to heal wounds.

UMN : Upper motor neuron.

UMNL : Upper motor neuron lesion.

UMP : Uridine monophosphate.

UN : Ulnar nerve.

UNC : Uncertain, unconscious, uncooperative.

Uncoating : The stage of viral replication at which structural proteins are lost and the virus genome is exposed to the replication machinery.

UND : Undetermined.

UNG : Ointment.

Ungulate : Hoof-shaped.

Unicellular : Single celled organism, such as bacteria. ***OR*** Having a body consisting of one cell.

Unidirectional replication : See bidirectional replication.

Unique manuka factor (UMF) : A substance found in manuka honey that can has antibacterial activity.

Uniserate : Phialides arising directly from a vesicle as in *Aspergillus*.

Universal veil : A membrane that initially entirely surrounds the fruiting body. It is broken as the cap expands the stem extends and the gills mature.

University of Maryland Cooperative Extension Service : The University of Maryland is the land grant institution in Maryland and is home for the Cooperative Extension Service. With extensive educational and scientific research programs the service provides growers and other citizens of Maryland current horticultural advice and training through such programs as Total Plant Management, Master Gardeners and the Home and Garden Information Center.

UNK : Unknown.

Unwinding proteins : Proteins that help to unwind double-stranded DNA during DNA replication.

UO : Undetermined origin; urinary output.

UOQ : Upper outer quadrant.

UP : Usual place of residence.

Upset : An upset digester does not decompose organic matter properly. The digester is characterized by low gas production, high volatile acid/alkalinity relationship, and poor liquid-solids separation. A digester in an upset condition is sometimes called a "sour" or "stuck" digester.

UQ : Upper quadrant.

UR : Unconditioned response; upper respiratory.

Uracil : The pyrimidine 2,4-dioxypyrimidine, which is found in nucleosides, nucleotides, and RNA.

URD : Upper respiratory disease.

Urea cycle : A metabolic pathway in the liver that leads to the synthesis of urea from amino groups and CO_2. The function of the pathway is to convert the ammonia resulting from catabolism to a nontoxic form, which is then secreted.

URI : Upper respiratory infection.

Uronic acid : Class of acidic compounds of the general formula HOOC (CHOH) nCHO that contain both carboxylic and aldehydic groups, are oxidation products of sugars, and occur in many polysaccharides; especially in the hemicelluloses.

UROQ : Upper right outer quadrant.

URQ : Upper right quadrant.

Urticaria : Urticaria is the technical term for hives, which are red, itchy skin welts usually brought on by an allergic reaction.

US : Ultrasound.

Usher : Person who, in1650,decided the Earth was only 6000 years old.

UT : Urinary tract; uterus.

UTI : Urinary tract infection.

Utilitarianism : The ethical framework which holds that the right course of action is that which maximises the amount of happiness or pleasure in the world. If you are a utilitarian it means that you while you might, for example, normally tell the truth, there could be occasions where you would believe that the right thing to do would be to lie.

UTOC : Upper thoracic outlet compression syndrome.

UTP : Uridine triphosphate.

UV : Ultraviolet; urinary volume.

UV irradiation : Electromagnetic radiation with a wavelength shorter than that of visible light (200-390 nm). Causes damage to DNA (mainly by forming pyrimidine dimers).

UVL : Ultraviolet light.

UVR : Ultraviolet radiation.

Uvula : A flap at the end of the palate which closes to block off the nose during swallowing.

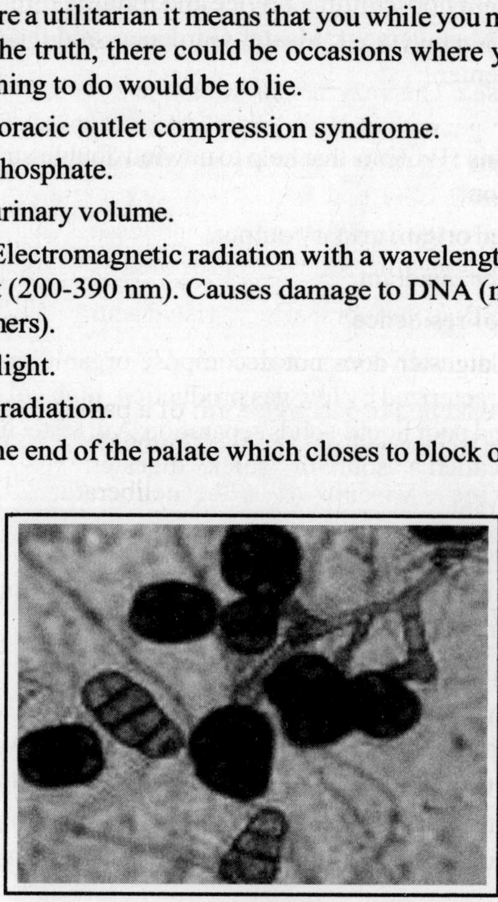

Ulocladium

V

V : Volt.

V domain : The variable region of the polypeptide chains of an immunoglobulin or T-cell receptor is composed of a single N-terminal V domain. Paired V domains form the antigen-binding sites of immunoglobulins and T-cell receptors.

V gene segments : The first 95 amino acids or so of immunoglobulin or T-cell receptor V domains are encoded in inherited V gene segments. There are multiple different V gene segments in the germline genome. To produce a complete exon encoding a V domain, one V gene segment must be rearranged to join up with a J or a rearranged DJ gene segment.

V&PS : Vibration and peripheral sensation.

V(D)J recombinase : The enzyme that joins the gene segments of B-cell and T-cell receptor genes is called the V(D)J recombinase. It is made up of several enzymes, but most important are the products of the recombinase activating genes RAG-1 and RAG-2 whose protein products RAG-1 and RAG-2 are expressed in developing lymphocytes and make up the only known lymphoid-specific components of the V(D)J recombinase.

VA : Vacuum aspiration; vertebral artery; visual acuity.

Vaccinate : Inoculate with a vaccine.

Vaccination : A weakened or inactive form of a pathogen given to enable the immune system to respond and produce immunity.

Vaccination, vaccine : Vaccination is the deliberate induction of adaptive immunity to a pathogen by injecting a vaccine, a dead or attenuated (nonpathogenic) form of the pathogen.

Vaccine : A preparation of killed or weakened microbes (or portions of them) that when taken into the body, causes the immune system to produce antibodies and makes the body resistant to the disease.OR A preparation of either killed microorganisms; living, weakened (attenuated) micro-organisms; or inactivated bacterial toxins (toxoids). It is administered to induce development of the immune response and protect the individual against a pathogen or a toxin. ***OR*** A molecule that causes an immune response in an organism without resulting in disease. A vaccine is normally used to 'train' the immune system to recognise a foreign substance, like a bacteria or virus, so that it can be destroyed quickly if it enters the body at a later stage.

Vaccinia : The first effective vaccine was vaccinia, a cowpox virus that causes a limited infection in humans that leads to immunity to the human smallpox virus.

Vaccinomics : The application of genomics and bioinformatics to vaccine development.

Vacuole : A larger storage area with in a cell.

Vadose zone : Unsaturated zone of soil above the groundwater, extending from the bottom of the capillary fringe all the way to the soil surface.

VAG : Vaginal.

VAH : Veteran's administration hospital.

Val : Valine.

VALE : Visual acuity, left eye.

Valence : The number of antigenic determinant sites on the surface of an antigen or the number of antigen-binding sites possessed by an antibody molecule. ***OR*** The valence of an antibody or antigen is the number of different molecules that it can combine with at one time.

Valid : Scientifically accurate.

Van der Waals forces : Refers to the combined effect of two types of interactions, one attractive and one repulsive. The attractive forces are due to favorable interactions among the induced instantaneous dipole moments that arise from fluctuations in the electron charge densities of neighboring nonbonded atoms. Repulsive forces arise when noncovalently bonded atoms come too close together.

VAOD : Visual acuity, right eye.

VAOS : Visual acuity, left eye.

VAOU : Visual acuity, both eyes.

VAR : Variety; variation.

Variability : The variability of a protein is a measure of the difference between the amino acid sequences of different variants of that protein. The most variable proteins known are antibodies and T-cell receptors.

Variability plot : See Wu and Kabat plot.

Variable gene segments : See V gene segments.

Variable region (VL and VH), V-region : The region at the N-terminal end of immunoglobulin heavy and light chains whose amino acid sequence varies between antibodies of different specificity. Variable regions form the antigen binding site. ***OR*** The variable region (V region) of an immunoglobulin or T-cell receptor is formed of the amino-terminal domains of its component polypeptide chains. These are called the variable domains (V domains) and are the most variable parts of the molecule. They contain the antigen-binding sites.

Varicose veins : Swollen, stretched, enlarged veins.

Vas deferens : The tubes which carry sperm from scrota to urethra (these are cut in a vasectomy).

Vascular : Relating to blood vessels, for example veins and arteries.

Vascular addressing : Vascular addressing are molecules on endothelial cells to which leukocyte adhesion molecules bind. They have a key role in selective homing of leukocytes to particular sites in the body.

Vascular system : The conductive system in plants, comprised of xylem and phloem, in which water and nutrients move.

Vasculitis : Inflammation of a blood vessel.

Vav : See adaptor proteins.

VBP : Venous blood pressure.

VC : Ventricular contraction; vital capacity; vocal cord.

VD : Venereal disease; venous dilation; ventrodorsal.

VDA : Visual discriminatory acuity.

VE : Vaginal examination.

Vector : Carrier of an infectious agent; capable of transmitting infection from one host to another; especially the animal that transfers an infectious agent from one host to another, usually an arthropodor An intermediate host in a parasitic life cycle, that may or may not (paratenic host) be essential for the development of the parasite. A vector will seek out the definitive host in the life cycle to prey on it, whereas other intermediate hosts will not. ***OR*** An organism that carries a pathogen to a plant. Aphids, nematodes, thrips and whiteflies are common vectors of plant viruses. Only thrips vector tospovirus. ***OR*** An insect or other organism capable of transmitting germs or other agents of disease.

Vector-borne transmission : The transmission of an infectious pathogen between hosts by means of a vector.

Vegetative : Actually growing state.

Vegetative cell : Growing or feeding form of a microbial cell, as opposed to a resting form such as a spore.

VEGF : Vascular endothelial growth factor.

Vehicle : An inanimate substance or medium involved in the transmission of a pathogen.

Veil : A thin membrane covering the gills during the development of the toadstool.

Venereal syphilis : A contagious, sexually transmitted disease caused by the spirochete *Treponema pallidum*.

Venereal warts : See anogenital condylomata.

VENT : Ventral.

Ventricose : Enlarged in the middle.

Vernalization : Exposing seeds to enough cold to trigger germination.

Verrucae vulgaris : The common wart; a raised, epidermal lesion with horny surface caused by an infection with a human papillomavirus.

Verrucose : Having many warts or covered with small rounded warts - used to describe rough spores.

VERT : Vertebra, vertical, vertigo.

Vertebrae : Bones that make up the spine.

Vertebrate : Animal with a back-bone.

Verticillate : Having branches arranged in verticils or whorls.

Verticillium (Hyphomycetes) : Mostly soilborne, root-inhabiting fungi that cause vascular wilt and other diseases on plants. Some species also infect mushrooms, rusts, and other fungi, as well as nematodes, ticks, mites and other insects. Other species can attack wool and textiles, or can decompose paper.

Very late antigens, VLA : The very late antigens (VLA) are members of the Pi family of integrins involved in cell-cell and cell-matrix interactions. Some VLAs are important in leukocyte and lymphocyte migration.

Vesicle : A swollen cell.

Vesicles : Spherical structures, formed intracellularly, by some arbuscular mycorrhizal fungi. ***OR*** Vesicles are small membrane-bounded compartments within the cytosol.

Vesicular arbuscular mycorrhiza : See arbuscular mycorrhiza.

VF : Ventricular fibrillation; visual field.

VFI : Visual fields intact.

VH : Vaginal hysterectomy.

VI : Viral infection.

Viability : Whether something is likely to work or not, e.g., the viability of automatic milking.

Viable : Alive; able to reproduce.

Viable but nonculturable : Organisms that are alive but cannot be cultured on laboratory media.

Viable count : Measurement of the number of live cells in a microbial population.

Vibrio : (i) Curved, rodshaped bacterial cell. (ii) Bacterium of the genus *Vibrio*. ***OR*** A rod-shaped bacterial cell that is curved to form a comma or an incomplete spiral.

VIGS : Virus-induced gene silencing.

Vinaceous : Pinkish.

Violaceae : Plant family commonly known as the violet family and having bilaterally symmetrical flowers.

Viral envelope : Many viruses that are produced by mammalian cells are enclosed in a viral envelope of host cell membrane lipid and proteins bound to the viral core by viral envelope proteins.

Viral hemagglutination : The clumping or agglutination of red blood cells caused by some viruses.

Viral neutralization : An antibody-mediated process in which IgA, IgM, and IgA antibodies bind to some viruses during their extracellular phase and inactivate or neutralize them.

Viremia : The presence of viruses in the blood stream.

Viricide : An agent that inactivates viruses so that they cannot reproduce within host cells.

Virion : Virus particle; the virus nucleic acid surrounded by protein coat and in some cases other material. ***OR*** A complete virus particle that represents the extracellular phase of the virus life cycle; at the simplest, it consists of a protein capsid surrounding a single nucleic acid molecule.

Virions : Structurally mature, extracellular virus particles. ***OR*** Virions are complete virus particles, the form in which viruses spread from cell to cell or from one individual to another.

Virioplankton : Viruses that occur in waters; high levels are found in marine and freshwater environments.

Viroid : An infectious agent of plants that is a single-stranded RNA not associated with any protein; the RNA does not code for any proteins and is not translated.

Viroids : Pathogenic agents, mostly of plants, that consist of short (usually circular) RNA molecules.

Virology : The branch of microbiology that is concerned with viruses and viral diseases.

Virtue : A good character trait to possess. Precisely what the virtues are is open to disagreement, and may vary from place to place and at different times in history. However, certain virtues, such as kindness and courage, are valued by most cultures.

Virulence : Degree of pathogenicity of a parasite. ***OR*** A measure of pathogenicity; a measurement of the degree of disease producing ability of a microorganism as indicated by the severity of the disease produced; a measure of the dosage required to caused a specific degree of pathogenicity one general standard is the LD_{50}. ***OR*** Ability to cause disease. ***OR*** Degree of pathogenicity of a parasite.

Virulence factor : A bacterial product, usually a protein or carbohydrate, that contributes to virulence or pathogenicity.

Virulent bacteriophages : Bacteriophages that lyse their host cells during the reproductive cycle.

Viruliferous : A term used to describe a vector that has acquired and carries a virus, and can transmit the virus to a healthy plant.

Virus : An infectious agent with a simple structure consisting of DNA or RNA and reproducing only within living host cells.OR Any of a large group of submicroscopic infective agents that typically contain a protein coat surrounding a nucleic acid core and are capable of growth only in a living cell. ***OR*** A microscopic particle that can only multiply inside the cells of living organisms; they are totally inactive outside of a living host cell. ***OR*** A complex of nucleic-acid and protein, that can infect and replicate inside a specific host cell to make more virus particles. ***OR*** Viruses are pathogens composed of a nucleic acid genome enclosed in a protein coat. They can replicate only in a living cell, as they do not possess the metabolic machinery for independent life.

Virus attachment protein : The protein on the surface of a virus particle responsible for binding the receptor.

Virus reservoir : Plants that may harbor a virus without obvious symptoms. Weeds are often virus reservoirs.

Visceral larva migrans : Nematode larvae migration in hosts that are suitable for long survival but are unsuitable for development to the adult stage. The larva worms wander for a time in the hosts tissues.

Viscid : Slimy.

Vitamin : A trace organic substance required in the diet of some species. Many vitamins are precursors of coenzymes. ***OR*** An organic compound required by organisms in minute quantities for growth and reproduction because it cannot be synthesized by the organism; vitamins often serve as enzyme cofactors or parts of cofactors. ***OR*** Organic components in food that are needed in very small amounts for growth and for maintaining good health. The vitamins include vitamin D, vitamin E, vitamin A, and vitamin K, or the fat-soluble vitamins, and folate (folic acid), vitamin B12, biotin, vitamin B6, niacin, thiamin, riboflavin, pantothenic acid, and vitamin C (ascorbic acid), or the water-soluble vitamins. Vitamins are required in the diet in only tiny amounts, in contrast to the energy components of the diet such as sugars, starches, fats, and oils.

Vitamin C : An antioxidant found in fruits and vegetables, such as oranges, kiwifruit, berries, tomatoes, and leafy greens.

Viviparous : Condition in which the embryo develops within the female and is nourished by the mother (no egg shell exists at any time) e.g. mammals.

Viviparous : Bearing live young.

VLDL : Very low-density lipoprotein.

VM : Vasomotor.

V_{max} **:** Maximal velocity.

VMR : Vasomotor rhinitis.

VMS : Visual memory span.

VMT : Vasomotor tone.

VNA : Visiting nurse's association.

VNR : Ventral nerve root.

VNTR : Variable number tandem repeat.

VO : Verbal order.

VOC : Vocal; vocational.

VOL : Volatile; volume; voluntary.

Volatile : A volatile substance is one that is capable of being evaporated or changed to a vapor at a relatively low temperature. Volatile substances also can be partially removed by air stripping.

Volutin granules : See metachromatic granules.

Volva : The expanded bulbous remains of the universal veil found at the base of the stipe in some toadstools e.g. Amanita and lepiota.

Volvox : Member of Kingdom Protista ; Division Chlorophyta that occurs in spherical colonies, often with daughter colonies inside.

Vomiting : (Oremesis). The expelling of undigested food through the mouth.

VP : Venipuncture; venous pressure.

VPB : Ventricular premature beats.

VPC : Ventricular premature contractions.

VpreB : See pre-B-cell receptor.

VPS : Vibrations per second.

VQ : Voice quality.

VR : Ventilation rate; ventricular rate; vocational rehabilitation.

VRC : Ventral root, cervical.

VRL : Ventral root, lumbar.

VRT : Ventral root, thoracic.

VS : Vesicular sounds; vital signs.

VS/L : Measure of volatile solids, usually expressed as g VS/L/day = grams volatile solids per liter per day.

VSBE : Very short below elbow (cast).

VSD : Ventral septal defect.

VSG : Variable surface glycoprotein.

VSO : Vital signs normal.

VSS : Vital signs stable.

VSTT : Ventral spinothalamic tract.

VT : Ventricular tachycardia.

VV : Varicose veins; veins.

Verticillium

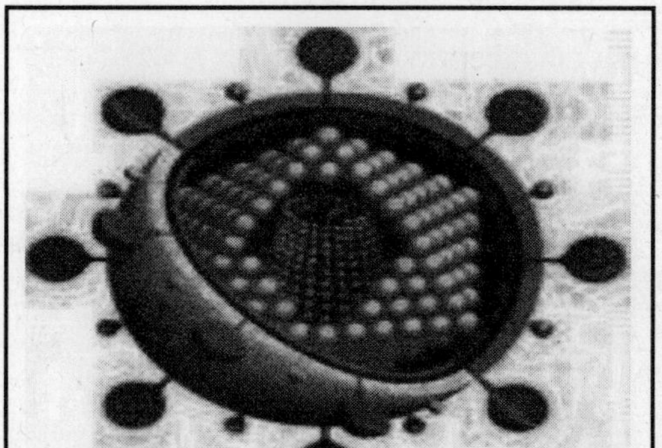

Virus

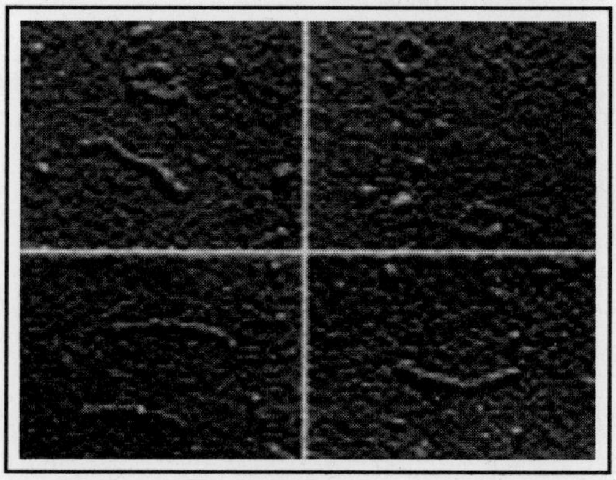

Viroid

W

W : Watt.

W&D : Warm and dry.

W/A : Weakness atrophy.

W/C : Wheelchair.

W/E : Wound of entry.

W/U : Workup.

WAIS : Wechsler's adult intelligence scale.

Wallemia (Hyphomycetes) : A very xerophilic fungus that has been isolated from soil, air, hay, textiles, and food such as jam, salted fish, and milk products. Can cause allergies.

Wart (wt) An epidermal tumor of viral : Origin.

WAS : Waste activated sludge, mg/l. The excess growth of microorganisms which must be removed from the process to keep the biological system in balance.

Wastewater : The used water and solids from a community that flow to a treatment plant. Storm water, surface water, and groundwater infiltration also may be included in the wastewater that enters a wastewater treatment plant. The term "sewage" usually refers to household wastes, but this word is being replaced by the term "wastewater".

Wastewater treatment : The use of physical and biological processes to remove particulate and dissolved material from sewage and to control pathogens.

Water activity (aw) : A quantitative measure of water availability in the habitat; the water activity of a solution is one-hundredth its relative humidity.

Water cluster : The clustering of different numbers of water molecules, which are held together by weak bonds called hydrogen bonds.

Water content : Water contained in a material expressed as the mass of water per unit mass of oven dry material.

Water mold : A common term for a member of the division Oomycota.

Water potential : See soil water potential.

Water retention curve : Graph showing soilwater content as a function of increasingly negative soil water potential.

Watson-Crick base pairs : The type of hydrogen-bonded base pairs found in DNA, or comparable base pairs found in RNA. The base pairs are A-T, G-C, and A-U.

Wavelength : The distance between two successive points of an electromagnetic waveform, usually measured in nanometers (nm).

WB : Weight bearing; whole blood.

WBAT : Weight bearing as tolerated.

WBC : White blood cell; white blood cell count.

WC : Will call; wound check; wheel chair.

WD : Well developed; wound.

WDWN : Well developed, well nourished.

Weathering : All physical and chemical changes produced in rock by atmospheric agents.

Wegener's granulomatosis : Antibodies against the neutrophil granule proteinase-3 are formed in Wegener's granulomatosis, an autoimmune disease in which there is severe necrotizing vasculitis. The presence of anti-neutrophil cytoplasmic antigen, or ANCA, helps in the diagnosis of this disease.

Weibel-Palade bodies : Weibel-Palade bodies are granules within endothelial cells that contain P-selectin. Activation of the endothelial cell by mediators such as histamine and C5a leads to rapid translocation of P-selectin to the cell surface.

Weil-Felix reaction : A test for the diagnosis of typhus and certain other rickettsial diseases. In this test, the blood serum of a patient with suspected rickettsial disease is tested against certain strains of *Proteus vulgaris* (OX-2, OX-19, OX-K). The agglutination reactions, based on antigens common to both organisms, determine the presence and type of rickettsial infection.

Weir : A wall or plate placed in an open channel and used to measure the flow of water.

Western blot : Similar in principle to a Southern blot, but where the species adsorbed to the nitrocellulose filter is a protein, and the detection makes use of specific antibodies.

Western blotting : In Western blotting, a mixture of proteins is separated, usually by gel electrophoresis and transferred by blotting to a nitrocellulose membrane; labeled antibodies are then used as probes to detect specific proteins.

WF : White female.

WFO : Will follow in office.

WGD : Whole genome duplication.

WGS : Whole genome shotgun.

WH : Walking heel cast; well healed; well hydrated.

Wheal-and-Flare reaction : When small amounts of allergen are injected into the dermis of an allergic individual, a wheal-and-flare reaction is observed. This consists of a raised area of skin containing fluid and a spreading, red, itchy circular reaction.

White piedra : A fungal infection caused by the yeast *Trichosporon beigelii* that forms light-colored nodules on the beard and mustache.

White pulp : The discrete areas of lymphoid tissue in the spleen are known as the white pulp.

White rot fungus : Fungus that attacks lignin, along with cellulose, and hemicellulose, leading to a marked lightening of the infected wood.

WHO : World health organization.

Whole-genome shotgun sequencing : An approach to genome sequencing in which the complete genome is broken into random fragments, which are then individually sequenced. Finally the fragments are placed in the proper order using sophisticated computer programs.

Whole-organism vaccine : A vaccine made from complete pathogens, which can be of four types: inactivated viruses; attenuated viruses; killed microorganisms; and live, attenuated microorganisms.

WHP : Whirlpool.

Widal test : A test involving agglutination of typhoid bacilli when they are mixed with serum containing typhoid antibodies from an individual having typhoid fever; used to detect the presence of *Salmonella typhi* and *S. paratyphi*.

Wild type : Strain of microorganism isolated from nature. The usual or native form of a gene or organism.

Wild-type gene : The form of a gene (allele) normally found in nature.

William : "Strata"Smith Person who lived from 1769 to 1839 and studied the distribution of fossils.

Wilt : A collapse or drooping of leaves and shoots that may be caused by lack of water, excess fertilization or disease.

Wilting point : See permanent wilting point.

Winogradsky column : Glass column with an anaerobic lower zone and an aerobic upper zone, which allows growth of microorganisms under conditions similar to those found in nutrient rich water and sediment. ***OR*** A glass column with an anaerobic lower zone and an aerobic upper zone, which allows growth of microorganisms under conditions similar to those found in a nutrient-rich lake.

Wiskott-Aldrich syndrome : The Wiskott-Aldrich syndrome is characterized by defects in the cytoskeleton of cells due to a mutation in the protein WASP. Patients with this disease are highly susceptible to pyogenic bacteria.

WK : Weak; week; work.

WL : Work load.

WM : White male.

WMF : White middle aged female.

WMM : White middle aged male.

WN : Well nourished.

WNL : Within normal limits.

WO : Written order.

Wobble : A proposed explanation for base pairing that is not of the Watson-Crick type and that often occurs between the 3' base in the codon and the 5' base in the anticodon.

WOE : Wound of entry.

Woolster's disease : See anthrax.

Worms : An infestation where worms live as parasites in the human body.

Woronin body : Spherical structure associated with the simple pore in the septa separating hyphal compartments of fungi in the phylum Ascomycota.

Wt : The filtrate of malted grains used as the substrate for the production of beer and ale by fermentation.

WOX : Wound of exit.

WP : Wet pack; whirlpool.

WPD : Whirlpool bath.

WPK : Wet pack.

WR : Weak response; wrist.

WT : Weight.

WTB : Weight bearing.

Wu and Kabat plot : A Wu and Kabat plot, or variability plot, is generated from the amino acid sequences of related proteins by plotting the variability of the sequence against amino acid residue number. Variability is the number of different amino acids observed at a position divided by the frequency of the most common amino acid.

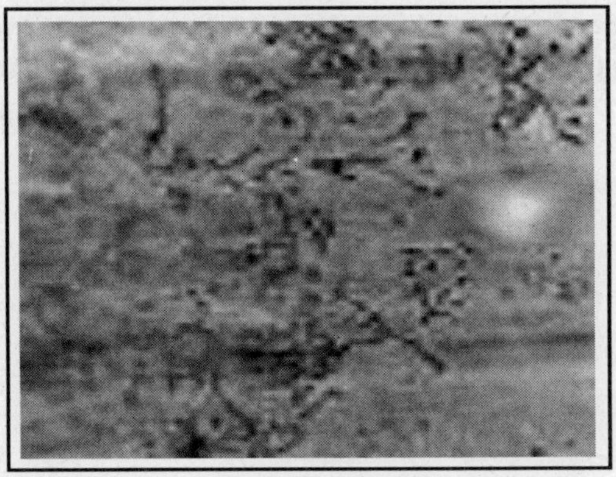

Wallemia

XYZ

X : Times; power; multiplied by.

Xenobiotic : Compound foreign to biological systems. Often refers to humanmade compounds that are resistant or recalcitrant to biodegradation and decomposition.

Xenogeneic : Animals of different species are xenogeneic.

Xenograft : A tissue graft between animals of different species.

Xenografts : The use of xenografts, organs from a different species, is being explored as a solution to the severe shortage of human organs for transplantation. The main problem with xenografting is the presence of natural antibodies against xenograft antigens; attempts are being made to modify these reactions by creating transgenic animals.

Xenophanes : Person who lived around 570 BC and was one of the first people to observe and write about fossils.

Xenophore : Group or structure built into a chemical that imparts xenobiotic character, which may hinder biodegradation of the chemical.

Xenotransplantation : An organ transplant where the organ from an individual of one species is transplanted into an individual from a different species.

Xerophile : Organism adapted to grow at low water potential, i.e., very dry habitats.

Xerophilic : Able to grow under dry conditions.

Xerophilic microganisms : Microorganisms that grow best under low aw conditions, and may not be able to grow at high aw values.

Xid : Mice with mutations in the btk gene have a deficiency in making antibodies, especially in primary responses. These mice are called xid, for X-linked immunodeficiency, the mouse equivalent of X-linked agammaglo-bulinemia, the human form of this disease.

XIP : X-ray in plaster; extra-ocular movements, muscles.

X-linked agammaglobulinemia, XLA : X-linked agammaglobulinemia (XLA) is a genetic disorder in which B-cell development is arrested at the pre-B-cell stage and no mature B cells or antibodies are formed. The disease is due to a defect in the gene encoding the protein tyrosine kinase btk.

X-linked hyper IgM syndrome : X-linked hyper IgM syndrome is a disease in which little or no IgG, IgE, or IgA antibody is produced and even IgM responses are deficient, but serum igm levels are normal to high. It is due to a defect in the gene encoding the CD40 ligand or CD154.

X-linked lymphoproliferative syndrome : X-linked lymphoproliferative syndrome is a rare immunodeficiency that results from mutations in a gene named SH2-domain containing gene 1A (SH2DlA). Boys with this deficiency typically develop overwhelming EBV infection during childhood, and sometimes lymphomas.

X-linked severe combined immunodeficiency, X-linked SCID : X-linked severe combined immuno-deficiency (X-linked SCID) is a disease in which T-cell development fails at an early intrathymic stage and no production of mature T cells or T-cell dependent antibody occurs. It is due to a defect in a gene that encodes the y, chain that is a component of the receptors for several different cytokines.

XMP : Xanthosine monophosphate.

XOOP : X-rays out of plaster.

XOP : X-rays out of plaster.

XR : X-ray.

X-ray : Crystallography: A method of seeing viruses by turning them into crystals, shooting them with X-rays, then recording the image on film.

X-ray crystallography : A technique for determining the structure of molecules from the X-ray diffraction patterns that are produced by crystalline arrays of the molecules.

XS : Excessive.

XT : Exotropia.

Xu5P : Xylulose-5-phosphate.

XYL : Xylocaine.

Y/A : Years of age.

YAC : Yeast artificial chromosome.

YADH : Yeast alcohol dehydrogenase.

YD : Yard.

Yeast : Fungus whose thallus consists of single cells that multiply by budding or fission. ***OR*** Any of a number of one-celled fungi that break down sugar into alcohol and carbon dioxide. ***OR*** Microorganisms that produce the enzymes which convert sugar to alcohol. ***OR*** Eukaryotic organisms larger than most bacteria, commonly divides by budding. ***OR*** Fungus whose thallus consists of single cells that multiply by budding or fission. ***OR*** A growth form exhibited by some fungi in which the fungus exists as single budding cells.

Yeast artificial chromosome (YAC) : A stretch of DNA that contains all the elements required to propagate a chromosome in yeast and which is used to clone foreign DNA fragments in yeast cells.

Yellow fever : An acute infectious disease caused by a flavivirus, which is transmitted to humans by mosquitoes. The liver is affected and the skin turns yellow in this disease.

Yellow rings : Circles of yellow in a green leaf. Also called chlorotic rings.

YFP : Yellow fluorescent protein.

Ylid : A compound in which adjacent, covalently-bonded atoms, both having an electronic octet, have opposite charges.

YM shift : The change in shape by dimorphic fungi when they shift from the yeast (Y) form in the animal body to the mold or mycelial form (M) in the environment.

YO : Tears old.

YOB : Year of birth.

Yoghurt : (Or yogurt). A custard-like food made from fermented milk.

YPD : Yeast proteome database.

YR : Year.

Z : Zero.

Z form : A duplex DNA structure in which there is the usual type of hydrogen bonding between the base pairs but in which the helix formed by the two polynucleotide chains is left-handed rather than right-handed.

Z value : The increase in temperature required to reduce the decimal reduction time to one-tenth of its initial value.

ZAP-70 : The protein called ZAP-70 is found in T cells and is a relative of Syk in B cells. It contains two SH2 domains that, when bound to the phosphorylated æ chain, leads to activation of kinase activity. The main cellular substrate of ZAP-70 is a large adaptor protein called LAT.

ZD : Zero defects.

Zonate : Marked with concentric bands of colour

Zooflagellates : Flagellate protozoa that do not have chlorophyll and are either holozoic, saprozoic, or symbiotic.

Zoogleal film : A complex population of organisms that form a "slime growth" on a trickling-filter media and break down the organic matter in wastewater.

Zoogleal mass : Jelly-like masses of bacteria found in both the trickling filter and activated sludge processes.

Zoology : The study of animals.

Zoomastigophora : Division in Kingdom Protista whose members have a whip-like flagellum ; some are parasitic and some are free-living.

Zoonosis : A disease of animals that can be transmitted to humans.

Zooplankton : A community of floating, aquatic, minute animals and nonphotosynthetic protists.

Zoospore : A motile, flagellated spore.OR An asexual spore formed by some fungi that usually can move in an aqueous environment via one or more flagella.

Zooxanthella : A dinoflagellate found living symbiotically within cnidarians and other invertebrates.

Z-plasty : Plastic surgery skin repair technique.

Zwitterion : A dipolar ion with spatially-separated positive and negative charges. For example, most amino acids are zwitterions, having a positive charge on the ?-amino group and a negative charge on the ?-carboxyl group but no net charge on the overall molecule.

ZYG : Zygoma.

Zygomorphic : A term which describes an irregular flower.

Zygomycetes : A class of fungi where the asexual spores are mostly formed endogenously in sporangia. The majority of the species are saprobes.

Zygospore : Thickwalled resting spore resulting from fusion of two gametangia of fungi in the phylum Zygomycota.

Zygospores : A thick-walled sexual spore formed by the fusion of two similar gametangia; characteristic of the Zygomycetes.

Zygote : In eukaryotes, the single diploid cell resulting from the union (fusion) of two haploid gametes. ***OR*** A cell that results from the union of haploid male and female sex cells. Zygotes are diploid.

Zygote : The first cell of the new 2n organism formed by fertilization.

Zymogen : An inactive precursor of an enzyme. For example, trypsin exists in the inactive form trypsinogen before it is converted to its active form, trypsin.

Zymogenous flora : Refers to microorganisms, often transient or alien, that respond rapidly by enzyme production and growth when simple organic substrates become available. Also called copiotrophs.

Zymogenous organism : Refers to an often transient or alien microorganism that grows rapidly when high energycontaining nutrients become available. Also called copiotroph.